RILEM State-of-the-Art Reports

# RILEM STATE-OF-THE-ART REPORTS

Volume 28

RILEM, The International Union of Laboratories and Experts in Construction Materials, Systems and Structures, founded in 1947, is a non-governmental scientific association whose goal is to contribute to progress in the construction sciences, techniques and industries, essentially by means of the communication it fosters between research and practice. RILEM's focus is on construction materials and their use in building and civil engineering structures, covering all phases of the building process from manufacture to use and recycling of materials. More information on RILEM and its previous publications can be found on www.RILEM.net.

The RILEM State-of-the-Art Reports (STAR) are produced by the Technical Committees. They represent one of the most important outputs that RILEM generates—high level scientific and engineering reports that provide cutting edge knowledge in a given field. The work of the TCs is one of RILEM's key functions.

Members of a TC are experts in their field and give their time freely to share their expertise. As a result, the broader scientific community benefits greatly from RILEM's activities.

RILEM's stated objective is to disseminate this information as widely as possible to the scientific community. RILEM therefore considers the STAR reports of its TCs as of highest importance, and encourages their publication whenever possible.

The information in this and similar reports is mostly pre-normative in the sense that it provides the underlying scientific fundamentals on which standards and codes of practice are based. Without such a solid scientific basis, construction practice will be less than efficient or economical.

It is RILEM's hope that this information will be of wide use to the scientific community.

More information about this series at http://www.springer.com/series/8780

William G. Buttlar · Armelle Chabot
Eshan V. Dave · Christophe Petit
Gabriele Tebaldi
Editors

# Mechanisms of Cracking and Debonding in Asphalt and Composite Pavements

## State-of-the-Art of the RILEM TC 241-MCD

rilem

Springer

*Editors*
William G. Buttlar
Department of Civil and Environmental Engineering
University of Missouri
Columbia, MO
USA

Armelle Chabot
Department of Materials and Structures (MAST)
IFSTTAR
Nantes
France

Eshan V. Dave
Department of Civil and Environmental Engineering
University of New Hampshire
Durham, NH
USA

Christophe Petit
Département de Génie, Institut Universitaire de Technologie
Université de Limoges
Egletons
France

Gabriele Tebaldi
Department of Civil, Environmental Engineering and Architecture
University of Parma
Parma
Italy

ISSN 2213-204X ISSN 2213-2031 (electronic)
RILEM State-of-the-Art Reports
ISBN 978-3-030-08308-3 ISBN 978-3-319-76849-6 (eBook)
https://doi.org/10.1007/978-3-319-76849-6



Printed on acid-free paper

This Springer imprint is published by the registered company Springer International Publishing AG part of Springer Nature
The registered company address is: Gewerbestrasse 11, 6330 Cham, Switzerland

# Preface

## History of the Previous RILEM Technical Committees on Pavement Cracking

Crack reflection through a road structure has been addressed since 1989 by the RILEM TC97-GCR *Application of Geotextiles to the prevention of Cracks in Roads* led by Louis Francken. It started a series of four international RILEM conferences on *Reflective Cracking* with the first two international RILEM RC conferences held in Liege (RC1989, RC1993) [1] and [2]. The RILEM TC 157-PRC *Prevention of Reflective Cracking in Pavements* (1993–1997) was established to further explore the potential for improving pavement crack resistance with a broader range of products and technical solutions. This led to the organization of the third and fourth RILEM RC conferences in Maastricht (RC1996) [3] and Ottawa (RC2000) [4]. Many of the innovative systems recommended at RC conferences have been implemented, and the four proceedings of these conferences remain unparalleled sources of information on the problem of Pavement Reflective Cracking. The information gathered by the RILEM Technical Committee 157-PRC was used to finalize a state-of-the-art report that was published in 1997 and presented in the RILEM 18 report with updated results [5].

To approach cracking problems in pavements in a more general sense, the scope of the state-of-the-art activities and the conferences was extended, after 2000, to other modes of cracking modes such as fatigue, ageing or top-down cracking. A successful series of three conferences on ‘Cracking in Pavement’ have presented these updates first in Limoges (CP2004) [6] and then in Chicago (CP2008) [7] and Delft (CP2012) [8]. Between 2004 and 2011, the RILEM Technical Committee 210-CAP Cracking in Asphalt Pavement led by Prof. Andre A. A. Molenaar has organized the last two conferences.

## Main Objectives of TC241-MCD (2011–2017)

The RILEM Technical Committee 241-MCD on *Mechanisms of Cracking and Debonding in Asphalt and Composite Pavements* has been conducting scholarly activities geared towards developing a deeper fundamental understanding of the mechanisms behind cracking and debonding in asphalt concrete and composite (e.g. asphalt overlays placed on PCC) pavement systems. Although most modern pavement design methods do not directly account for the presence of cracking and debonding in asphalt and composite pavements, many of the critical modes of pavement deterioration involve one or both of these mechanisms.

In order to achieve its goals, the TC 241-MCD (Chair: Bill Buttlar, University of Missouri, USA; Secretary: Armelle Chabot, IFSTTAR/MAST, France) has been performing its work within three interrelated technical groups (TGs) (see Fig. 1). Each TG has involved an integration of testing, modelling and connection to related field studies:

- *TG1 Cracking in Asphalt Material* (Convener: Eshan Dave, University of New Hampshire, USA)
- *TG2 Interface Debonding Behaviour* (Convener: Christophe Petit, Université de Limoges, France)
- *TG3 Advanced Measurement Systems for Crack Characterization* (Convener: Gabriele Tebaldi, Université of Parma, Italy—University of Florida, USA)

The main advances have been presented during the 8th RILEM International Conference on Mechanisms of cracking and Debonding in Pavements in Nantes (MCD2016) [9].

> Cracking in asphalt and concrete pavements has been a research topic for decades and many conferences like the RILEM conferences on Cracking in Asphalt Pavements have been dedicated to it. In time we have acquired a vast amount of knowledge about the factors affecting cracking in pavements as well as about the factors that control the cracking resistance of pavement materials like asphalt and cement concrete and stabilized materials. However the factors influencing de-bonding between layers only received limited attention. It almost looks like we are ignoring the importance of a good bond between successive layers. This is perhaps best reflected by the following example When designing the

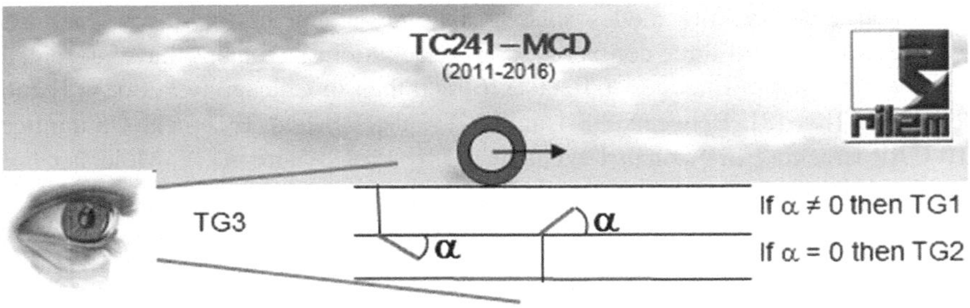


**Fig. 1** Logo of the RILEM TC241-MCD

thickness of an asphalt pavement we mostly assume that the total asphalt thickness acts as one layer. The fact that the total thickness is made up of multiple single layers is simply ignored although it is known for decades that the shear strength of the interface between two asphalt layers as determined by means of e.g. the Leutner test is approximately 50% of the shear strength of the asphalt mixtures above and below the interface. This indicates that the layer interfaces are the weakest part of the asphalt layer! Therefore it is no surprise that many of the cracks we observe in our pavements are affected by the poor bond between layers. Poor bonding between the wearing course and layer below it is one of the reasons of pothole formation.

This first conference, MCD2016, on "de-bonding" is a very important one because it is the first time that the weakest part in pavement structures being the interface between layers gets the full attention it deserves. I am convinced that the outcomes of this conference, MCD2016, will contribute to a better understanding of how interfaces perform, to improved design methodologies that do take into account the initiation and propagation of de-bonding and its effect on the overall performance of pavements.

Prof. A. A. A. Molenaar (January 29th, 2014)

This book is authored and edited under the auspices of RILEM Technical Committee TC 241-MCD. Contributions were collected during the first half of the committee's planned five-year scope, followed by compilation, discussion and final editing. It presents technical contributions from RILEM members and international partners that provide an outline of the state of the art in field investigations, laboratory testing and modelling devoted to the understanding of pavement cracking mechanisms, organized according to the theme areas of the TGs (i.e. fracture in the asphalt bulk material, interface debonding behaviour and advanced measurement systems). This book also draws from the contributions of the TC241-MCD members emanating from their work conducted in the latter stages of the TC 210-CAP (Cracking in Asphalt Pavements). The TC 241-MCD membership list is given below.

Airey Gordon, University of Nottingham, UK
Apeagyei Alex, University of Nottingham Ph.D.→ University of East London, UK
Behzad Behnia, University of Illinois Ph.D.→ Clarkson University, USA
Bodin Didier, ARRB, Australia
Botella Ramón, Universidad Politécnica de Cataluña, Espagne
Buttlar Bill, University of Illinois, USA
Cannone-Falchetto A., TU Braunschweig, Germany
Chabot Armelle, IFSTTAR, France
Chupin Olivier, IFSTTAR, France
Collop Andy, De Montfort University, UK
Dave Eshan, University of New-Hampshire, USA
Destree Alexandra, BRRC, Belgium
Di Benedetto Hervé, Université de Lyon, ENTPE, France
Dubois Frédéric, Université de Limoges, France
Graziani Andrea, Univresita Politecnica Delle Marche, Italy
Grenfell James, University of Nottingham, UK
Hammoum Ferhat, IFSTTAR, France

Hornych Pierre, IFSTTAR, France
Hun Manitou, IFSTTAR Ph.D.→ ASBIC Construction Co., Ltd, Cambogia
Jelagin Denis, KTH Royal Institute of Technology, Sweden
Kringos Niki, KTH Royal Institute of Technology, Sweden
Loizos Andreas/Plati, University of Athens, Grèce
Marsac Paul, IFSTTAR, France
Millien Anne, Université de Limoges, France
Moreno Fernando, Universidad de Granada, Espagne
Nguyen Mai Lan, IFSTTAR, France
Pais Jorge, University of Minho, Portugal
Petit Christophe, Université de Limoges, France
Raab Christiane, EMPA, Switzerland
Sauzeat Cedric, Université de Lyon, ENTPE, France
Simonin Jean Michel, IFSTTAR, France
Tebaldi Gabriele, University of Parma, Italy
Toumi Ahmed, INSA Toulouse, France
Turatsinze Anaclet, INSA Toulouse, France
Zofka Adam, Road and Bridge Research Institute, Poland

## Acknowledgements

Membership and participation in the RILEM TC 241-MCD committee and events have evolved substantially over the active years of the committee (2011–2017). The editors thank all of who have actively contributed to this report through meeting attendance, discussions and participation in the direct input to the various chapters as listed below.

Abraham Odile, IFSTTAR, France
Canestrari Francesco, Universita Politecnica Delle Marche, Italy
Fouchal Fazia, Université de Limoges, France
Guarin Alvaro, KTH Royal Institute of Technology, Sweden
Hakimzadeh Khoee Salman, Ph.D. University of Illinois, USA
Hill Brian, Ph.D. University of Illinois, USA
Lebon Frédéric, Université de Marseille, France
Miro Rodrigo, Universidad Politécnica de Cataluña, Espagne
Pérez-Jiménez Félix, Universidad Politécnica de Cataluña, Espagne
Piau Jean-Michel, IFSTTAR, France
Picoux Benoit, Université de Limoges, France
Plati Christina, University of Athens, Grèce
Pop Octavian, Université de Limoges, France
Roque Rey, University of Florida, USA
Sullivan Chris, Institute of Asphalt Technology, UK
Takarli Mokhfi, Université de Limoges, France

In particular, the different chapters have been produced with different contributions coming specifically from:

Contributors to Chap. 1: *Introduction*: W. G. Buttlar, A. Chabot, A de Bondt, S. Guthrie, P. Hornych, and C. Sullivan

Contributors to Chap. 2: *Cracking in Asphalt Material*: W. G. Buttlar, A. Collop, B. Hill, B. Behnia, F. Perez-Jimenez, G. Valdes, R. Miro, A. Martinez, H. Di Benedetto, C. Sauzeat, Q. T. Nguyen, M.-L. Nguyen, N. Tapsoba, R. Roque, D. Hernando, J. Zou, G. Tebaldi, R. Kim, A. Chabot, L. Wendling, F. Hammoum, P. Marsac, F. Dubois, B. Picoux, C. Petit, D. Bodin, C. Maggiore, and A. Collop

Contributors to Chap. 3: *Interface Debonding Behaviour*: W. G. Buttlar, A. Chabot, A. Destree, F. Fouchal, F. Hammoum, F. Lebon, N. M. Louay, A. Millien, C. Petit, O. Pop, C. Raab, G. Tebaldi, A. Toumi, A. Turatsinze, T. Takarli, R. Roque, A. Zofka

Contributors to Chap. 4: *Advanced Measurement Systems for Crack Characterization*: G. Tebaldi, M. E. Kutay, O. Pop, A. Millien, C. Petit, M. Hun, A. Chabot, F. Hammoum, A. Montepara, E. Romeo, R. Roncella, J. M. Simonin, B. Picoux, V. Balthazard, X. Dérobert, B. Behnia, W. G. Buttlar, H. Reis, M. Takarli, N. Angellier, A. Guarin, D. Jelagin, N. Kringos, B. Birgisson

Columbia, USA William G. Buttlar
Nantes, France Armelle Chabot
Durham, USA Eshan V. Dave
Egletons, France Christophe Petit
Parma, Italy Gabriele Tebaldi

## References

1. Rigo JM, Degeimbre R (eds) (1989) Reflective cracking in pavements—assessment and control. In: 1st international RILEM conference on Reflective Cracking in Pavements, Liège, Mardaga Press
2. Rigo JM, Degeimbre R, Francken L (eds) (1993) Reflective cracking in pavements: state of the art and design recommendations. In: 2nd international RILEM conference on Reflective Cracking in Pavements—Assessment and Control, Liège. RILEM proceedings, 20, CRC Press, ISBN-10: 0419182209, ISBN-13: 978-0419182207
3. Francken L, Beuving E, Molenaar AAA (eds) (1996) Reflective cracking in pavements: design and performance of overlay systems. In: 3rd international RILEM conference on Reflective Cracking in Pavements, Maastricht. In: RILEM proceedings 13, Spon Press (Verlag), ISBN: 978-0-419-22260-6
4. Abd El Halim AO, Taylor DA, Mohamed El HH (eds) (2000) Reflective cracking in pavements—research in Practice. In: 4th international RILEM conference on Reflective Cracking in Pavements, Ottawa. In: RILEM proceedings 011, ISBN: 2-912143-14-4, e-ISBN: 2351580265
5. Vanelstraete A, Francken L (eds) (1997) State-of-the-art report of RILEM technical committee 157 PRC, systems to prevent reflective cracking in pavements, RILEM Report 18
6. Petit C, Al Qadi I, Millien A (eds) (2004) 5th RILEM international conference on Cracking in Pavements—Mitigation, Risk, Assessment and Prevention, Limoges. In: RILEM proceedings 37, ISBN 2-912143-47-0

7. Loizos A, Scarpas T, Al-Qadi I (eds) (2008) 6th RILEM international conference on Cracking in Pavements—Pavement Cracking Mechanisms, Modeling, Detection, Testing and Case Histories, Chicago. CRC Press, Print ISBN: 978-0-415-47575-4, Doi: 10.1201/9780203882191
8. Scarpas T, Kringos N, Al-Qadi I, Loizos A (eds) (2012) 7th RILEM international conference on Cracking in Pavements, Delft. Springer Series: RILEM Bookseries, Vol 4, ISBN 978-94-007-4565-0, Doi: 10.1007/978-94-007-4566-7
9. Chabot A, Buttlar B, Dave E, Petit C, Tebaldi G (eds) (2016) 8th RILEM international conference on Mechanisms of Cracking and Debonding in Pavements, Nantes. Springer Series: RILEM Bookseries, Vol 13, ISBN 978-94-024-0866-9, Doi: 10.1007/978-94-024-0867-6

# RILEM Publications

The following list is presenting the global offer of RILEM Publications, sorted by series. Each publication is available in printed version and/or in online version.

## RILEM PROCEEDINGS (PRO)

**PRO 1**: Durability of High Performance Concrete (ISBN: 2-912143-03-9; e-ISBN: 2-351580-12-5; e-ISBN: 2351580125); *Ed. H. Sommer*

**PRO 2**: Chloride Penetration into Concrete (ISBN: 2-912143-00-04; e-ISBN: 2912143454); *Eds. L.-O. Nilsson and J.-P. Ollivier*

**PRO 3**: Evaluation and Strengthening of Existing Masonry Structures (ISBN: 2-912143-02-0; e-ISBN: 2351580141); *Eds. L. Binda and C. Modena*

**PRO 4**: Concrete: From Material to Structure (ISBN: 2-912143-04-7; e-ISBN: 2351580206); *Eds. J.-P. Bournazel and Y. Malier*

**PRO 5**: The Role of Admixtures in High Performance Concrete (ISBN: 2-912143-05-5; e-ISBN: 2351580214); *Eds. J. G. Cabrera and R. Rivera-Villarreal*

**PRO 6**: High Performance Fiber Reinforced Cement Composites—HPFRCC 3 (ISBN: 2-912143-06-3; e-ISBN: 2351580222); *Eds. H. W. Reinhardt and A. E. Naaman*

**PRO 7**: 1st International RILEM Symposium on Self-Compacting Concrete (ISBN: 2-912143-09-8; e-ISBN: 2912143721); *Eds. Å. Skarendahl and Ö. Petersson*

**PRO 8**: International RILEM Symposium on Timber Engineering (ISBN: 2-912143-10-1; e-ISBN: 2351580230); *Ed. L. Boström*

**PRO 9**: 2nd International RILEM Symposium on Adhesion between Polymers and Concrete ISAP '99 (ISBN: 2-912143-11-X; e-ISBN: 2351580249); *Eds. Y. Ohama and M. Puterman*

**PRO 10**: 3rd International RILEM Symposium on Durability of Building and Construction Sealants (ISBN: 2-912143-13-6; e-ISBN: 2351580257); *Ed. A. T. Wolf*

**PRO 11**: 4th International RILEM Conference on Reflective Cracking in Pavements (ISBN: 2-912143-14-4; e-ISBN: 2351580265); *Eds. A. O. Abd El Halim, D. A. Taylor and El H. H. Mohamed*

**PRO 12**: International RILEM Workshop on Historic Mortars: Characteristics and Tests (ISBN: 2-912143-15-2; e-ISBN: 2351580273); *Eds. P. Bartos, C. Groot and J. J. Hughes*

**PRO 13**: 2nd International RILEM Symposium on Hydration and Setting (ISBN: 2-912143-16-0; e-ISBN: 2351580281); *Ed. A. Nonat*

**PRO 14**: Integrated Life-Cycle Design of Materials and Structures—ILCDES 2000 (ISBN: 951-758-408-3; e-ISBN: 235158029X); (ISSN: 0356-9403); *Ed. S. Sarja*

**PRO 15**: Fifth RILEM Symposium on Fibre-Reinforced Concretes (FRC)—BEFIB'2000 (ISBN: 2-912143-18-7; e-ISBN: 291214373X); *Eds. P. Rossi and G. Chanvillard*

**PRO 16**: Life Prediction and Management of Concrete Structures (ISBN: 2-912143-19-5; e-ISBN: 2351580303); *Ed. D. Naus*

**PRO 17**: Shrinkage of Concrete—Shrinkage 2000 (ISBN: 2-912143-20-9; e-ISBN: 2351580311); *Eds. V. Baroghel-Bouny and P.-C. Aïtcin*

**PRO 18**: Measurement and Interpretation of the On-Site Corrosion Rate (ISBN: 2-912143-21-7; e-ISBN: 235158032X); *Eds. C. Andrade, C. Alonso, J. Fullea, J. Polimon and J. Rodriguez*

**PRO 19**: Testing and Modelling the Chloride Ingress into Concrete (ISBN: 2-912143-22-5; e-ISBN: 2351580338); *Eds. C. Andrade and J. Kropp*

**PRO 20**: 1st International RILEM Workshop on Microbial Impacts on Building Materials (CD 02) (e-ISBN 978-2-35158-013-4); *Ed. M. Ribas Silva*

**PRO 21**: International RILEM Symposium on Connections between Steel and Concrete (ISBN: 2-912143-25-X; e-ISBN: 2351580346); *Ed. R. Eligehausen*

**PRO 22**: International RILEM Symposium on Joints in Timber Structures (ISBN: 2-912143-28-4; e-ISBN: 2351580354); *Eds. S. Aicher and H.-W. Reinhardt*

**PRO 23**: International RILEM Conference on Early Age Cracking in Cementitious Systems (ISBN: 2-912143-29-2; e-ISBN: 2351580362); *Eds. K. Kovler and A. Bentur*

**PRO 24**: 2nd International RILEM Workshop on Frost Resistance of Concrete (ISBN: 2-912143-30-6; e-ISBN: 2351580370); *Eds. M. J. Setzer, R. Auberg and H.-J. Keck*

**PRO 25**: International RILEM Workshop on Frost Damage in Concrete (ISBN: 2-912143-31-4; e-ISBN: 2351580389); *Eds. D. J. Janssen, M. J. Setzer and M. B. Snyder*

**PRO 26**: International RILEM Workshop on On-Site Control and Evaluation of Masonry Structures (ISBN: 2-912143-34-9; e-ISBN: 2351580141); *Eds. L. Binda and R. C. de Vekey*

**PRO 27**: International RILEM Symposium on Building Joint Sealants (CD03; e-ISBN: 235158015X); *Ed. A. T. Wolf*

**PRO 28**: 6th International RILEM Symposium on Performance Testing and Evaluation of Bituminous Materials—PTEBM'03 (ISBN: 2-912143-35-7; e-ISBN: 978-2-912143-77-8); *Ed. M. N. Partl*

**PRO 29**: 2nd International RILEM Workshop on Life Prediction and Ageing Management of Concrete Structures (ISBN: 2-912143-36-5; e-ISBN: 2912143780); *Ed. D. J. Naus*

**PRO 30**: 4th International RILEM Workshop on High Performance Fiber Reinforced Cement Composites—HPFRCC 4 (ISBN: 2-912143-37-3; e-ISBN: 2912143799); *Eds. A. E. Naaman and H. W. Reinhardt*

**PRO 31**: International RILEM Workshop on Test and Design Methods for Steel Fibre Reinforced Concrete: Background and Experiences (ISBN: 2-912143-38-1; e-ISBN: 2351580168); *Eds. B. Schnütgen and L. Vandewalle*

**PRO 32**: International Conference on Advances in Concrete and Structures 2 vol. (ISBN (set): 2-912143-41-1; e-ISBN: 2351580176); *Eds. Ying-shu Yuan, Surendra P. Shah and Heng-lin Lü*

**PRO 33**: 3rd International Symposium on Self-Compacting Concrete (ISBN: 2-912143-42-X; e-ISBN: 2912143713); *Eds. Ó. Wallevik and I. Níelsson*

**PRO 34**: International RILEM Conference on Microbial Impact on Building Materials (ISBN: 2-912143-43-8; e-ISBN: 2351580184); *Ed. M. Ribas Silva*

**PRO 35**: International RILEM TC 186-ISA on Internal Sulfate Attack and Delayed Ettringite Formation (ISBN: 2-912143-44-6; e-ISBN: 2912143802); *Eds. K. Scrivener and J. Skalny*

**PRO 36**: International RILEM Symposium on Concrete Science and Engineering—A Tribute to Arnon Bentur (ISBN: 2-912143-46-2; e-ISBN: 2912143586); *Eds. K. Kovler, J. Marchand, S. Mindess and J. Weiss*

**PRO 37**: 5th International RILEM Conference on Cracking in Pavements—Mitigation, Risk Assessment and Prevention (ISBN: 2-912143-47-0; e-ISBN: 2912143764); *Eds. C. Petit, I. Al-Qadi and A. Millien*

**PRO 38**: 3rd International RILEM Workshop on Testing and Modelling the Chloride Ingress into Concrete (ISBN: 2-912143-48-9; e-ISBN: 2912143578); *Eds. C. Andrade and J. Kropp*

**PRO 39**: 6th International RILEM Symposium on Fibre-Reinforced Concretes—BEFIB 2004 (ISBN: 2-912143-51-9; e-ISBN: 2912143748); *Eds. M. Di Prisco, R. Felicetti and G. A. Plizzari*

**PRO 40**: International RILEM Conference on the Use of Recycled Materials in Buildings and Structures (ISBN: 2-912143-52-7; e-ISBN: 2912143756); *Eds. E. Vázquez, Ch. F. Hendriks and G. M. T. Janssen*

**PRO 41**: RILEM International Symposium on Environment-Conscious Materials and Systems for Sustainable Development (ISBN: 2-912143-55-1; e-ISBN: 2912143640); *Eds. N. Kashino and Y. Ohama*

**PRO 42**: SCC'2005—China: 1st International Symposium on Design, Performance and Use of Self-Consolidating Concrete (ISBN: 2-912143-61-6; e-ISBN: 2912143624); *Eds. Zhiwu Yu, Caijun Shi, Kamal Henri Khayat and Youjun Xie*

**PRO 43**: International RILEM Workshop on Bonded Concrete Overlays (e-ISBN: 2-912143-83-7); *Eds. J. L. Granju and J. Silfwerbrand*

**PRO 44**: 2nd International RILEM Workshop on Microbial Impacts on Building Materials (CD11) (e-ISBN: 2-912143-84-5); *Ed. M. Ribas Silva*

**PRO 45**: 2nd International Symposium on Nanotechnology in Construction, Bilbao (ISBN: 2-912143-87-X; e-ISBN: 2912143888); *Eds. Peter J. M. Bartos, Yolanda de Miguel and Antonio Porro*

**PRO 46**: ConcreteLife'06—International RILEM-JCI Seminar on Concrete Durability and Service Life Planning: Curing, Crack Control, Performance in Harsh Environments (ISBN: 2-912143-89-6; e-ISBN: 291214390X); *Ed. K. Kovler*

**PRO 47**: International RILEM Workshop on Performance Based Evaluation and Indicators for Concrete Durability (ISBN: 978-2-912143-95-2; e-ISBN: 978291 2143969); *Eds. V. Baroghel-Bouny, C. Andrade, R. Torrent and K. Scrivener*

**PRO 48**: 1st International RILEM Symposium on Advances in Concrete through Science and Engineering (e-ISBN: 2-912143-92-6); *Eds. J. Weiss, K. Kovler, J. Marchand, and S. Mindess*

**PRO 49**: International RILEM Workshop on High Performance Fiber Reinforced Cementitious Composites in Structural Applications (ISBN: 2-912143-93-4; e-ISBN: 2912143942); *Eds. G. Fischer and V. C. Li*

**PRO 50**: 1st International RILEM Symposium on Textile Reinforced Concrete (ISBN: 2-912143-97-7; e-ISBN: 2351580087); *Eds. Josef Hegger, Wolfgang Brameshuber and Norbert Will*

**PRO 51**: 2nd International Symposium on Advances in Concrete through Science and Engineering (ISBN: 2-35158-003-6; e-ISBN: 2-35158-002-8); *Eds. J. Marchand, B. Bissonnette, R. Gagné, M. Jolin and F. Paradis*

**PRO 52**: Volume Changes of Hardening Concrete: Testing and Mitigation (ISBN: 2-35158-004-4; e-ISBN: 2-35158-005-2); *Eds. O. M. Jensen, P. Lura and K. Kovler*

**PRO 53**: High Performance Fiber Reinforced Cement Composites—HPFRCC5 (ISBN: 978-2-35158-046-2; e-ISBN: 978-2-35158-089-9); *Eds. H. W. Reinhardt and A. E. Naaman*

**PRO 54**: 5th International RILEM Symposium on Self-Compacting Concrete (ISBN: 978-2-35158-047-9; e-ISBN: 978-2-35158-088-2); *Eds. G. De Schutter and V. Boel*

**PRO 55**: International RILEM Symposium Photocatalysis, Environment and Construction Materials (ISBN: 978-2-35158-056-1; e-ISBN: 978-2-35158-057-8); *Eds. P. Baglioni and L. Cassar*

**PRO 56**: International RILEM Workshop on Integral Service Life Modelling of Concrete Structures (ISBN 978-2-35158-058-5; e-ISBN: 978-2-35158-090-5); *Eds. R. M. Ferreira, J. Gulikers and C. Andrade*

**PRO 57**: RILEM Workshop on Performance of cement-based materials in aggressive aqueous environments (e-ISBN: 978-2-35158-059-2); *Ed. N. De Belie*

**PRO 58**: International RILEM Symposium on Concrete Modelling—CONMOD'08 (ISBN: 978-2-35158-060-8; e-ISBN: 978-2-35158-076-9); *Eds. E. Schlangen and G. De Schutter*

**PRO 59**: International RILEM Conference on On Site Assessment of Concrete, Masonry and Timber Structures—SACoMaTiS 2008 (ISBN set: 978-2-35158-061-5; e-ISBN: 978-2-35158-075-2); *Eds. L. Binda, M. di Prisco and R. Felicetti*

**PRO 60**: Seventh RILEM International Symposium on Fibre Reinforced Concrete: Design and Applications—BEFIB 2008 (ISBN: 978-2-35158-064-6; e-ISBN: 978-2-35158-086-8); *Ed. R. Gettu*

**PRO 61**: 1st International Conference on Microstructure Related Durability of Cementitious Composites 2 vol., (ISBN: 978-2-35158-065-3; e-ISBN: 978-2-35158-084-4); *Eds. W. Sun, K. van Breugel, C. Miao, G. Ye and H. Chen*

**PRO 62**: NSF/ RILEM Workshop: In-situ Evaluation of Historic Wood and Masonry Structures (e-ISBN: 978-2-35158-068-4); *Eds. B. Kasal, R. Anthony and M. Drdácký*

**PRO 63**: Concrete in Aggressive Aqueous Environments: Performance, Testing and Modelling, 2 vol., (ISBN: 978-2-35158-071-4; e-ISBN: 978-2-35158-082-0); *Eds. M. G. Alexander and A. Bertron*

**PRO 64**: Long Term Performance of Cementitious Barriers and Reinforced Concrete in Nuclear Power Plants and Waste Management—NUCPERF 2009 (ISBN: 978-2-35158-072-1; e-ISBN: 978-2-35158-087-5); *Eds. V. L'Hostis, R. Gens, C. Gallé*

**PRO 65**: Design Performance and Use of Self-consolidating Concrete—SCC'2009 (ISBN: 978-2-35158-073-8; e-ISBN: 978-2-35158-093-6); *Eds. C. Shi, Z. Yu, K. H. Khayat and P. Yan*

**PRO 66**: 2nd International RILEM Workshop on Concrete Durability and Service Life Planning—ConcreteLife'09 (ISBN: 978-2-35158-074-5); *Ed. K. Kovler*

**PRO 67**: Repairs Mortars for Historic Masonry (e-ISBN: 978-2-35158-083-7); *Ed. C. Groot*

**PRO 68**: Proceedings of the 3rd International RILEM Symposium on 'Rheology of Cement Suspensions such as Fresh Concrete (ISBN 978-2-35158-091-2; e-ISBN: 978-2-35158-092-9); *Eds. O. H. Wallevik, S. Kubens and S. Oesterheld*

**PRO 69**: 3rd International PhD Student Workshop on 'Modelling the Durability of Reinforced Concrete (ISBN: 978-2-35158-095-0); *Eds. R. M. Ferreira, J. Gulikers and C. Andrade*

**PRO 70**: 2nd International Conference on 'Service Life Design for Infrastructure' (ISBN set: 978-2-35158-096-7; e-ISBN: 978-2-35158-097-4); *Ed. K. van Breugel, G. Ye and Y. Yuan*

**PRO 71**: Advances in Civil Engineering Materials—The 50-year Teaching Anniversary of Prof. Sun Wei' (ISBN: 978-2-35158-098-1; e-ISBN: 978-2-35158-099-8); *Eds. C. Miao, G. Ye, and H. Chen*

**PRO 72**: First International Conference on 'Advances in Chemically-Activated Materials—CAM'2010' (2010), 264 pp., ISBN: 978-2-35158-101-8; e-ISBN: 978-2-35158-115-5; *Eds. Caijun Shi and Xiaodong Shen*

**PRO 73**: 2nd International Conference on 'Waste Engineering and Management—ICWEM 2010' (2010), 894 pp., ISBN: 978-2-35158-102-5; e-ISBN: 978-2-35158-103-2; *Eds. J. Zh. Xiao, Y. Zhang, M. S. Cheung and R. Chu*

**PRO 74**: International RILEM Conference on 'Use of Superabsorsorbent Polymers and Other New Addditives in Concrete' (2010) 374 pp., ISBN: 978-2-35158-104-9; e-ISBN: 978-2-35158-105-6; *Eds. O. M. Jensen, M. T. Hasholt, and S. Laustsen*

**PRO 75**: International Conference on 'Material Science—2nd ICTRC—Textile Reinforced Concrete—Theme 1' (2010) 436 pp., ISBN: 978-2-35158-106-3; e-ISBN: 978-2-35158-107-0; *Ed. W. Brameshuber*

**PRO 76**: International Conference on 'Material Science—HetMat—Modelling of Heterogeneous Materials—Theme 2' (2010) 255 pp., ISBN: 978-2-35158-108-7; e-ISBN: 978-2-35158-109-4; *Ed. W. Brameshuber*

**PRO 77**: International Conference on 'Material Science—AdIPoC—Additions Improving Properties of Concrete—Theme 3' (2010) 459 pp., ISBN: 978-2-35158-110-0; e-ISBN: 978-2-35158-111-7; *Ed. W. Brameshuber*

**PRO 78**: 2nd Historic Mortars Conference and RILEM TC 203-RHM Final Workshop—HMC2010 (2010) 1416 pp., e-ISBN: 978-2-35158-112-4; *Eds. J. Válek, C. Groot, and J. J. Hughes*

**PRO 79**: International RILEM Conference on Advances in Construction Materials Through Science and Engineering (2011) 213 pp., ISBN: 978-2-35158-116-2; e-ISBN: 978-2-35158-117-9; *Eds. Christopher Leung and K. T. Wan*

**PRO 80**: 2nd International RILEM Conference on Concrete Spalling due to Fire Exposure (2011) 453 pp., ISBN: 978-2-35158-118-6; e-ISBN: 978-2-35158-119-3; *Eds. E. A. B. Koenders and F. Dehn*

**PRO 81**: 2nd International RILEM Conference on Strain Hardening Cementitious Composites (SHCC2-Rio) (2011) 451 pp., ISBN: 978-2-35158-120-9; e-ISBN: 978-2-35158-121-6; *Eds. R. D. Toledo Filho, F. A. Silva, E. A. B. Koenders and E. M. R. Fairbairn*

**PRO 82**: 2nd International RILEM Conference on Progress of Recycling in the Built Environment (2011) 507 pp., e-ISBN: 978-2-35158-122-3; *Eds. V. M. John, E. Vazquez, S. C. Angulo and C. Ulsen*

**PRO 83**: 2nd International Conference on Microstructural-related Durability of Cementitious Composites (2012) 250 pp., ISBN: 978-2-35158-129-2; e-ISBN: 978-2-35158-123-0; *Eds. G. Ye, K. van Breugel, W. Sun and C. Miao*

**PRO 84**: CONSEC13—Seventh International Conference on Concrete under Severe Conditions—Environment and Loading (2013) 1930 pp., ISBN: 978-2-35158-124-7; e-ISBN: 978-2- 35158-134-6; *Eds Z. J. Li, W. Sun, C. W. Miao, K. Sakai, O. E. Gjorv & N. Banthia*

**PRO 85**: RILEM-JCI International Workshop on Crack Control of Mass Concrete and Related issues concerning Early-Age of Concrete Structures—ConCrack 3—Control of Cracking in Concrete Structures 3 (2012) 237 pp., ISBN: 978-2-35158-125-4; e-ISBN: 978-2-35158-126-1; *Eds. F. Toutlemonde and J.-M. Torrenti*

**PRO 86**: International Symposium on Life Cycle Assessment and Construction (2012) 414 pp., ISBN: 978-2-35158-127-8; e-ISBN: 978-2-35158-128-5; *Eds. A. Ventura and C. de la Roche*

**PRO 87**: UHPFRC 2013—RILEM-fib-AFGC International Symposium on Ultra-High Performance Fibre-Reinforced Concrete (2013), ISBN: 978-2-35158-130-8; e-ISBN: 978-2-35158-131-5; *Eds. F. Toutlemonde*

**PRO 88**: 8th RILEM International Symposium on Fibre Reinforced Concrete (2012) 344 pp., ISBN: 978-2-35158-132-2; e-ISBN: 978-2-35158-133-9; *Eds. Joaquim A. O. Barros*

**PRO 89**: RILEM International workshop on performance-based specification and control of concrete durability (2014) 678 pp., ISBN: 978-2-35158-135-3; e-ISBN: 978-2-35158-136-0; *Eds. D. Bjegović, H. Beushausen and M. Serdar*

**PRO 90**: 7th RILEM International Conference on Self-Compacting Concrete and of the 1st RILEM International Conference on Rheology and Processing of Construction Materials (2013) 396 pp., ISBN: 978-2-35158-137-7; e-ISBN: 978-2-35158-138-4; *Eds. Nicolas Roussel and Hela Bessaies-Bey*

**PRO 91**: CONMOD 2014—RILEM International Symposium on Concrete Modelling (2014), ISBN: 978-2-35158-139-1; e-ISBN: 978-2-35158-140-7; *Eds. Kefei Li, Peiyu Yan and Rongwei Yang*

**PRO 92**: CAM 2014—2nd International Conference on advances in chemically-activated materials (2014) 392 pp., ISBN: 978-2-35158-141-4; e-ISBN: 978-2-35158-142-1; *Eds. Caijun Shi and Xiadong Shen*

**PRO 93**: SCC 2014—3rd International Symposium on Design, Performance and Use of Self-Consolidating Concrete (2014) 438 pp., ISBN: 978-2-35158-143-8; e-ISBN: 978-2-35158-144-5; *Eds. Caijun Shi, Zhihua Ou, Kamal H. Khayat*

**PRO 94 (online version)**: HPFRCC-7—7th RILEM conference on High performance fiber reinforced cement composites (2015), e-ISBN: 978-2-35158-146-9; *Eds. H. W. Reinhardt, G. J. Parra-Montesinos, H. Garrecht*

**PRO 95**: International RILEM Conference on Application of superabsorbent polymers and other new admixtures in concrete construction (2014), ISBN: 978-2-35158-147-6; e-ISBN: 978-2-35158-148-3; *Eds. Viktor Mechtcherine, Christof Schroefl*

**PRO 96 (online version)**: XIII DBMC: XIII International Conference on Durability of Building Materials and Components (2015), e-ISBN: 978-2-35158-149-0; *Eds. M. Quattrone, V. M. John*

**PRO 97**: SHCC3—3rd International RILEM Conference on Strain Hardening Cementitious Composites (2014), ISBN: 978-2-35158-150-6; e-ISBN: 978-2-35158-151-3; *Eds. E. Schlangen, M. G. Sierra Beltran, M. Lukovic, G. Ye*

**PRO 98**: FERRO-11—11th International Symposium on Ferrocement and 3rd ICTRC—International Conference on Textile Reinforced Concrete (2015), ISBN: 978-2-35158-152-0; e-ISBN: 978-2-35158-153-7; *Ed. W. Brameshuber*

**PRO 99 (online version)**: ICBBM 2015—1st International Conference on Bio-Based Building Materials (2015), e-ISBN: 978-2-35158-154-4; *Eds. S. Amziane, M. Sonebi*

**PRO 100**: SCC16—RILEM Self-Consolidating Concrete Conference (2016), ISBN: 978-2-35158-156-8; e-ISBN: 978-2-35158-157-5; *Ed. Kamal H. Kayat*

**PRO 101 (online version)**: III Progress of Recycling in the Built Environment (2015), e-ISBN: 978-2-35158-158-2; *Eds. M. Quattrone, V. M. John*

**PRO 102 (online version)**: RILEM Conference on Microorganisms-Cementitious Materials Interactions (2016), e-ISBN: 978-2-35158-160-5; *Eds. Alexandra Bertron, Henk Jonkers, Virginie Wiktor*

**PRO 103 (online version)**: ACESC'16—Advances in Civil Engineering and Sustainable Construction (2016), e-ISBN: 978-2-35158-161-2; *Eds. T. Ch. Madhavi, G. Prabhakar, Santhosh Ram, and P. M. Rameshwaran*

**PRO 104 (online version)**: SSCS'2015—Numerical Modeling—Strategies for Sustainable Concrete Structures (2015), e-ISBN: 978-2-35158-162-9

**PRO 105**: 1st International Conference on UHPC Materials and Structures (2016), ISBN: 978-2-35158-164-3; e-ISBN: 978-2-35158-165-0

**PRO 106**: AFGC-ACI-fib-RILEM International Conference on Ultra-High-Performance Fibre-Reinforced Concrete—UHPFRC 2017 (2017), ISBN: 978-2-35158-166-7; e-ISBN: 978-2-35158-167-4; *Eds. François Toutlemonde & Jacques Resplendino*

**PRO 107 (online version)**: XIV DBMC—14th International Conference on Durability of Building Materials and Components (2017), e-ISBN: 978-2-35158-159-9; *Eds. Geert De Schutter, Nele De Belie, Arnold Janssens, Nathan Van Den Bossche*

**PRO 108**: MSSCE 2016—Innovation of Teaching in Materials and Structures (2016), ISBN: 978-2-35158-178-0; e-ISBN: 978-2-35158-179-7; *Ed. Per Goltermann*

**PRO 109 (2 volumes)**: MSSCE 2016—Service Life of Cement-Based Materials and Structures (2016), ISBN Vol. 1: 978-2-35158-170-4, Vol. 2: 978-2-35158-171-4, Set Vol. 1&2: 978-2-35158-172-8; e-ISBN: 978-2-35158-173-5; *Eds. Miguel Azenha, Ivan Gabrijel, Dirk Schlicke, Terje Kanstad and Ole Mejlhede Jensen*

**PRO 110**: MSSCE 2016—Historical Masonry (2016), ISBN: 978-2-35158-178-0; e-ISBN: 978-2-35158-179-7; *Eds. Inge Rörig-Dalgaard and Ioannis Ioannou*

**PRO 111**: MSSCE 2016—Electrochemistry in Civil Engineering (2016), ISBN: 978-2-35158-176-6; e-ISBN: 978-2-35158-177-3; *Ed. Lisbeth M. Ottosen*

**PRO 112**: MSSCE 2016—Moisture in Materials and Structures (2016), ISBN: 978-2-35158-178-0; e-ISBN: 978-2-35158-179-7; *Eds. Kurt Kielsgaard Hansen, Carsten Rode and Lars-Olof Nilsson*

**PRO 113**: MSSCE 2016—Concrete with Supplementary Cementitious Materials (2016), ISBN: 978-2-35158-178-0; e-ISBN: 978-2-35158-179-7; *Eds. Ole Mejlhede Jensen, Konstantin Kovler and Nele De Belie*

**PRO 114**: MSSCE 2016—Frost Action in Concrete (2016), ISBN: 978-2-35158-182-7; e-ISBN: 978-2-35158-183-4; *Eds. Marianne Tange Hasholt, Katja Fridh and R. Doug Hooton*

**PRO 115**: MSSCE 2016—Fresh Concrete (2016), ISBN: 978-2-35158-184-1; e-ISBN: 978-2-35158-185-8; *Eds. Lars N. Thrane, Claus Pade, Oldrich Svec and Nicolas Roussel*

**PRO 116**: BEFIB 2016–9th RILEM International Symposium on Fiber Reinforced Concrete (2016), ISBN: 978-2-35158-187-2; e-ISBN: 978-2-35158-186-5; *Eds. N. Banthia, M. di Prisco and S. Soleimani-Dashtaki*

**PRO 117**: 3rd International RILEM Conference on Microstructure Related Durability of Cementitious Composites (2016), ISBN: 978-2-35158-188-9; e-ISBN: 978-2-35158-189-6; *Eds. Changwen Miao, Wei Sun, Jiaping Liu, Huisu Chen, Guang Ye and Klaas van Breugel*

**PRO 118 (4 volumes)**: International Conference on Advances in Construction Materials and Systems (2017), ISBN Set: 978-2-35158-190-2, Vol. 1: 978-2-35158-193-3, Vol. 2: 978-2-35158-194-0, Vol. 3: ISBN:978-2-35158-195-7, Vol. 4: ISBN:978-2-35158-196-4; e-ISBN: 978-2-35158-191-9; *Ed. Manu Santhanam*

**PRO 119 (online version)**: ICBBM 2017—Second International RILEM Conference on Bio-based Building Materials, (2017), e-ISBN: 978-2-35158-192-6; *Ed. Sofiane Amziane*

**PRO 120**: 2nd International RILEM/COST Conference on Early Age Cracking and Serviceability in Cement-based Materials and Structures (EAC-02), 2017, ISBN: 978-2-35158-197-1; e-ISBN: 978-2-35158-198-8; *Eds. Dimitrios Aggelis and Stéphanie Staquet*

**PRO 121**: SynerCrete18, Interdisciplinary Approaches for Cement-based Materials and Structural Concrete: Synergizing Expertise and Bridging Scales of Space and Time, (2018), ISBN: 978-2-35158-202-2; e-ISBN: 978-2-35158-203-9; *Eds. Miguel Azenha, Dirk Schlicke, Farid Benboudjema, Agnieszka Knoppik*

## RILEM REPORTS (REP)

**Report 19**: Considerations for Use in Managing the Aging of Nuclear Power Plant Concrete Structures (ISBN: 2-912143-07-1); *Ed. D. J. Naus*

**Report 20**: Engineering and Transport Properties of the Interfacial Transition Zone in Cementitious Composites (ISBN: 2-912143-08-X); *Eds. M. G. Alexander, G. Arliguie, G. Ballivy, A. Bentur and J. Marchand*

**Report 21**: Durability of Building Sealants (ISBN: 2-912143-12-8); *Ed. A. T. Wolf*

**Report 22**: Sustainable Raw Materials—Construction and Demolition Waste (ISBN: 2-912143-17-9); *Eds. C. F. Hendriks and H. S. Pietersen*

**Report 23**: Self-Compacting Concrete state-of-the-art report (ISBN: 2-912143-23-3); *Eds. Å. Skarendahl and Ö. Petersson*

**Report 24**: Workability and Rheology of Fresh Concrete: Compendium of Tests (ISBN: 2-912143-32-2); *Eds. P. J. M. Bartos, M. Sonebi and A. K. Tamimi*

**Report 25**: Early Age Cracking in Cementitious Systems (ISBN: 2-912143-33-0); *Ed. A. Bentur*

**Report 26**: Towards Sustainable Roofing (Joint Committee CIB/RILEM) (CD 07) (e-ISBN 978-2-912143-65-5); *Eds. Thomas W. Hutchinson and Keith Roberts*

**Report 27**: Condition Assessment of Roofs (Joint Committee CIB/RILEM) (CD 08) (e-ISBN 978-2-912143-66-2); *Ed. CIB W 83/RILEM TC166-RMS*

**Report 28**: Final report of RILEM TC 167-COM 'Characterisation of Old Mortars with Respect to Their Repair (ISBN: 978-2-912143-56-3); *Eds. C. Groot, G. Ashall and J. Hughes*

**Report 29**: Pavement Performance Prediction and Evaluation (PPPE): Interlaboratory Tests (e-ISBN: 2-912143-68-3); *Eds. M. Partl and H. Piber*

**Report 30**: Final Report of RILEM TC 198-URM 'Use of Recycled Materials' (ISBN: 2-912143-82-9; e-ISBN: 2-912143-69-1); *Eds. Ch. F. Hendriks, G. M. T. Janssen and E. Vázquez*

**Report 31**: Final Report of RILEM TC 185-ATC 'Advanced testing of cement-based materials during setting and hardening' (ISBN: 2-912143-81-0; e-ISBN: 2-912143-70-5); *Eds. H. W. Reinhardt and C. U. Grosse*

**Report 32**: Probabilistic Assessment of Existing Structures. A JCSS publication (ISBN 2-912143-24-1); *Ed. D. Diamantidis*

**Report 33**: State-of-the-Art Report of RILEM Technical Committee TC 184-IFE 'Industrial Floors' (ISBN 2-35158-006-0); *Ed. P. Seidler*

**Report 34**: Report of RILEM Technical Committee TC 147-FMB 'Fracture mechanics applications to anchorage and bond' Tension of Reinforced Concrete Prisms—Round Robin Analysis and Tests on Bond (e-ISBN 2-912143-91-8); *Eds. L. Elfgren and K. Noghabai*

**Report 35**: Final Report of RILEM Technical Committee TC 188-CSC 'Casting of Self Compacting Concrete' (ISBN 2-35158-001-X; e-ISBN: 2-912143-98-5); *Eds. Å. Skarendahl and P. Billberg*

**Report 36**: State-of-the-Art Report of RILEM Technical Committee TC 201-TRC 'Textile Reinforced Concrete' (ISBN 2-912143-99-3); *Ed. W. Brameshuber*

**Report 37**: State-of-the-Art Report of RILEM Technical Committee TC 192-ECM 'Environment-conscious construction materials and systems' (ISBN: 978-2-35158-053-0); *Eds. N. Kashino, D. Van Gemert and K. Imamoto*

**Report 38**: State-of-the-Art Report of RILEM Technical Committee TC 205-DSC 'Durability of Self-Compacting Concrete' (ISBN: 978-2-35158-048-6); *Eds. G. De Schutter and K. Audenaert*

**Report 39**: Final Report of RILEM Technical Committee TC 187-SOC 'Experimental determination of the stress-crack opening curve for concrete in tension' (ISBN 978-2-35158-049-3); *Ed. J. Planas*

**Report 40**: State-of-the-Art Report of RILEM Technical Committee TC 189-NEC 'Non-Destructive Evaluation of the Penetrability and Thickness of the Concrete Cover' (ISBN 978-2-35158-054-7); *Eds. R. Torrent and L. Fernández Luco*

**Report 41**: State-of-the-Art Report of RILEM Technical Committee TC 196-ICC 'Internal Curing of Concrete' (ISBN 978-2-35158-009-7); *Eds. K. Kovler and O. M. Jensen*

**Report 42**: 'Acoustic Emission and Related Non-destructive Evaluation Techniques for Crack Detection and Damage Evaluation in Concrete'—Final Report of RILEM Technical Committee 212-ACD (e-ISBN: 978-2-35158-100-1); *Ed. M. Ohtsu*

**Report 45**: Repair Mortars for Historic Masonry—State-of-the-Art Report of RILEM Technical Committee TC 203-RHM (e-ISBN: 978-2-35158-163-6); *Eds. Paul Maurenbrecher and Caspar Groot*

**Report 46**: Surface delamination of concrete industrial floors and other durability related aspects guide—Report of RILEM Technical Committee TC 268-SIF (e-ISBN: 978-2-35158-201-5); *Ed. Valerie Pollet*

# Contents

# Chapter 1
# Introduction

**William G. Buttlar and Armelle Chabot**

## 1.1 Field and Accelerated Pavement Testing Studies

This section provides selected results from field observations and full-scale accelerated pavement tests involving cracking studies that have been conducted worldwide. The pictures, observations, and measurements from these studies help in 'qualifying' the engineering problem of asphalt and composite pavement cracking. Qualification is a crucial first step in understanding the pavement cracking problem, and provides motivation and guidance in the subsequent activities of lab testing and crack modeling.

### *1.1.1 French Motorways*

In the 2010s, IFSTTAR has been involved in a detailed survey of 16 old French toll motorway sections (from 2 to about 20 km long), in order to identify their failure mechanisms and compare their real behaviour with French design hypotheses [1, 2]. The study concerned thick bituminous and composite pavements, built more than 20 years ago, from different parts of France (except from the south). This section summarizes the main outcomes found for two types of French pavement roads.

W. G. Buttlar (✉)
University of Missouri, Columbia, USA
e-mail: buttlarw@missouri.edu

A. Chabot
Department of materials and Structures (MAST), IFSTTAR, Nantes, France
e-mail: armelle.chabot@ifsttar.fr



W. G. Buttlar et al. (eds.), *Mechanisms of Cracking and Debonding in Asphalt and Composite Pavements*, RILEM State-of-the-Art Reports 28,
https://doi.org/10.1007/978-3-319-76849-6_1

#### 1.1.1.1 Thick Bituminous Pavements

The survey included 11 thick bituminous pavements, with thicknesses of bituminous layers varying between 25 and 40 cm. The study showed that only two of the 11 bituminous pavement presented typical fatigue cracking, starting at the base of the bound layers. These two pavements had both received a level of traffic exceeding largely their design life. Two sections were affected by thermal cracking, associated with high aging of the upper pavement layers. These pavements were both located in mountainous regions, with more severe climatic conditions (cold winters). The main deterioration mode (5 sections) was debonding, which could affect surface layers (binder course), or deeper layers, which were in fact old wearing courses, which had been overlayed by several new layers. This debonding was mainly identified by coring, but could also be detected by analysis of deflection measurements. Sections with debonding presented consistently higher deflections, and especially higher values of radius of curvature than sections with well-bonded layers. These sections were analysed using the French pavement design method, to try to estimate their remaining life, and these calculations confirmed that debonding can reduce very significantly the pavement life, and in some cases lead to very high tensile stresses at the bottom of the debonded layers, leading to rapid deterioration of these upper layers (Fig. 1.1).

**Fig. 1.1** Examples of cores from thick bituminous pavements, showing debonding of surface or internal layers

#### 1.1.1.2 Semi Rigid Pavements

The 5 remaining sections of the study were semi rigid sections, including cement bound base or subbase, and covered by a variable thickness of bituminous materials. These sections were all more than 25 years old. The study showed that on these sections, the main mode of deterioration is reflective cracking, propagating upward form the cement treated layers. When a significant level of deterioration was attained, a debonding between the cement treated and bituminous layers was also observed on most sections (Fig. 1.2). Four of the 5 semi rigid sections evaluated in the study required an important rehabilitation, after 20–25 years, consisting in removing totally the old bituminous layers placed on top of the hydraulic treated base, and replacing them by new layers. However, the investigations made during these rehabilitation works indicated that, on all 4 sections, the cement treated base was still in very good condition, and presented no fatigue deterioration.

One of the conclusions that this study was the important role played by layer debonding in the deterioration of heavy traffic pavements (both bituminous and

**Fig. 1.2** Example of core from semi-rigid pavement, showing typical debonding between bituminous and cement-treated layers

semi-rigid), and the need for improved non-destructive testing (NDT) methods to evaluate the extent and severity of such debonding. Some studies have been initiated at IFSTTAR along this direction, which have shown that ground penetrating radar (GPR) and mechanical wave propagation methods [3] as well as the use of new sensors [4] can provide accurate information on the bond conditions between layers. More realistic models, taking into account variable interface conditions, influence of climatic conditions and real distribution of traffic also need to be developed.

### *1.1.2 United States Roadways*

The predominant form of cracking on US roadways varies by geographic and climatic region. Cracks in asphalt pavements are expensive and difficult to properly treat, which motivates studies to understand, prevent and treat crack-related distress forms. In southern states where temperatures are warmer, fatigue cracking, as shown in Fig. 1.3 (including top-down cracking) and debonding are of primary concern. Block cracking is also sometimes noted in mid-continental and/or desert locations, where aged surfaces and large temperature swings from day to night are present. Reflective cracking is also prevalent in composite pavement structures (such as jointed Portland cement concrete pavements with asphalt overlays). Reflective, block, and slippage cracking also occur in northern, colder climates, as shown in , 1.4, 1.5, 1.6.

**Fig. 1.3** Fatigue cracking in asphalt pavement in Savoy, Illinois. *Photo* W. Buttlar

**Fig. 1.4** Reflective cracking in asphalt pavement in Chicago, Illinois. *Photo* W. Buttlar

**Fig. 1.5** Debonding in asphalt pavement in Chicago, Illinois. *Photo* W. Buttlar

**Fig. 1.6** Block cracking in asphalt pavement in Urbana, Illinois. *Photo* W. Buttlar

In cold climates, thermal cracking is a prevalent cause of asphalt pavement premature deterioration. It is especially problematic when this distress form develops early in the pavement life, as it contributes to a significant loss of pavement serviceability long before the fatigue-based design life has been reached. A typical thermal cracking pattern occurring in asphalt pavements is shown in Fig. 1.7. If left untreated, thermal cracks will continue to deteriorate and widen over time, allowing moisture to readily infiltrate the pavement system. Detrimental effects of low-temperature cracking have motivated a number of studies in an effort to experimentally design and control asphalt properties related to the low temperature performance of asphalt pavements. Two major Pooled Fund studies on low-temperature cracking were conducted in the past decade, which provide new research results and a comprehensive summary of past research on this subject [5, 6]. These studies recognized the need to better understand and predict thermal, block, and reflective cracking mechanisms via discrete fracture testing and modelling approaches, which led to fracture-mechanics inspired tests and models. For instance, a disk-shaped compact tension test (DC(T)), which is now specified in ASTM D7313, and a cohesive zone fracture model were implemented in a thermal cracking specification, involving fracture energy requirements and an optional finite element based thermal cracking simulation model, called ILLI-TC.

**Fig. 1.7** Thermal Cracking in Asphalt Pavement in Savoy, Illinois. *Photo* W. Buttlar

## *1.1.3 Other Pavement Configurations*

### 1.1.3.1 Example of Typical UTW Degradations for Urban Use

Due to the structural heterogeneity of multilayer systems, debonding between pavement layers can occur during its service life. This phenomenon happens especially near the edges or vertical cracks through one layer as observed for instance during an accelerated loading test made on a specific "composite" pavement composed of a cement concrete overlay on a bituminous material layer [7] or on ultra-thin white topping pavements that is often used on French urban roads such as in Nantes city bus stop areas (Fig. 1.8). Due to existing joints of the cement

**Fig. 1.8** Corner crack of UTW in Nantes, France. *Photo* A. Chabot

concrete overlay, combined to environmental conditions and heavy loads, cracks appear quite rapidly in such pavements [8].

In order to investigate the mechanisms of debonding of bilayer material interfaces, concrete-asphalt-interface, several bending tests emanating from Ph.D. studies, have been developed at IFSTTAR [9, 10]. These tests allow the generation of interface failure in mixed mode (mode I and II), which is described later in Chaps. 3 and 4.

#### 1.1.3.2 Example of Polymer Modified Stress Absorbing Membrane Interlayer Trails

A demonstration installation of a polymer modified binder as a Stress Absorbing Membrane Interlayer (SAMI) was constructed on a 200 m section of the A9 at Cairnleith Moss, north of Bankfoot, near Perth. The object of the demonstration project was to investigate the improvement of the bond between asphalt and concrete layers and also to prevent, or at least reduce reflective cracking within the asphalt layers. Constructed between 1975 and 1985, the existing pavement consists of approximately 135 mm of asphalt over 180 mm of lean mix cement bound material (CBM).

The transverse cracking found during the original investigation was attributed to either normal shrinkage or thermally induced cracking within the CBM. The coring report, carried out for the original site investigation, shows that the pavement is between 310 and 356 mm thick and is in a generally good structural condition. All cores show delamination at depths between 86 and 140 mm, while in situ testing showed the pavement foundation to be strong. The proposed treatment was to place a 100 mm inlay, consisting of 60 mm of AC 20 dense binder 40/60 and 40 mm of thin surfacing, through 472 m of the site. Also proposed was a 200 m section of a 140 mm inlay, consisting of 100 mm of AC 20 Dense Binder 40/60 with a 40 mm thin surfacing with a Stress Absorbing Membrane Interlayer (SAMI) on both sides of the carriageway and a conventional bond coat on another 200 m section to serve as a control section. The results from Ooms (that used the Leutner shear test described in SHW clause 954 [11] and Transerv labs had conflicting conclusions regarding the use of tack coat under the bond coat. The effect of different thicknesses and types of tack coats and bond coats including polymer modified bitumen products on bonding of lean concrete and asphalt layers is not clear due to the variation in site conditions, cores taken and testing performed [12]. A second example of a SAMI, tested via accelerated pavement testing in the US, is provided later in this chapter.

## 1.1.4 Accelerated Pavement Cracking Studies Examples

Full scale observation of damage and failure mechanisms of pavements due to traffic is essential for the improvement of road materials and construction methods, as well as for the development and the validation of modelling tools and pavement design methods. As an alternative to the monitoring of in-service roads over long periods of time, accelerated pavement tests (APT) have been used. The aim of this section is to provide an overview of experiments that have contributed to the understanding of crack initiation and propagation under moving loads and varying climatic conditions.

### 1.1.4.1 IFSTTAR Cracking and Debonding APT Examples

The studies presented in this section have been performed at the French IFSTTAR (ex. LCPC) facilities, including the pavement fatigue Carrousel [13] and the 'FABAC' machines [14]. Observations reported herein deal primarily with publications written during the time frame of the RILEM TC-CAP committee and specifically for the RILEM CP2008 conference in Chicago. This includes deterioration mechanisms observed in APT experiments [15], in situ performance of a pavement interface between a thin cement concrete layer and an asphalt layer [7] and observations of reflective cracking over a new asphalt overlay [16]. As these experiments lead to similar deterioration mechanisms as on real pavements, these works may contribute to the understanding of where cracks travel in a pavement under moving loads and varying climatic conditions. In particular, the APT results obtained demonstrate the necessity to account for the crack initiation and propagation phenomenon in the bulk asphalt material as well as at interface locations between adjacent layers. Specific mechanical information coming from different sensors of the reflective cracking study done by Pérez et al. [16] was particularly useful for validation of research modelling tools of the RILEM Technical committees dealing with Cracking and Debonding in Pavements.

### 1.1.4.2 Cracking Mechanism Observed on the IFSTTAR Pavement Fatigue Carousel

As of 2008, the LCPC organization had more than 20 years of experience in APT applied to the study of pavement deterioration and cracking. The pavement fatigue carrousel (Fig. 1.9) of IFSTTAR (ex. LCPC) is a large scale circular outdoor test facility, unique in Europe according to its size (120 m long) and loading capabilities (maximum loading speed 100 km/h, loading rate 1 million cycles per month). Contrary to most ALT equipment, it is able to test pavements up to failure in just a few weeks. The machine is comprised of a central drive motor and 4 arms that can be equipped with different wheel configurations. The circular test track can be

**Fig. 1.9** View of the LCPC fatigue test track [15]

divided into several different test sections, which can be loaded simultaneously. The width of the test track (6 m) allows the application of traffic loads on the same track at two different radii.

Hornych et al. [15] presents deterioration mechanisms observed in APT experiments on the LCPC fatigue carrousel for low traffic pavement structures with unbound granular bases. The main deterioration mechanism observed depends on the thickness of the bituminous surfacing (less than 6 cm, or between 6 and 12 cm). For pavements with very thin bituminous layers (less than 6 cm), the main deterioration mechanism is rutting due to permanent deformation of the unbound layers, where significant cracking appeared only after 10–15 mm of rutting and was mainly longitudinal in orientation (Fig. 1.10a). The experiments appeared to elicit similar deterioration mechanisms as observed on real pavements. For the pavements with thicker bituminous layers (6–12 cm), the main deterioration mode observed was cracking, with minor rutting (3–7 mm). In this case, the cracking was mainly transverse in orientation, whereas longitudinal cracking is generally observed on real pavements (Fig. 1.10b). This discrepancy may be attributable to the circular moving loads.

For the two types of structures, design calculations have been performed with the French design method [2]. For the very thin pavements, reasonable predictions of the service life are obtained, despite the simplifications of the model (subgrade rutting criterion only). For the thicker pavements, where bituminous layer fatigue is the design criterion, the design method leads to a large under-estimation of the observed pavement life, showing the limits of the conventional fatigue criterion. These results suggest the necessity to:

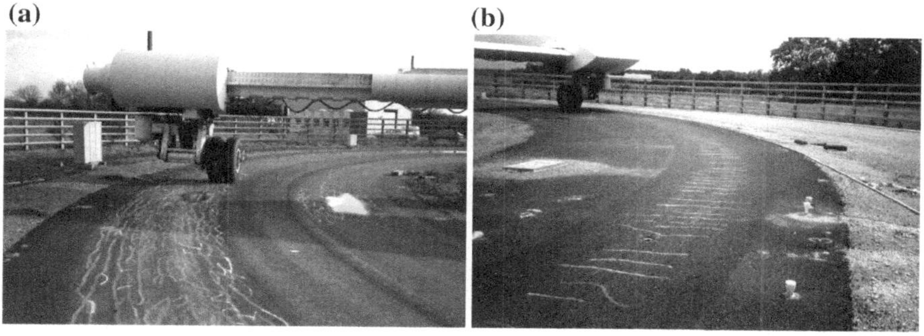


**Fig. 1.10** Typical cracks observed on the LCPC carrousel for flexible pavement with thin (**a**) and thick (**b**) asphalt layer [15]

- improve modelling accuracy of fatigue damage in pavements. Research at LCPC was carried out to develop a damage model for bituminous materials in pavement structures [17, 18], to model pavements with vertical discontinuities, and to understand crack propagation phenomena [19–21].
- study the limits of accelerated testing, which does not take into account aging of materials, climatic effects and variability of loads as they pertain to fatigue cracking mechanisms.
- develop specific sensors to investigate the development of cracking in pavements.

Chabot et al. [7] presented the results of an accelerated loading test to investigate the bond between a Portland cement concrete (PCC) overlay and asphalt layer, based on the Pouteau [9] Ph.D. dissertation study. With the support of the French cement industry, a pavement section similar to an ultra-thin white-topping structure was constructed (Fig. 1.11). Vertical joints of plain concrete were loaded up to one million cycles. The traffic loading was simulated by means of two IFSTTAR linear Accelerated Loading Facilities, called 'FABAC' devices. The FABAC devices simulate heavy traffic by applying a 6.5 ton, twin-axle repeated load. Each machine has four twin wheels, which are driven by a chain to apply the load across 2 m of the test track (Fig. 1.11b).

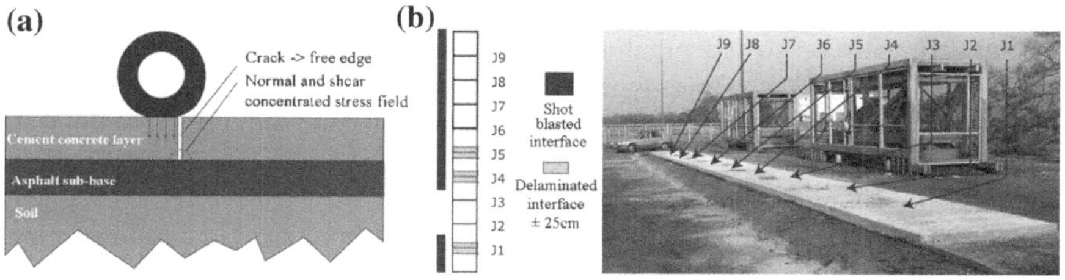


**Fig. 1.11** Location of stress singularities at the interface near a crack on a loaded composite pavement structure **a** composite pavement test section and the two FABAC devices (**b**) [9]

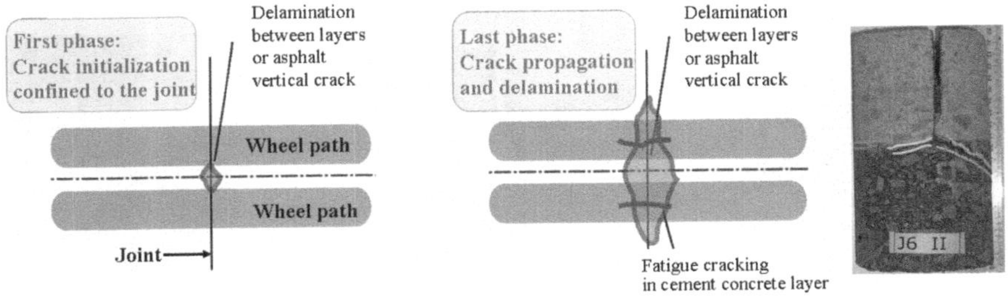


**Fig. 1.12** Damage scenario and core sample after 1 million loadings for joint number 6 [9]

After one million passes of the reference axle on different pavement joints, fatigue damage was observed. Cracking was also observed at the interface between PCC and asphalt layers. Transversal strain measurements at the bottom of the BC layer, non-destructive Colibri measurements combined with core sampling after the tests, make it possible to reconstitute a structural deterioration scenario [9, 22] (Fig. 1.12).

These tests demonstrated that pavement temperature and position of load relative to the joint are the most critical factors. For high temperature environmental conditions, it was also shown that a shot blasted interface treatment significantly delayed the delamination between the PCC overlay and asphalt layer. During lab testing of an associated bi-material specimen (at lower temperatures), the macro-texture of the top asphalt surface was also shown to significantly deter the reflective cracking rate [23].

Towards developing a simple delamination criteria for use in routine design, research was proposed to gain a deeper fundamental understanding of the mechanisms behind this deterioration mode through modeling [19–21] and laboratory static and fatigue tests on bi-material specimens [9]. Connected to reflective cracking ALT experiments conducted by Pérez [24] and Florence et al. [25], these studies led to the development of laboratory equipment to realistically evaluate crack propagation patterns observed in the ALT experiments. These investigations were continued at IFSTTAR in Ph.D. studies by Hun [10, 26–28] (see also details in Chap. 3) and Nasser [29–31].

In the domain of pavement reinforcement, one of the major problems is the frequent incidence of reflective cracking in new asphalt overlays. The full-scale test track experiments presented previously [9] was tested using the aforementioned Fabac devices [16, 32]. Three types of surface layers used in the former ALT experiments were investigated (Fig. 1.13). Experimental and numerical results were compared in order to better understand the reflective cracking process (debonding, damage and cracking) in actual pavements.

In this test section structure, it was possible to record and to visualize cinematically, the reflective cracking propagation, using very simple instrumentation by optimizing its position on the pavement structure. The deflections and the displacements have been measured at several levels of the structure, as well as the

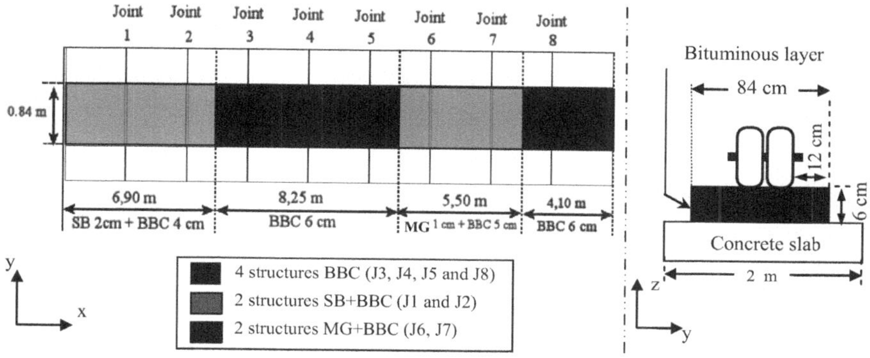


**Fig. 1.13** Position of the different surface layers on the tested track [16]

temperature and the lateral edge cracking, which were monitored by cracking sensors on both sides of the track. The resulting data presented new insights regarding the strain fields occurring on full-scale pavements in the vicinity of reflective cracking sites. With this data the initial strain and stress field could be deduced. The impact of bonding interface properties in the reflective cracking evolution were then confirmed using the FEM code of calculation CESAR and its contact module (TACT) simulating the structure test pavement including the bonding interface properties [16, 33]. The calculations results match particularly well to the strain data obtained from the test (see Fig. 1.14 for instance).

Preliminary results on two joints were used to determine the mechanisms of rupture of the structure (Fig. 1.15). Four more tests with improved sensors have completed the information.

Furthermore, theoretical types of cracking evolutions depending on bonding interface conditions were also confirmed by crossing experimental failure data (test recorded data and failure observation conclusions) and strain field calculation data.

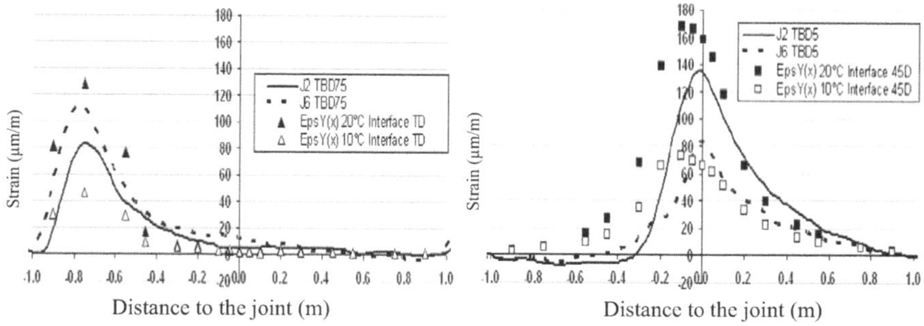


**Fig. 1.14** Comparison between an APT strain gage acquisition data placed 75 and 5 cm before the joint for one passage of the load (Temperature 17.5 °C) and CESAR simulation results for 10 and 20 °C [16]

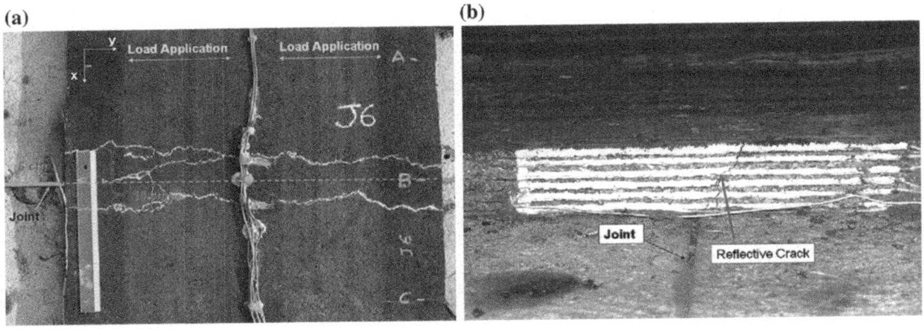


**Fig. 1.15** **a** Top view of double reflective cracking over the tested joint no [33]; **b** lateral view of single reflective cracking over the tested joint no 5 [33]

Based on this work, it was possible to develop a schema illustrating mechanisms behind reflective crack propagation (Fig. 1.16). Namely when single or double reflective cracking occurs, the specific causes are detected and their relationship is coherent in the experiment and the numerical simulations.

Among the three papers [7, 15, 16] presented at the Sixth International RILEM Conference on Cracking in Pavements at Chicago in 2008, experiments lead to similar deterioration mechanisms as on real pavements. The advantage to be able to put the instrumentation very close to the loaded road is an advantage only given by the APT facility and it would not be feasible to install in a real pavement. These studies related in this chapter here show the necessity to study the cracks initiation and propagation phenomenon in asphalt material as well as in interface between layers. The last one, from research work [16], offers interesting detail measurements for the validation of pavement modelling tool if one is able to simulate properly crack initiation and propagation at the interface between two layers as well as the asphalt material.

#### 1.1.4.3 Evaluation of Interface Behaviour During APT Test

Following the studies showing the importance of debonding in the deterioration of old pavements, since 2011, investigations have been carried out on the IFSTTAR Accelerated pavement testing facility, with bituminous sections including artificial discontinuities, (internal cracks and debonding between bituminous layers). These sections are used to evaluate different non-destructive methods, for detection of debonding: the GPR, and a mechanical impact method [3, 35] (see Chap. 4). In addition, in 2012, a full scale experiment was also conducted on the IFSTTAR APT facility, to evaluate different systems of instrumentation, for the evaluation of strains in upper pavement layers, and at their interfaces [36]. As part of this study, a system for local strain measurements, using optical fiber sensors was evaluated on the IFSTTAR fatigue carrousel. This system uses plates equipped with small fiber optic sensors, placed at different levels horizontally and vertically, used to measure

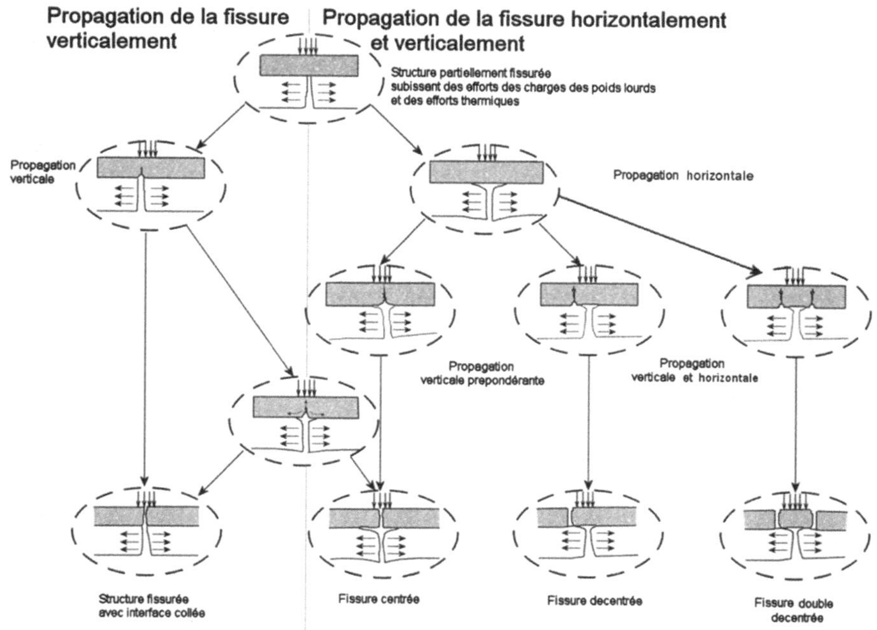


**Fig. 1.16** Reflective crack propagation scenario [24] after [34]

horizontal and vertical strains in the asphalt layers. All the results presented in these different papers show how important is to study the mechanisms of debonding between layers that mostly affect the durability of pavements.

#### 1.1.4.4 Information Coming from Fiber Grid Use for Pavement Reinforcement

A study has also been conducted recently on the IFSTTAR APT facility [37, 38], to evaluate the effect of a fiberglass grid on the resistance to fatigue and cracking of a low traffic pavement. The tests were carried out on typical French low traffic pavement structures consisting of an 80 mm thick bituminous wearing course, over a granular subbase (300 mm thick), and a sandy subgrade soil, with a bearing capacity of about 95 MPa. Two structures were compared, one without grid, and one reinforced with a high strength fiber-glass grid. The grid tested was an open fiberglass grid coated with a patent-pending elastomeric polymer and self-adhesive glue (Tensile Strength: 100 kN/m × 100 kN/m). The mesh of the grid was (25 mm$^2$ × 25 mm$^2$). The grid was placed in the lower part of the bituminous layer, 2 cm above the interface with the subbase. Each experimental pavement structure was about 10 m long and 3 m wide. The pavement structures were built on the existing subgrade of the test track, which is a sand with 10% fines, sensitive to water. The granular base consisted of 30 cm of 0/31.5 mm unbound granular material (UGM). A 2 cm thick bituminous layer was first laid and compacted on the 2 sections, with and without reinforcement. Then a tack coat with 300 g/m$^2$ of residual bitumen was applied. The final 6 cm thick bituminous layer was laid on the 2 sections, and compacted with a rubber-tired roller. The loading program started in April 2011. Until September 2011, 1 million 65 kN loads (standard French equivalent axle load) were applied. Between January 2012 and March 2012, 200,000 additional loads at 70 kN were applied. Measurements made during the experiment included deflection measurements, rut depth measurements, crack monitoring, and measurement of strains and temperatures in pavement layers by internal sensors. Concerning deflection and rutting, no significant difference was observed between the behavior of the reinforced and unreinforced pavements, throughout the experiment. After 1,200,000 cycles, the final average rut depths were 14.6 mm for the reinforced section and 16 mm for the unreinforced section. Therefore, the grid did not seem to improve the resistance to rutting of the pavement.

Cracking was monitored approximately every 100,000 load cycles. Crack percentage was determined by the ratio between the length of pavement with cracks and the initial length. The first cracks were observed on the unreinforced section after 800,000 load cycles. Until 1.2 million of loads, the extent of cracking increased, reaching 70% at the end of the test. On this section, very fine isolated transverse cracks appeared first. Then, under traffic, these cracks started to open, and crushed fines started to come out due to a pumping effect. Other thin transverse cracks developed nearby. The transverse orientation of the cracks is typical of fatigue cracking observed on the carrousel, for pavements with relatively thin bituminous layers (about 8 cm) [15]. Figure 1.17a presents the evolution of the extent of cracking, as a function of the level of traffic. It corresponds to the percentage of the length of the pavement affected by cracks (for a transverse crack, the affected length is considered, arbitrarily, to be 500 mm). On the grid-reinforced section, no crack was observed until the end of the test. A view of the two sections at the end of the experiment is shown on Fig. 1.17b and c. In conclusion, the test confirmed that the grid improved very significantly the resistance to cracking of the pavement, and therefore its design life.

A second experiment on pavements reinforced by geogrids was performed at IFSTTAR using the FABAC traffic simulators presented previously [39]. The objective was to study the behaviour of flexible pavements (8 cm thick bituminous wearing course, unbound granular base) reinforced by two types of Tensar geogrids, AR-G (rectangular mesh) and Tri-Ax (triangular mesh), and to compare their performance with that of a reference non-reinforced pavement. About 500,000 loads were applied on each structure, using dual wheels, loaded at 65 kN. The tests were performed in two phases: The first phase consisted in testing a non-reinforced reference section (P1) and a section reinforced by a tri-Ax grid (P3). Concerning the resistance to rutting, section P3 presented a better performance, with a maximum rut depth of 10 mm after 200,000 load cycles, compared with 21 mm for the non-reinforced section (P1). Concerning the resistance to cracking, the behaviour of

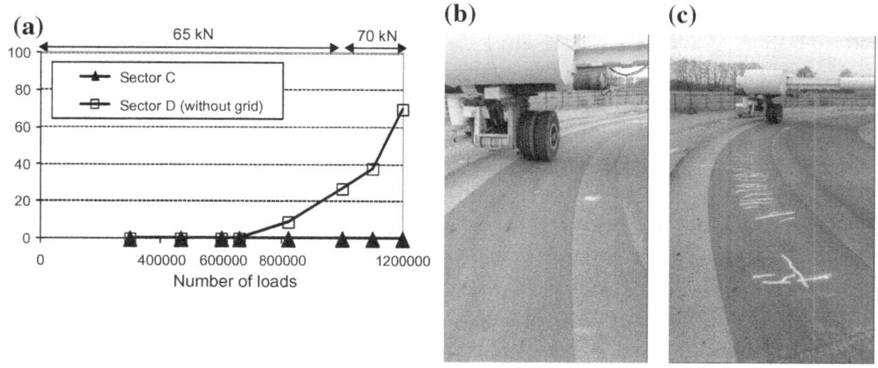


**Fig. 1.17** **a** Extent of cracking, in percent, on the grid reinforced and unreinforced sections view of reinforced section (**b**) and unreinforced section (**c**) at end of experiment [37]

the two sections was similar. Figure 1.18 shows the cracking obtained on the two sections. Due to the channelized traffic applied by the FABAC machines, the cracks developed longitudinally, at the edges of the wheel paths.

The two sections presented similar crack patterns, consisting of longitudinal cracks adjacent to the trafficked area. The first cracks appeared on the two sections after 220,000 and 180,000 cycles respectively. Investigations made at the end of the tests indicated that these cracks started from the surface of the bituminous layer, which can explain why the grids, placed at the bottom of the bituminous layer, did not slow down the development of these cracks. In the second phase, two tests sections, reinforced respectively with a tri-Ax grid (section P2) and a rectangular AR-G grid (section P4), were tested. These two sections were tested after 2 months of rainy weather, and they presented approximately 20% higher deflection values

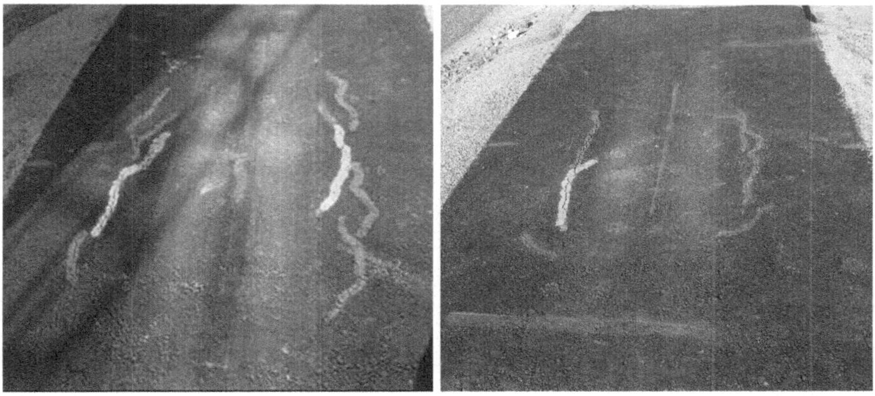

**Fig. 1.18** Final condition of structure P1 without grid (left) and structure P3 with Tri-Ax grid (right) [39]

than the first two sections, due to a decrease of the bearing capacity of the subgrade and unbound granular layers. This loss of bearing capacity can probably be explained by the infiltration of water in the structures, facilitated by the use of a relatively permeable road base asphalt layer. In these conditions, section P2 (tri-Ax grid), deteriorated relatively rapidly: After 200,000 load cycles, the rut depth attained 27 mm on this section, and the first cracks appeared after 95,000 cycles. By comparison, section P4 presented a significantly better performance, with 17 mm of rut depth after 200,000 load cycles, and the observation of the first cracks after 200,000 cycles. The crack patterns were similar to those observed in the first phase of tests, namely longitudinal cracks located near the edge of the trafficked area (see Fig. 1.19). The final investigations also indicated that the bond between the grid and the bituminous material was better for the rectangular AR-G grid. The investigations made at the end of the tests also indicated, on the reinforced sections, a deterioration of the interface between the granular material and the road base asphalt layer in the trafficked area, leading to a debonding of the grids.

#### 1.1.4.5 UIUC Reflective Cracking APT Interlayer Study

A study was conducted at the University of Illinois at Urbana-Champaign (UIUC) to investigate the mechanisms of reflective cracking and the mitigation thereof using single- and multi-layered reflective crack relief interlayer (RCRI) systems. The work was sponsored by the National Science Foundation and an industry partner through the GOALI program (Grant Opportunity for Academic Liaison with Industry) [40]. Testing was carried out using the Accelerated Loading and Testing Assembly (ATLAS).

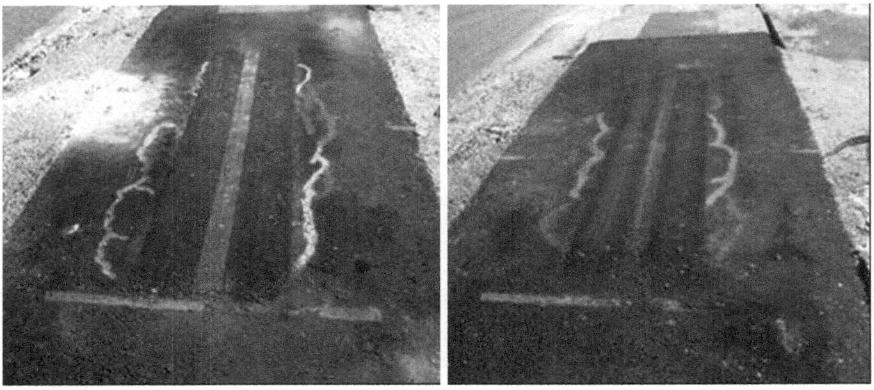

**Fig. 1.19** Final condition of structure P2 with Tri-Ax grid (left) and section P4 with AR-G grid (right) [39]

#### 1.1.4.6 Accelerated Pavement Test Loading System

The ATLAS device, as shown in Fig. 1.20, was used to apply 65,300 passes of varying load magnitudes to the pavement test lane during the winter of 2007–2008 at ATREL. This system is capable of applying moving wheel loads up to 80 kips (350 kN) using either highway or aircraft tires in single or dual tire configurations. The system can run tire loadings up to a speed of 15 mph (25 kph) and apply uni- or multi- directional loading. ATLAS operates year round and can be moved around the Advanced Transportation Research and Engineering Laboratory (ATREL) facility in Rantoul, IL using its crawler track supports. Finite element modeling using a cohesive zone fracture model and fracture energy results from the disk-shaped compact tension test were used to design several innovate, single- and multi-layer RCRI systems. More details regarding this integrated modeling and testing approach is described in Chap. 2.

#### 1.1.4.7 Cracking Mechanisms in Control Section

A load sequence starting with typical highway load levels of 9 kips (40 kN) and stepping upwards to a heavy overload of 30 kips (133 kN) was used in the accelerated reflective cracking study. Three doweled and four undoweled (full-depth saw cut) PCC joints were located in each of the test sections. Figure 1.21 shows the typical overlay cracking that was observed in Section-1

**Fig. 1.20** Accelerated Loading and Testing Assembly (ATLAS)

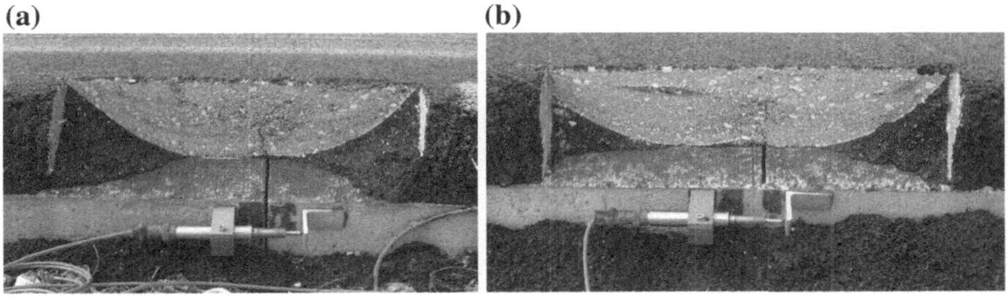

**Fig. 1.21** Section-1 typical overlay response to APT loading **a** Saw-Cut joint, **b** doweled joint

(untreated control section) after the completion of accelerated testing. At the edge of the pavement, the region of the overlay within 2 ft (0.6 m) of each PCC joint was saw-cut [4 in. (0.1 m) from edge of PCC] in a longitudinal direction to provide a planar face for observation of reflective crack formation and propagation. All seven joint locations in Section-1 exhibited reflective cracking in the overlay. Figure 1.21a shows a typical joint location for the full depth saw-cut joint and Fig. 1.21b shows a typical doweled joint location.

In the case of saw cut joints, the initial reflective crack developed in a vertical direction indicating mode-I dominance; however, the cracks started to deviate at an angle as the top of the overlay was reached, indicating a mixed-mode response. In general, the direction of crack deviation seemed to be governed by the type of slab faulting experienced. In the majority of cases, the slab faulted in such a manner that the trailing slab (i.e., the last slab loaded by the moving wheel, or 'down-traffic' slab) had a lower elevation than the leading or 'up-traffic' slab. In this case, the crack deviated in the direction of 'down-traffic', or to the right as shown in Fig. 1.22a. Faint surface reflective cracks also became visible in the spring following testing (Fig. 1.22a). The doweled PCC joints exhibited a vertical crack indicating a minimization of joint shearing which would be expected due to high load transfer efficiency. Significant interface debonding was observed to occur at a load level of around 16 kips (71 kN).

#### 1.1.4.8 Cracking Mechanisms in Single RCRI Overlay System

In the case of Section-2, at the end of the loading sequence, none of the joint locations exhibited visible macro-cracking. This section had a 1 in. (25 mm) thick SAMI layer (also known as an RCRI system) at the bottom of the 4 in. (101 mm) overlay system. The RCRI material was designed to have very high fracture energy, over 2000 J/m$^2$ at a test temperature of −12 °C. However, after a few months of loading, at two joint locations without dowels, the overlay started to exhibit visible macro-cracks. Both of these locations experienced significant joint faulting during

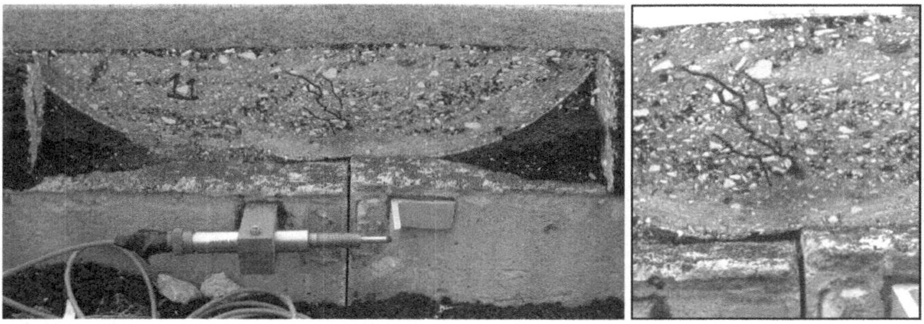

**Fig. 1.22** Section-2 crack jumping and slab faulting

the course of APT loading. The delayed formation of reflective cracks might have been attributed to PCC joint movement under thermal cycling. However, in both locations, the RCRI material did not crack. Rather, "crack jumping" was observed (Fig. 1.22). Notice that in these cases the cracks were manifested as distributed (smeared) as compared to the localized reflective cracks observed in the control section and with fewer crack branches. Similar to Section-1, interface debonding was observed during overloading sequences. This phenomenon combined with crack jumping and mixed mode (angled) cracking may explain why significant crack deviations and distributed, hairline surface cracks are often observed in RCRI treated pavements in the field, as opposed to straighter, more localized cracks directly over joints that are often observed in untreated overlay systems in the field.

#### 1.1.4.9 Cracking Mitigation in Double RCRI Overlay System

In Section-3, a double RCRI system was used. This consisted of the same RCRI system used in Section-2, also placed over the PCC pavement, topped with a second RCRI layer of similar thickness but with less fracture energy. This layer was designed to serve as an transitional RCRI layer, based on finite element simulation results, which showed less severe bending strain at this position of the overlay. The overlay did not crack at any of the seven joint locations in Section-3, even after the full overload sequence. One of the joints underwent faulting; however, unlike Section-2, no macro cracks formed into the overlay or the transitional layer (see Fig. 1.23). Thus, the dual-interlayer system appears to have achieved its objective of mitigating the crack jumping phenomenon, even under severe overloading and harsh environmental conditions.

**Fig. 1.23** Section-3 slab faulting and intact overlay system

#### 1.1.4.10 APT Conferences

Additional information on completed accelerated pavement studies can be found in the proceedings of the series of International Conferences on Accelerated Pavement Testing, including:

- 1st International Conference on Accelerated Pavement Testing (Nevada—1999) https://sites.google.com/site/afd40web/apt-conferences/reno-nevada-october-18-20-1999
- 2nd International Conference on Accelerated Pavement Testing (Minnesota—2004) https://sites.google.com/site/afd40web/apt-conferences/test
- 3rd International Conference on Accelerated Pavement Testing (Spain—2008) https://sites.google.com/site/afd40web/apt-conferences/Madrid-2008
- 4th International Conference on Accelerated Pavement Testing (California—2012) http://www.ucprc.ucdavis.edu/apt2012/
- 5th International Conference on Accelerated Pavement Testing (Costa Rica—Sept. 19/21, 2016) http://www.lanamme.ucr.ac.cr/APT2016/
- 6th International Conference on Accelerated Pavement Testing (Nantes—2020) https://sites.google.com/site/afd40web/apt-conferences, http://apt2020.sciencesconf.org.

## 1.2 Studies with Integrated Fracture Testing and Modelling Approaches

This section presents a brief summary of previous studies conducted by current TC241-MCD committee members and their colleagues, where fracture tests and models have been integrated. The section starts with a summary of the work done under TC-CAP on the extension of verification and validation methods for asphalt concrete, leading to the QVCV approach (qualification, verification, calibration, and validation). A case is then made for the need to develop mechanics-based models in order to advance the field of pavement engineering, particularly with regards to mitigating cracking in asphalt pavements.

### *1.2.1 QVCV as a Rational Framework for Advancing Cracking Research*

Engineered systems by definition involve approximation of all key system inputs and characteristics, i.e., loads, climate/environment, geometry, interface/boundary conditions, material properties, etc. pavements in particular involve a number of poorly known inputs and stochastic variables (asphalt and other geomaterial layer properties, future traffic, future environmental conditions), and therefore, pavement models will at best be mechanistic-empirical and will often require numerical approaches rather than closed form solutions.

Due to the complexity of numerical modelling approaches and the inherent risks of a 'black box' solution, the field of verification and validation (a.k.a., V & V) has emerged in the engineering/mechanics community. Verification refers to the process of checking whether or not the computational solution provides the correct solution when compared against known solutions (i.e., did the model get the computations correct?). This is often done by checking against selected problems with known closed-form solutions, and sometimes checking against other numerical models when closed-form solutions do not exist. Validation refers to checking whether or not the model adequately captures the intended physical phenomena, possibility with or without the need for model calibration. More background on verification and validation studies in the field of mechanics are provided in the bibliography [41–58]. Another important step, and a great starting point for model development is model qualification. Model qualification involves the process of understanding the salient physical features of a system to be modelled and the selection of a model that can potentially capture those features. Figure 1.24 presents the classic V & V model of Oberkampf et al. [53].

In the pavement-engineering field, model calibration is almost always required. It is important to delineate between model calibration and model validation, and these points are often confused in pavement research investigations. Model calibration involves the tuning of model parameters to aid in fitting the model to experimentally obtained data (lab test or field). After model calibration, a necessary but insufficient check on model qualification and validation is obtained. In other words, it is necessary to achieve a good fit between a calibrated model and the data it intends to fit, but not sufficient to consider the model validated. It is only when the model is put to the test outside of the calibrated data set that it can be truly validated. Models that continue to fit data outside of the calibration set, and particularly when such validation, is likewise achieved by other researchers, can be

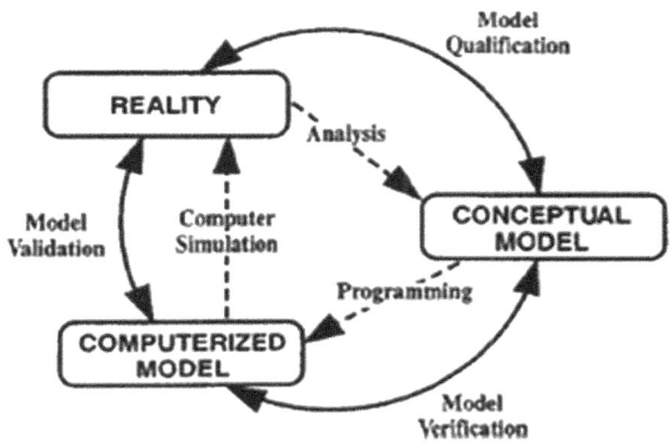


**Fig. 1.24** Classic V & V model [53]

considered as truly validated. Models that do not fit data outside of the calibration set must be used with caution (restricted to cases within the calibrated parameters). That notwithstanding, models that fail the validation stage provide considerable insight towards model qualification in subsequent iterations.

Thus, for pavement engineering, a variant on V&V was proposed to recognize the importance of model qualification and calibration; QVCV, or Qualification, Verification, Calibration, and Validation. Figure 1.25 illustrates the process of QVCV in the context of pavement engineering, as developed by the TG-02 group in TC-CAP. In this example, the four key steps involved the following activities:

- **Qualification**: Preliminary fracture tests revealed the general load-crack mouth opening behavior, suggesting the use of a cohesive zone model.
- **Verification**: Double-cantilever beam simulations were verified against closed-form solutions, to establish the veracity of the cohesive zone fracture modeling in the finite element model used.
- **Calibration**: Cohesive zone model parameters were inversely obtained by bringing simulation results into agreement with DC(T) test results. This was accomplished by trial-and-error approach. A rigorous optimization scheme has been subsequently developed [59].
- **Validation**: The calibrated model was used to simulate fracture in a notched beam specimen and results were compared to experiment. At this stage, the model is either considered as validated or another QVCV iteration can be started by re-qualifying the model with additional physical considerations/parameters. Iterating through the QVCV process can lead to improved physical understanding, better ability to accurately test, and to accurately model the material and structural phenomena of interest.

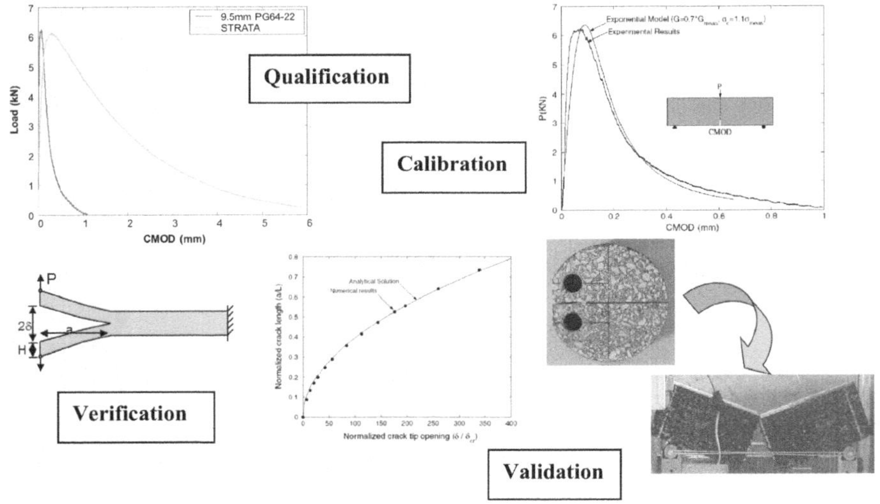


**Fig. 1.25** Example of proposed QVCV model for modelling of asphalt-based systems

### 1.2.2 Modeling and Testing Approaches Used in Past Research: Shortcomings of Empirical Approaches

Considerable empiricism was necessary in the design of pavements in the previous century, as testing and modelling capabilities were limited prior to the advent of the modern computer and modern sensors, test devices, and numerical modelling methods. Some examples of highly empirical approaches and their shortcomings are:

- Penetration test: needle penetration of asphalt binder provided a simple measure of asphalt binder consistency, but extension of test results beyond the penetration index value was hampered by the difficulty in assessing the actual stress and strain fields present in the test.
- Structural number and layer coefficient: The pavement design method resulting from the AASHO road test in the late 1950s in Ottawa, Illinois USA was based on the determination of a pavement structural number, based on an empirical layer coefficient multiplied by thickness and a drainage coefficient for each pavement layer [60]. The most recent version of the guide was release in 1998, but this version was based largely on the 1993 version of the design guide [60]. Because the approach does not consider actual stresses, strains, and deflections in the pavement system, and because empirical transfer functions are used in place of more physically and mechanistically appropriate deterioration (distress) models, the resulting design method was highly approximate. In addition, most design require extrapolation of the Ottawa, Illinois test pavement results to account for differences in subgrade soils, climate, and loading types, magnitudes, and repetitions.

In France, a mechanistic-empirical (ME) design method has been used for pavement design since 1980, and was published as an official design guide in 1994 [1, 2]. A very similar method has also been published recently for airfield pavements [61]. This method, similar to other design method used in Europe, is based on the following principles:

- Calculation of the response of the pavement to loading using the Burmister‘s linear elastic multi-layer model [62] (pavement design software ALIZE-LCPC) [63]
- Application of 3 main design criteria depending on the type of pavement system to be designed:
  - For bituminous materials: a fatigue criterion, based on the maximum tensile strain $\varepsilon_t$ at the bottom of the bituminous layers.
  - For cement-treated materials: a fatigue criterion, based on the maximum tensile stress $\sigma_t$ at the bottom of the cement treated layers.
  - For unbound granular materials and subgrades, a rutting criterion, based on the maximum vertical strain $\varepsilon_{zz}$ at the top of the unbound granular layers, or at the top of the subgrade.

This method can also be used for the design of reinforcing layers for existing pavements, which has been implemented recently in France [64]. The new approach, however, has several limitations:

- Only a relatively limited number of deterioration mechanisms are considered (fatigue, permanent deformation).
- Other important modes of deterioration such as layer partial debonding, or reflective cracking, and thermal cracking are not considered. Calculations can only be performed for completely bonded or sliding interfaces.

These limitations are particularly important when dealing with the evaluation of the residual life of existing pavements, or in the design of reinforcement layers; namely, existing cracks and existing deteriorated interfaces cannot be considered in the modelling scheme used. For cracked, existing pavement layers, the empirical work-around is to use a reduced value of elastic modulus for the layer. This is not an ideal approach, since the actual failure mechanisms (propagation of the existing cracks and discontinuities are not directly considered.

To address the shortcomings in the AASHTO 1993 guide of design of pavement structures, a major research program was undertaken under the National Cooperative Highway Research Program (NCHRP), project 1-37A, in the late 1990s and early 2000s [65]. The resulting guide was originally planned to be termed as the 'AASHTO 2002' or 'mechanistic-empirical pavement design guide' (MEPDG). However, delays in finalization of the guide and evolution of the implementation approach has resulted in the AASHTOWare Pavement ME design software system [66], which can now be downloaded from the AASHTOWare website. Due to improvements in material testing capabilities, modelling, and desktop computer processing speed, a comprehensive, simulation-based design program was developed. Some of the key advances made possible through the NCHRP 1-37A project and Pavement ME software in the area of flexible pavements are:

- Extremely detailed pavement loading information can be considered, leading to a load spectrum approach rather than an Equivalent Single Axle Load (ESAL) approach. For instance, various axle types, axle and dual tire spacing, wander patterns, seasonal and even hourly truck distributions, etc., can be considered.
- Detailed temperature, moisture, and freezing calculations are included in the design procedure.
- The empirical structural number and layer coefficient approach has been replaced by calibrated, mechanistic-empirical models, similar in concept to those used in the French ME design procedure. An asphalt concrete complex (dynamic) modulus master curve is a key input to the ME fatigue and rutting models used in the software.
- In addition to ME models for pavement fatigue and rutting, an ME thermal cracking model based primarily on hourly pavement temperatures, the creep compliance master curve, mixture tensile strength and mixture thermal coefficient was included, called TCModel [67–69].

Thus, the last 30 years has ushered in the widespread usage of more fundamental material tests and mechanics-based modelling in pavement design. However, a number of key shortcomings in modern, ME pavement design still exists, which has motivated recent research. These shortcomings include the inability to accurately capture the following effects: (1) capturing material embrittlement with time, and associated material gradients in all distress models; (2) capturing the amount of aging, or the aging rate, as a function of crude source, refining techniques, additives, type and amount of recycled materials used, climate, and volumetric characteristics of the mixture; (3) capturing loading time and sequence, damage, healing, complexity of interface morphology and associated mechanics, and bulk material nonlinearity; (4) capturing complex interface behavior such as debonding, sliding and variable contact between surfaces; (5) capturing the complexities of rehabilitated, composite pavement structures with cracks and other discontinuities; (6) accounting for healing effects, and; (7) capturing moisture effects on pavement distress. Obviously, many of these shortcomings are related to accurately predicting various cracking modes in new and rehabilitated pavements, such as thermal, block, reflective, interface and fatigue cracking.

## 1.3 Summary

Asphalt pavement is often an attractive option for new construction, and moreover, for pavement rehabilitation due to its ease and speed of constructability and for economic reasons. However, asphalt pavement longevity and serviceability can be lessened as a result of a number of different cracking forms that can develop. Avoiding or delaying cracks can be best achieved as a result of a thorough understanding of their underlying mechanisms. Cracking tests and models are useful tools in reaching this goal; however, the qualification stage is generally first required, involving field and/or experimental observations to identify the key physical parameters needed to be measured and modelled. Typical cracking types and conventional thinking on their underlying mechanisms were presented in this chapter. A brief review of cracking tests and models developed in recent RILEM committees on cracking in asphalt materials/pavements leading into the current RILEM TC241-MCD committee was also presented.

The following chapters provide recent or current tests, models, and field studies being developed by TC-241MCD members and their research teams. Chapter 2 focuses on cracking in 'the bulk material,' or within the asphalt concrete itself; Chap. 3 deals with cracking along material interfaces, and; Chap. 4 presents advanced testing methods evolving in the discipline of advanced pavement cracking studies.

## References


1. SETRA-LCPC (1997) French design manual for pavement structures. SETRA-LCPC, LCPC-IST Editions, Paris, 248p
2. Corte JF, Goux MT (1996) Design of pavement structures: the French technical guide. Transp Res Rep 1539:116–124
3. Simonin JM, Kerzreho JK, Hornych P, Gouy T (2013) Comparison of NDT methods to detect debonded interfaces. In: 9th international conference on the bearing capacity of roads railways and airfields, Trndheim, Norway
4. Grellet D, Doré G, Kerzrého JP, Piau JM, Chabot A, Hornych P (2012) Experimental and theoretical investigation of three dimensional strain occurring near the surface in asphalt concrete layers. In: Proceedings of the 7th rilem international conference on cracking in pavements, 20–22 June 2012, Delft, The Netherlands. RILEM Bookseries, vol 4, pp 1017–1027, https://doi.org/10.1007/978-94-007-4566-7_97, Print ISBN: 978-94-007-4565-0
5. Marasteanu M, Zofka A, Turos M, Li X, Velasquez R, Li X, Buttlar W, Paulino G, Braham A, Dave E, Ojo J, Bahia H, Williams C, Bausano J, Gallistel A, McGraw J (2007) Investigation of low temperature cracking in asphalt pavements national pooled fund study 776. Minnesota Department of Transportation, pp. 1–338
6. Marasteanu M, Moon KH, Teshale EZ, Falchetto AC, Turos M, Buttlar W, Dave E, Paulino G, Ahmed S, Leon S, Bahia H, Arshadi A, Tabatabaee H, Ojo J, Velasquez R, Mangiafico S, Williams C, Buss A (2012) Investigation of low temperature cracking in asphalt pavements national pooled fund study—phase II. Final Report, Minnesota Department of Transportation, St. Paul, MN
7. Chabot A, Pouteau B, Balay JM, De Larrard F (2008) FABAC accelerated loading test of bond between cement overlay and asphalt layers. In: Sixth international RILEM conference on cracking in pavements, 16–18 June, Chicago, US, pp 13–23. Taylor & Francis Group Proceedings, ISBN 978-0-415-47575-4
8. Vandenbossche J, Barman M, Mu F, Gatti K (2011) Development of design guide for thin and ultra-thin concrete overlays of existing asphalt pavements, task 1 report: compilation and review of existing performance data and information. Technical report, University of Pittsburgh, Department of Civil and Environmental Engineering, Swanson School of Engineering
9. Pouteau B (2004) Durabilité mécanique du collage blanc sur noir dans les chaussées. Ph.D. thesis of the Ecole Centrale de Nantes. http://media.lcpc.fr/ext/pdf/theses/rou/throu_pouteau.pdf
10. Hun M (2012) Influence de leau dans le décollement par flexion de bicouches de chaussée urbaine. Ph.D. thesis of ED SPIGA, ECN. http://hal.archives-ouvertes.fr/tel-00777011/
11. Leutner R (1979) Undersuchung des Schichtenverbundes beim bituminosen Oberbau. Bitumen Heft 3, ARBIT, Hamburg Germany
12. Sullivan C, Guthrie S, de Bondt A (2012) Polymer modified stress absorbing membrane interlayer trails, A9 Cairleith moss, Scotland. Rilem TC241-MCD Star contribution, 5 pages
13. Autret P, de Boissoudy AB, Gramsammer JC (1987) The circular test track of the Laboratoire Central des Ponts et Chaussées—first results. In: Proceeding of 6th international conference on structural design of asphalt pavements, Ann Arbor, 13–17 June 1997, vol 1, pp 550–561
14. Aunis J, Balay JM (1998) An applied research programme on continuous reinforced concrete pavements: the FABAC project. In: Proceedings of the 8th international symposium on concrete roads, Lisbon, 13–16 Sep
15. Hornych P, Kerzreho JP, Chabot A, Bodin D, Balay JM, Deloffre L (2008) The LCPCs ALT facility contribution to pavement cracking knowledge. In: Sixth international RILEM conference on cracking in pavements, 16–18 June, Chicago, US, pp 671–681. Taylor & Francis Group Proceedings, ISBN 978-0-415-47575-4

16. Pérez SA, Balay JM, Tamagny P, Chabot A, Petit C, Millien A, Wendling L (2008) Accelerated pavement cracking testing. Experimental and finite element analysis. In: Sixth international RILEM conference cracking in pavements, 16–18 June, Chicago, US, pp 55–65. Taylor & Francis Group Proceedings, ISBN 978-0-415-47575-4
17. Bodin D, Belzung F, de La Roche C, Chabot A (2004) Thermomechanical modelling for fatigue damage of bituminous materials. In: Proceedings of the fifth international RILEM conference cracking in pavements, 5–8 May, Limoges, pp 675–682. Rilem Proceedings, Print ISBN 2-912143-47-0
18. Bodin D, Pijaudier-Cabot G, de La Roche C, Piau JM, Chabot A (2004) Continuum damage approach to asphalt concrete fatigue modeling. J Eng Mech (ASCE) 130(6):700–708
19. Tran QD, Chabot A, Ehrlacher A, Tamagny P (2004) A simplified modelling for cracking in pavements. In: Rilem proceedings (ISBN 2-912143-47-0), 5th international RILEM conference cracking in pavements, 5–8 May, Limoges, pp 299–306
20. Chabot A, Tran QD, Ehrlacher A (2005) A simplified modeling for cracked pavements. Bulletin des Laboratoires des Ponts et chaussées, (258–259), pp 105–120
21. Chabot A, Tran QD, Ehrlacher A (2007) A modeling to understand where a vertical crack can propagate in pavements. In: Taylor & Francis Group Proceedings (CRC Press ISBN 10: 0415448824 ISBN 13: 9780415448826), international conference on advanced characterization of pavement and soil engineering materials, 20–22 June, Athens, Greece, vol 1, pp 431–440
22. Pouteau B, Chabot A, De Larrard F, Balay JM (mars) (2006) Mécanique des chaussées Béton sur grave-bitume, étude de la tenue du collage entre béton et enrobé sur chaussée expérimentale (1re partie). Revue Générale des Routes et des Aérodromes (RGRA), issue 847, pp 85–90
23. Pouteau B, Balay JM, Chabot A, De Larrard F (2004) Fatigue test and mechanical study of adhesion between concrete and asphalt. In: 9th international symposium on concrete roads, 3–6 April, Istanbul, Turkey
24. Pérez SA (2008) Approche expérimentale et numérique de la fissuration réflective des chaussées. Ph.D. thesis of Université de Limoges
25. Florence C, Foret G, Tamagny P, Sener JY, Ehrlacher (2004) A design of a new laboratory test simulating the reflective cracking in pavements with cement treated bases. In: RILEM proceedings PRO 37, fifth international RILEM conference on cracking in pavements, Limoges, 5–8 May 2004, pp 147–154
26. Chabot A, Hun M, Hammoum F (2013) Mechanical analysis of a mixed mode debonding test for composite pavements. Constr Build Mater 40:1076–1087
27. Chabot A, Hammoum F, Hun M (2016) Mixed-mode debonding approach to evaluate water sensibility in bi-layer composite pavements. In: Chabot A et al (eds) 8th international conference on mechanisms of cracking and debonding in pavements (MCD2016), Nantes, France, 7–9 June 2016. Springer RILEM Bookseries, vol 13, pp 613–618. ISBN: 978-94-024-0867-6. https://doi.org/10.1007/978-94-024-0867-6_86
28. Chabot A, Hammoum F, Hun M (2017) A 4pt bending bond test approach to evaluate water effect in a composite beam. Eur J Environ Civil Eng 21(supply1):54–69
29. Nasser H (2016) Solving M4-5n by a mixed finite element method for the analysis of pavements with discontinuities. ECN Ph.D. thesis, Université Bretagne Loire, 2016. Retrieved from https://hal.archives-ouvertes.fr/tel-01529735
30. Nasser H, Piau JM, Chupin O, Chabot, A (2016) M4-5n numerical solution using the Mixed FEM, validation against the Finite Difference Method. In: Chabot A et al (eds) 8th international conference on mechanisms of cracking and debonding in pavements (MCD2016), Nantes, France, 7–9 June 2016. Springer RILEM Bookseries, vol 13, pp 363–369. ISBN: 978-94-024-0867-6. https://doi.org/10.1007/978-94-024-0867-6_51
31. Nasser H, Chabot A (2017) A half-analytical elastic solution for 2D analysis of cracked pavements. Adv Eng Softw. https://doi.org/10.1016/j.advengsoft.2017.06.008
32. Florence C (2005) Etude expérimentale de la fissuration réflective et modélisation de la résistance de structures cellulaires. Ph.D. thesis of ENPC (ParisTech)

33. Pérez SA, Balay JM, Tamagny P, Petit C (2007) Accelerated pavement testing and modelling of reflective cracking in pavements. J Eng Fail Anal Sci Dir 14(8):1526–1537
34. Gouacolou Honore, Marchand Jean-Pierre Et Mouratidis Anastassios (1983) La méthode des éléments finis: application à la fissuration des chaussées et au calcul du temps de remontée des fissures. Laboratoire Central des Ponts et Chaussées
35. Simonin JM, Balthazard V, Hornych P, Dérobert X, Kerzrého JP, Trichet S (2012) Survey of debonding and vertical cracks evolution with non destructive techniques during accelerated loading tests. In: 4th international conference on accelerated pavement testing, September 2012, Davis, California, USA
36. Grellet D, Doré G, Kerzrého JP, Piau JM, Chabot A, Hornych P (2012) Experimental and theoretical investigation of three dimensional strain occurring near the surface in asphalt concrete layers. In: Proceedings of the 7th Rilem international conference on cracking in pavements, 20–22 June 2012, Delft, The Netherlands. RILEM Bookseries, vol 4, pp 1017–1027. https://doi.org/10.1007/978-94-007-4566-7_97, Print ISBN: 978-94-007-4565-0
37. Hornych P, Kerzrého JP, Sohm J, Chabot A, Trichet S, Joutang JL, Bastard N (2012) Full scale tests on grid reinforced flexible pavements on the French fatigue carrousel. In: Proceedings of the 7th Rilem international conference on cracking in pavements, 20–22 June 2012, Delft, The Netherlands. RILEM Bookseries, vol 4, pp 1251–1260. https://doi.org/10.1007/978-94-007-4566-7_119, Print ISBN: 978-94-007-4565-0
38. Nguyen ML, Sohm J, Kerzrého JP, Hornych P (2013) Review of glass fiber grid use for pavement reinforcement and APT experiments at IFSTTAR. Intern J Road Mater Pavement Design (Special edition EATA Conference)
39. Kerzrého JP, Hornych P, Sohm J (2011) Accelerated loading test on structures reinforced by geogrids with the FABAC machines. In: Technical Report for Tensar International, LTD, IFSTTAR, May 2011, 67p
40. Dave EV, Ahmed SA, Buttlar WG, Bausano J, Lynn T (2010) Investigation of strain tolerant mixture reflective crack relief systems: an integrated approach. J Assoc Asphalt Paving Technol 79:119–156
41. ASME (2006) Guide for verification and validation in computational solid mechanics. Am Soc Mech Eng. V & V 10, ISBN: 079183042X
42. Balci O (1989) How to assess the acceptability and credibility of simulation results. In: 1989 winter simulation conference proceedings (Cat. No.89CH2778–9), SCS, Washington, DC, USA, pp 62–71
43. Balci O (1997) Principles of simulation model validation, verification, and testing. Trans Soc Comput Simul 14(1):3–12
44. Barlas Y, Carpenter S (1990) Philosophical roots of model validation: two paradigms. Syst Dyn Rev 6(2):148–166
45. Caughlin D (1999) Ensuring thorough comparison of modeling and simulation (M&S) results with experimental and test observations. In: Enabling technology for simulation science III, SPIE-International Society for Optics and Photonics, Orlando, FL, USA, pp 271–282
46. Gong M, Murray-Smith DJ (1998) A practical exercise in simulation model validation. Math Comput Model Dyn Syst 4(1):100–117
47. Kleijnen JP (1995) Verification and validation of simulation models. Eur J Oper Res 82 (1):145–162
48. Oberkampf WL (2001) Feature: computational validation series: part 2. What are validation experiments? Exp Tech 25(3):35–40
49. Oberkampf WL (1998) Bibliography for verification and validation in computational simulations. In: Report: SAND98–2041, Sandia National Labs, Albuquerque, NM, United States
50. Oberkampf WL, Barone MF (2005) Measures of agreement between computation and experiment: validation metrics. In: Report: SAND2005-4302, Sandia National Labs, Albuquerque, NM, United States
51. Oberkampf WL, Barone MF (2006) Measures of agreement between computation and experiment: validation metrics. J Comput Phys 217(1):5–36

52. Oberkampf WL, Trucano TG (2008) Verification and validation benchmarks. Nucl Eng Des 238(3):716–743
53. Oberkampf WL, Trucano TG, Hirsch C (2004) Verification, validation, and predictive capability in computational engineering and physics. Appl Mech Rev 57(5):345–384
54. Ripley BD (1988) Uses and abuses of statistical simulation. Martin Beale memorial symposium on mathematical models and their solutions, London, UK, pp 53–68
55. Robinson S (1999) Simulation verification, validation and confidence: a tutorial. Transac Soc Comput Simul Intern 16(2):63–69
56. Sargent RG (2005) Verification and validation of simulation models. In: Proceedings of the 2005 winter simulation conference, IEEE, Orlando, FL, USA, p 14
57. Trucano TG, Swiler LP, Igusa T, Oberkampf WL, Pilch M (2006) Calibration, validation, and sensitivity analysis: what's what. Reliab Eng Syst Saf 91(10–11):1331–1357
58. Velasco FRD (1987) Method for test data selection. J Syst Softw 7(2):89–97
59. Hill BC, Giraldo-Londoño O, Paulino GH, Buttlar WG (2017) Inverse estimation of cohesive fracture properties of asphalt mixtures using an optimization approach experimental mechanics, vol 57, issue 4, pp 637–648. https://doi.org/10.1007/s11340-017-0257-3
60. AASHTO (1993) Guide for design of pavement structures. American Association of State Highway and Transportation Officials
61. STAC (2nd edn) (2016) Rational design method for flexible airfield pavements/Méthode rationnelle de dimensionnement des chaussées aéronautiques souples. Guide technique STAC, (http://www.stac.aviation-civile.gouv.fr/publications/documents/dim_chau_souple-v2-gb.pdf)
62. IFSTTAR (2017) Alizé-Lcpc software for the design of pavements according to the French rational method. http://www.lcpc.fr/en/produits/alize/index.dml
63. Burmister DM (1943) The theory of stresses and displacements in layered systems and applications of the design of airport run ways. Proc Highway Res Board 23:126–148
64. CEREMA-IDRRIM (2016) Diagnostic et Conception des Renforcements de Chaussées. Guide technique, 259p. http://www.cerema.fr/guide-diagnostic-et-conception-des-renforcements-a1543.html
65. Applied Research Associates, Inc (2004) Development of the 2002 guide for the design of new and rehabilitated pavement structures, final report and software (version 0.70) NCHRP Project 1-37A. Transportation Research Board, Washington, D.C
66. AASHTO (2017) AASHTOWare pavement. http://www.aashtoware.org/Pavement/Pages/default.aspx
67. Buttlar WG, Roque R (1994) Development and evaluation of the new SHRP measurement and analysis system for indirect tensile testing of asphalt mixtures at low temperatures. Transportation Research Record, No. 1454, National Research Council, National Academy Press, Washington, D.C., pp 163–171
68. Roque R, Hiltunen DR, Buttlar WG (1995) Thermal cracking performance and design of mixtures using SUPERPAVE™, Symposium for SUPERPAVETM Implementation, Portland Oregon. J Assoc Asphalt Paving Technol 64:718–735
69. Buttlar WG, Roque R, Reid B (1998) An automated procedure for generation of the creep compliance master curve for asphalt mixtures. In: Journal of the Transportation Research Board, No. 1630, National Research Council, National Academy Press, Washington, D.C., pp 28–36

# Chapter 2
# Cracking in Asphalt Materials

**Eshan Dave, Ramon Botella, Paul Marsac, Didier Bodin, Cédric Sauzeat and Mai Lan Nguyen**

## 2.1 Introduction

As described in the first chapter of this report, asphalt mixtures are heterogeneous composites with temperature, rate and hereditary dependencies amongst other complexities such as oxidative aging and non-uniformities due to construction practices. Characterization of fracture in this highly complex, viscoelastic particulate composite is a challenging task. Cracking related failures in asphalt pavements are often associated with fractures occurring within the asphalt materials or debonding occurring between asphalt layers or between asphalt and other pavement layers. Thermal and fatigue cracking in asphalt pavements as well as certain aspects of reflective cracking are examples of fracturing within 'bulk' asphalt concrete materials. In order to understand the mechanisms of cracking in asphalt pavements it is critical to be able to adequately characterize and understand the physical nature of fracture in the bulk material, particularly within near-surface layers, where damage and cracking tend to appear most often. Such characterization includes the

E. Dave (✉)
University of New Hampshire, New Hampshire, USA
e-mail: eshan.dave@unh.edu

R. Botella
University of Catalonia, Catalonia, Spain

P. Marsac · M. L. Nguyen
Department of Materials and Structures (MAST), IFSTTAR, Nantes, France
e-mail: paul.marsac@ifsttar.fr

D. Bodin
ARRB Group, Vermont, Australia

C. Sauzeat
ENTPE - University of Lyon, Vaulx-en-Velin, France



W. G. Buttlar et al. (eds.), *Mechanisms of Cracking and Debonding in Asphalt and Composite Pavements*, RILEM State-of-the-Art Reports 28,
https://doi.org/10.1007/978-3-319-76849-6_2

ability to conduct and interpret bulk fracture tests and the availability of analytical and computational models that can capture key physical processes associated with failure of asphalt paving mixtures.

Significant progress has been made in both laboratory characterization and modelling of bulk fracture in asphalt materials in recent years. For example, the use of notched asphalt specimens for localizing crack initiation and propagation has become very popular over the last ten years. Similarly, a large number of computational and analytical models have emerged that are capable of capturing the fracture processes within the asphalt macro and micro-structure. Models that are capable of accurately simulating asphalt materials over wider ranges of temperatures, loading rates, aging conditions and other effects have also emerged. When applied in practice, these advances will lead to significant improvements over empirical test procedures and purely phenomenological modelling approaches that mainly consist of regression equations developed through data-fitting approaches using very specimen-centric or site-specific cracking observations, which are inherently biased by prevailing boundary conditions.

The titles and authors of the original contributions that were submitted for the characterization of fracture mechanisms in asphalt materials is as follows:

### Laboratory Tests

*Static Cracking and Damage*

1. Fracture Energy Measurements Using Disk-Shaped Compact Tension Test (Buttlar et al.)
2. Fracture Energy Measurements Using Fenix Test (Perez-Jimenez et al.)
3. Validity of Time–Temperature Superposition Principle for Cracking in Bitumenous Mixtures (Di Benedetto et al.)
4. Dissipated Creep Strain Energy Density Measurements (Hernanado et al.).

*Cyclic Cracking and Damage*

1. 2 Point Bending Test (Maggiore, Marsac)
2. Complex Modulus and Fatigue Tests on Cantilever Trapezoidal Beam (Chabot, Hammoum)
3. Complex Modulus Testing Using Uniaxial Cylindrical Test (Di Benedetto et al.)
4. Uniaxial Test for determining Viscoelastic Continuum Damage model parameters (Kim et al.).

### Constitutive Models

*Crack Initiation and Propagation Models*

1. Crack Growth Initiation Model (Dubois et al.)
2. Cohesive Zone Fracture Model (Behnia et al.)
3. Hot Mix Asphalt Fracture Mechanics Model (Hernando et al.).

*Cyclic Degradation Models*

1. Dissipated Energy Concept (Maggiore et al.)
2. Non-local Modeling of Fatigue Microcracking with Application to Specimen Size Effects (Marsac and Bodin)
3. Viscoelastic Continuum Damage Model (Kim).

## 2.2 Lab Tests: Static Cracking and Damage

This section describes three of the most recent static lab tests developed to evaluate cracking resistance in bituminous mixtures. Each subsection contains details regarding the theoretical background, specimen geometry and data interpretation methods.

### *2.2.1 Fracture Energy Measurements Using Disk-Shaped Compact Tension Test and Compact Tension Test*

The University of Illinois at Urbana-Champaign (UIUC) has developed a variety of tools to characterize the fracture properties of asphalt mixtures. This research has predominantly focused on the mechanisms related to reflective and thermal cracking. The experimental tests created at UIUC measure the fracture energy of mixtures to describe the work required to propagate an existing crack through the asphalt mastic and aggregate. The disk-shaped compact tension test (DC(T)), shown in Fig. 2.1a, was developed by Wagoner et al. [1] to characterize the Mode I fracture behavior of conventional intermediate and surface layer asphalt mixtures. In addition, Ahmed et al. [2] adapted the compact tension (C(T)) test from the DC (T) geometry to evaluate the fracture properties of thin-bonded overlay (TBO) systems. Thin-bonded overlays are thin lifts of gap or dense graded asphalt concrete (less than 25 mm thick) placed immediately after the application of a

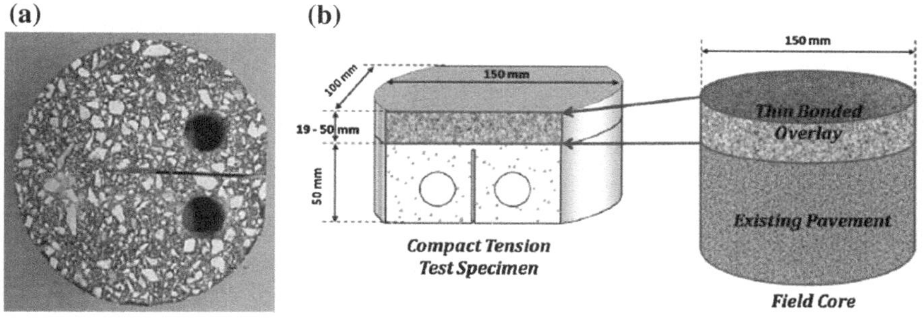


**Fig. 2.1** **a** DC(T) specimen **b** C(T) specimen (Ahmed et al. [3])

heavy tack coat to promote greater adhesion between the overlay and underlying layers. The C(T) test determines the Mode I fracture properties of TBO systems by driving a crack vertically through the overlay as shown in Fig. 2.1b.

#### 2.2.1.1 Test Setup, Procedures, and Analysis

Wagoner et al. [1] proposed an adapted DC(T) geometry to characterize conventional asphalt mixtures, using the ASTM E399 standard as a starting point. This geometry, shown in Fig. 2.2, can be fabricated from 150 mm diameter field core or laboratory produced gyratory specimens and is specified in ASTM D7313-14. The C(T) geometry, provided in Fig. 2.1a, has been proposed by Ahmed et al. [4] for fracture characterization of thin-bonded asphalt overlays. This geometry can also be generated from 150 mm dia. field cores or gyratory specimens. However, the ligament length of the specimen varies from 19 to 50 mm depending on the thickness of the TBO system. Additionally, the C(T) specimen notch spans the thickness of the entire underlying layer such that the fracture energy of the specimen relates to the TBO system only.

The DC(T) and C(T) tests are controlled using a crack mouth opening displacement (CMOD) gauge. A constant CMOD rate of 1.0 mm/min is used to initiate and propagate a crack through the asphalt material, generally at temperatures less than 0 °C. According to ASTM D7313-14, the testing temperature is recommended to be 10 °C higher than the low temperature performance grade (PG) of the asphalt binder. The primary outputs of the DC(T) and C(T) tests are load-CMOD plots, CMOD fracture energy, and peak load. A plot of four Load-CMOD curves is provided in Fig. 2.3. The dissipated energy during cracking, i.e. global fracture energy ($G_f$), is calculated as:

$$G_f = \frac{W_D}{h \cdot l} \tag{2.1}$$

where $G_f$ = dissipated energy during test application, (typically in units of J/m$^2$); $W_D$ = dissipated work during test application; h = specimen thickness, l = initial

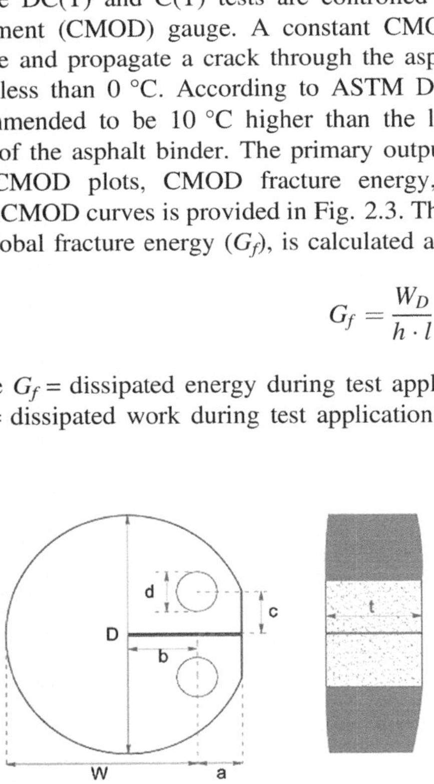


| DC(T), Dimensions (mm) | |
|---|---|
| D | 150 |
| W | 110 |
| a | 35 |
| b | 27.5 |
| c | 25 |
| d | 25 |
| t | 50 |

**Fig. 2.2** Asphalt concrete DC(T) geometry [5]

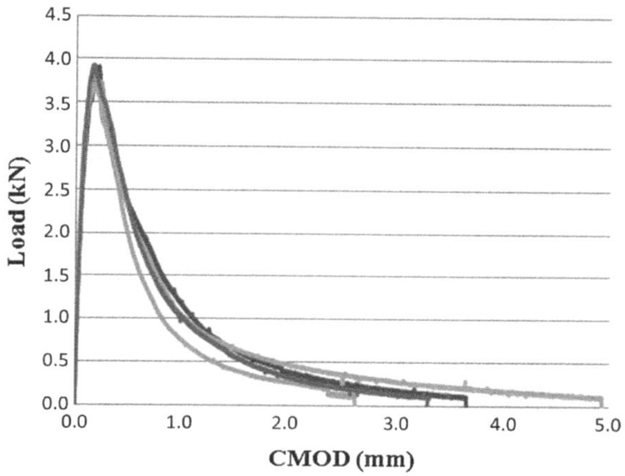


**Fig. 2.3** Example load versus CMOD plots for asphalt mixtures

ligament length. The dissipated work is calculated by measuring the area under the average Load-CMOD curve.

#### 2.2.1.2 Examples of Test Results

Initial DC(T) studies evaluated the effects of testing rate, temperature, aging, and specimen size. Wagoner et al. [1] examined the effect of four rates, 10, 1, 0.5, and 0.1 mm/min, and three temperatures, 0, −10, and −20 °C on CMOD fracture energy. Fracture energy was found to increase with temperature from 233 J/m$^2$ at −20 °C to 470 J/m$^2$ at 0 °C. In addition, fracture energy decreased with CMOD opening rate from 352 at 0.1 mm/min to 276 J/m$^2$ at 10 mm/min.

The increase in fracture energy due to the temperature increase likely occurred as a result of a combination of factors, including the decreasing brittleness (increasing ductility) of the binder with temperature, the corresponding increase in the fracture process zone (and decreased structural brittleness), and increased proportion of viscous dissipation entering into the energy measurement. In addition, the inverse relationship between the fracture energy and loading rate was likely related to a reduced ability to relax stress during the shorter testing period. Wagoner et al. [6] assessed the effect of specimen size on fracture resistance. Varying specimen diameters and thicknesses were used. CMOD fracture energies increased from 400 to 600 J/m$^2$ and 250 to 550 J/m$^2$ with respective increases in diameter from 100 to 450 mm and thickness from 25 to 150 mm, despite normalizing for the size of the fractured area in the energy calculation (Eq. 2.1).

Braham et al. [7] considered the effect of laboratory oven aging of asphalt mixtures on DC(T) fracture energy. Researchers placed loose asphalt mixtures at temperatures of 135 °C for 8 different oven aging periods ranging from 2 to 48 h. Then, specimens were compacted and tested. Braham and his colleagues found CMOD-based total fracture energy to increase for oven aging times up to about 8 h.

This result was counter-intuitive at first, but closer examination showed a rise in peak load, and slight increase in the slope of the post-peak softening curve (decreasing Load-CMOD curve in the post-peak region). Creep compliance values obtained from testing in AASHTO T-322 indicated a decrease in creep compliance with aging level. Up to the 8 h aging level, the energy increase associated with the peak load increase seems to outweigh the loss in energy due to the increased severity (slope) of the post-peak softening curve. After 8 h of aging at 135 °C, the post-peak softening behavior became quite severe (steep slope), more than offsetting the increased peak load, and leading to overall reduced fracture energy.

The DC(T) test has also been applied to sustainable materials such as reclaimed asphalt pavement (RAP) and warm mix asphalt (WMA) mixtures. Behnia et al. [5] found increased levels of RAP generally yielded decreased fracture energy. This result agreed with Braham et al. [7] as greater levels of aging or amounts of aged inclusions reduced the ability of the asphalt concrete to resist crack propagation above a certain load threshold. Furthermore, Hill et al. [8] found that the DC(T) test was able to distinguish among foaming, organic, and chemical WMA additives. In particular, this study found fracture energy to be more sensitive to additive type as compared to the reduced production temperatures of WMA.

Finally, Buttlar et al. [9] considered the relationship between transverse cracking and CMOD fracture energy. Researchers extracted field core samples at locations in which crack count surveys were conducted. The results are shown in Fig. 2.4 and produced a definitive correlation between fracture energy and transverse cracking. CMOD fracture energies in excess of 400 J/m$^2$ were determined to represent a threshold associated with negligible levels of transverse cracking. Fracture energies below this value related to sections containing significant transverse cracking. A study by Dave et al. [10] used 26 field pavement sections to further validate efficacy of DC(T) fracture energy to distinguish between poor and good transverse cracking performance in asphalt pavements. The limit of 400 J/m$^2$ has also been further validated by Dave and Hoplin [11] through use of computational modeling of asphalt pavement sections.

The C(T) test has been successfully used to characterize gap and dense graded TBOs. Ahmed et al. [2] determined that C(T) tests conducted on gap graded TBO's showed significantly greater fracture energies when compared to dense graded overlays. This result likely occurred due to the upward migration of the tack coat

**Fig. 2.4** Transverse cracking and DC(T) fracture energy relationship (Buttlar et al. [9])

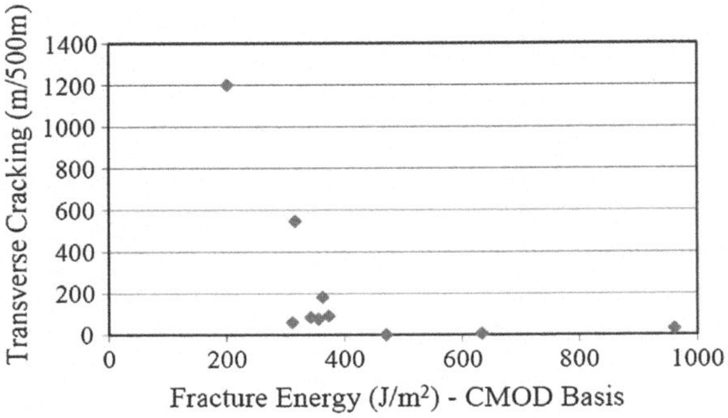

into the gap graded mixture during compaction. Ahmed et al. [2] also evaluated dense grade TBO's constructed with conventional and spray paver construction techniques using the C(T) test. Results showed that sections constructed using the spray paver construction technique yielded fracture energies 39% greater than those constructed using traditional paving techniques. Testing also showed that samples obtained from sections constructed using a polymer modified asphalt binder as a tack coat had 40% higher fracture energies as compared to those constructed using unmodified asphalt emulsion. Finally, Ahmed [12] compared laboratory produced and field core based C(T) test specimens in terms of fracture energy. Results showed that samples obtained from field cores exhibited higher fracture energies than those manufactured in the laboratory.

### 2.2.2 *Fracture Energy Measurements Using Fenix Test*

The Fenix test is a tensile test to evaluate cracking resistance of asphalt concrete mixtures by calculation of the dissipated energy during the cracking process. The test is named after the research project that funded the investigation: FENIX Project (www.proyectofenix.es).

The Fenix test represents a mode I type of fracture. It can be performed on laboratory samples or field cores. The typical test temperature applied is 20°C, however it has been applied at lower temperatures such as, 5, −5, −10, −15 and −25 °C. The displacement velocity its usually 1 mm/min but tests have been carried out at 10, 0.1 and 0.03 mm/min.

By this definition, at below 0 °C temperatures the dissipated energy is equal or very close to the fracture energy, however when the test is performed at higher temperatures an important part of this energy is employed in deformation of the material and therefore it is not all used in the fracture of the specimen. The dissipated energy is calculated in the same way as for the DC(T) and C(T) tests (refer to Eq. 2.1).

Other mechanical parameters like peak load, $F_{max}$, displacement at peak load, $\Delta_{Fmax}$, and displacement at 50% of post-peak load, $\Delta_{mdp}$, are determined from the load-displacement curve (refer to Fig. 2.5). In order to obtain a stiffness related parameter from the test, the tensile stiffness index, IRT, is defined by Eq. 2.2:

$$IRT = \frac{1/2 \cdot F_{\max}}{\Delta_m} \tag{2.2}$$

where $IRT$ = tensile stiffness index (kN/mm); $\Delta_m$ = displacement before peak load at ½ $F_{max}$, (mm).

#### 2.2.2.1 Test Setup, Procedures, and Analysis

The test procedure consists of subjecting one half of a 63.5 mm thick cylindrical specimen of a 101.6 mm diameter prepared by Marshall or gyratory compaction to

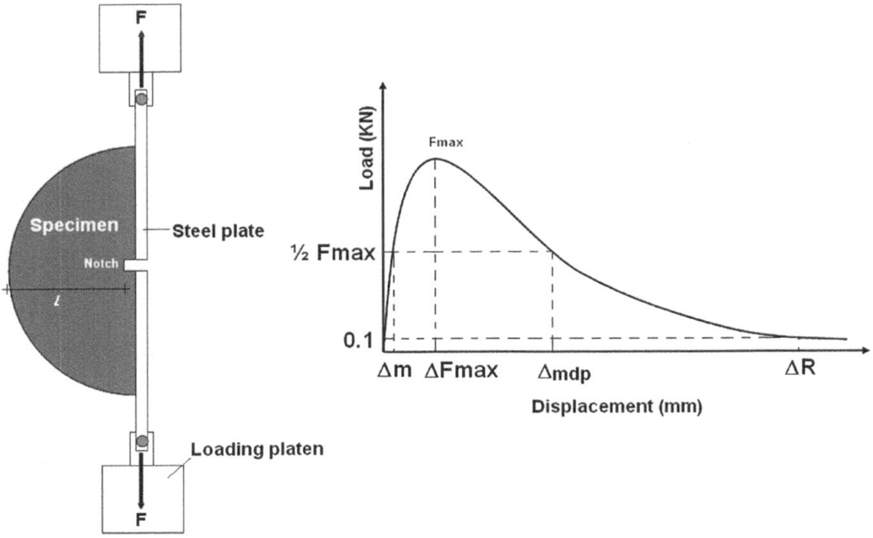


**Fig. 2.5** Fénix test set up and load-displacement output curve

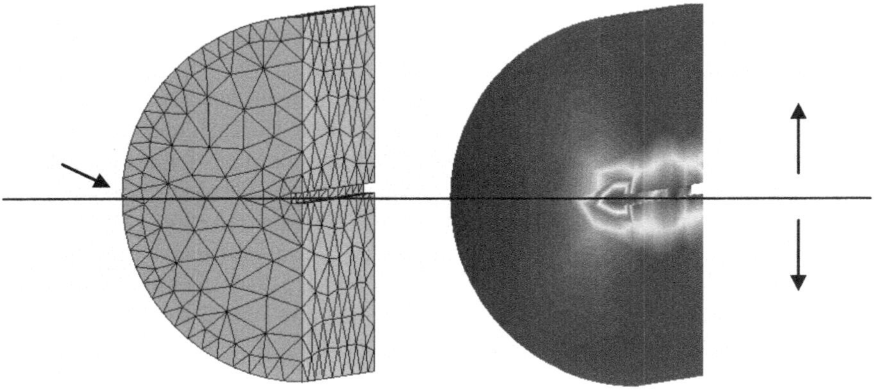

**Fig. 2.6** Finite element simulation of Fénix specimen (using GID® software)

a tensile stress at a constant displacement velocity and specific temperature. A 6 mm deep notch is made in the middle of its flat side where two steel plates are fixed. The specimen is glued to the steel plates with an adhesive mortar containing epoxy resins. Each plate is attached to a loading platen so that they can rotate about fixing points. Load and displacement data are recorded throughout the test to calculate the parameters involved in the cracking process. Figure 2.6 shows finite element simulation of Fénix specimen.

#### 2.2.2.2 Examples of Test Results

Figures 2.7, 2.8, and 2.9 summarize the results of the effect of binder type and test temperature for a S20 mixture (AC22S). At a low temperature (−10 °C), this asphalt mixture exhibited a brittle behaviour with high peak load and low displacement values, Fig. 2.7. At 20 °C, a more ductile behaviour with higher displacement values, $\Delta_{mdp}$, and low peak load, $F_{max}$, and stiffness, $I_{RT}$, values, can be observed (Fig. 2.9). At intermediate temperatures, 5 °C, the behaviour was also intermediate as seen in Fig. 2.8.

The results of dissipated energy during cracking, $G_f$, which is related to binder type and test temperature, reveal that BM3c mixture had the highest dissipated energy value at all temperatures, and therefore it is expected to have superior

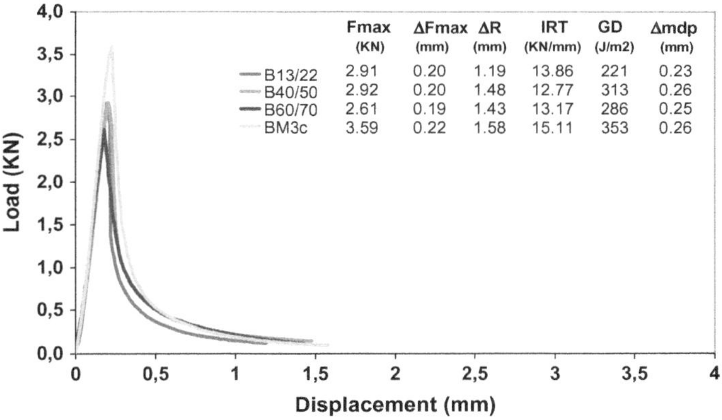


**Fig. 2.7** Fénix test with different bitumen types, S-20 mixture at −10 °C

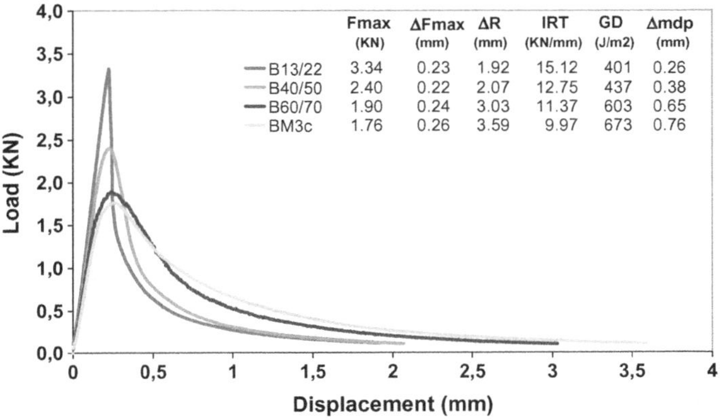


**Fig. 2.8** Fénix test with different bitumen types, S-20 mixture at 5 °C

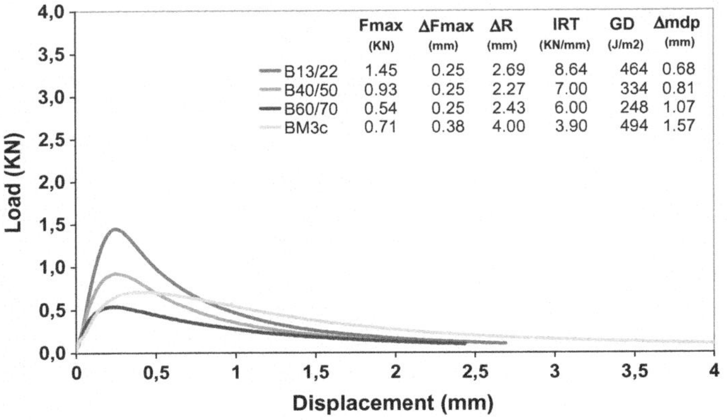


**Fig. 2.9** Fénix test with different bitumen types, S-20 mixture at 20 °C

cracking behaviour. Mixtures obtained the largest dissipated energy value at 5 °C irrespective of the binder used, except for B13/22 mixture that had the highest dissipated energy value at 20 °C due to the higher stiffness of the binder. At lower temperatures a more brittle behaviour can be observed.

#### 2.2.2.3 Repeatability and Sensitivity

Repeatability of the Fénix test was assessed by computing the coefficient of variation (COV) for each series of specimens tested. For dissipated energy, $G_f$, and maximum tensile load, $F_{max}$, COV mean values of 15 and 8.5% were obtained, respectively. Based on COV values for dissipated energy, the Fénix test seems to have good repeatability that is comparable to other fracture tests for asphalt mixtures. On basis of typical published results, the COVB ranges for other tests are 3–18% for SE(B), 4–25% for DC(T) and 15–34% for SCB tests (13).

Sensitivity of the Fénix test was evaluated by a variance analysis (ANOVA) of the results for dissipated energy obtained in the experimental tests. With ANOVA it is possible to find out whether the values of a set of numerical data are significantly different from those of other sets. Thus, test sensitivity for a specific parameter can be determined.

In the first stage, test sensitivity of S-20 mixture to binder type and test temperature was analyzed from the results obtained for dissipated energy during cracking in the Fénix test. A variance analysis of two factors, i.e. binder type and test temperature, was performed with several samples per group. Binder type levels were B60/70, B40/50, B13/22 and BM3c, with 9 data for each level; test temperature levels were −10, 5 and 20 °C, with 12 data for each level.

As can be seen in Table 2.1, the F ratio for binder type and test temperature factors, like for the interaction between both factors, is greater than its critical value

**Table 2.1** ANOVA comparison of dissipated energy, $G_F$, with bitumen type and test temperature. S-20 mixture (AC22 S)

| Source of variation | SS | DF | MS | F-value | P-value | Critical F-value |
|---|---|---|---|---|---|---|
| Bitumen type | 132414.178 | 3 | 44138.059 | 20.42 | $8.59 \times 10^{-7}$ | 3.0088 |
| Test temperature | 337020.377 | 2 | 168510.188 | 77.95 | $3.18 \times 10^{-11}$ | 3.4028 |
| Interaction | 167925.278 | 6 | 27987.546 | 12.95 | $1.66 \times 10^{-6}$ | 2.5082 |

determined for a significance level of 0.05. This means that for each evaluated factor, as well as in the interaction between them, significantly different values are obtained for dissipated energy, $G_f$. Thus, test sensitivity to factors analyzed at this stage is determined.

Analogously, in the second stage test sensitivity of G-20 mixture to binder content and test temperature was tested. As in the previous case, a variance analysis of two factors, i.e. binder content and test temperature, was conducted with several samples per group for each binder type (B60/70, B40/50 and B13/22). Binder content levels were 3.5, 4.5 and 5.5%, with 9 data for each level; test temperature levels were −10, 5 and 20 °C, with 9 data for each level.

From Table 2.2, it can be concluded that, for binder content and test temperature, significantly different values of dissipated energy during cracking, $G_f$, are obtained in the Fénix test. Thus, test sensitivity to factors analyzed at this stage is determined.

**Table 2.2** ANOVA comparison of dissipated energy, $G_F$, with bitumen content and test temperature. G-20 mixture (AC22 G)

| Bitumen type | Source of variation | SS | DF | MS | F-value | P-value | Critical F-value |
|---|---|---|---|---|---|---|---|
| B60/70 | Bitumen content | 112047.73 | 2 | 56023.865 | 5.2324 | $1.62 \times 10^{-2}$ | 3.5546 |
| | Test temperature | 1465290.22 | 2 | 732645.111 | 68.4263 | $3.87 \times 10^{-9}$ | 3.5546 |
| | Interaction | 103244.73 | 4 | 25811.183 | 2.4107 | $8.71 \times 10^{-2}$ | 2.9277 |
| B40/50 | Bitumen content | 181021.95 | 2 | 90510.974 | 6.9033 | $5.95 \times 10^{-3}$ | 3.5546 |
| | Test temperature | 581451.24 | 2 | 290725.619 | 22.1736 | $1.39 \times 10^{-5}$ | 3.5546 |
| | Interaction | 154442.81 | 4 | 38610.701 | 2.9448 | $4.91 \times 10^{-2}$ | 2.9277 |
| B13/22 | Bitumen content | 588307.52 | 2 | 294153.761 | 45.0689 | $9.81 \times 10^{-8}$ | 3.5546 |
| | Test temperature | 1104949.06 | 2 | 552474.530 | 84.6477 | $6.99 \times 10^{-10}$ | 3.5546 |
| | Interaction | 173596.07 | 4 | 43399.019 | 6.6494 | $1.81 \times 10^{-3}$ | 2.9277 |

### 2.2.3 *Four Point Bending Notch Fracture*

Crack propagation in bituminous mixtures using four point bending notch has been designed at University of Lyon/ENTPE. One of the main features of this test is that the middle one third of the beam is subjected to pure bending (constant momentum) without any shear deformation.

#### 2.2.3.1 Test Setup, Procedures, and Analysis

The test set up together with the tested pre-notched prismatic beams dimensions are shown in Fig. 2.10.

An initial notch of height "$a_0$" is made in the middle of the beam. This height assures stable crack propagation. The same ligament size ($W$-$a_0$) of about 5 cm (value used in previous study by Nguyen et al. [13]) is obtained for all specimens. As the energy needed for crack initiation is dependent of the initial notch edge shape, great care is taken to create it. It is performed in two steps. First, a circular saw is used to obtain a $a_0$_0.5 cm deep and 5 mm thick notch. Then the last 0.5 cm is made with a hacksaw to obtain a thinner crack tip of 1 mm thickness. Table 2.3 provides initial notch values ($a_0$) as well as ligament ($W$-$a_0$) of each considered specimen.

A servo-hydraulic machine is used to perform the tests. The displacement of the piston ($u$) (measured by an integrated transducer) is used as the feedback signal for the considered tests. However, any other sensors may be used to monitor the test

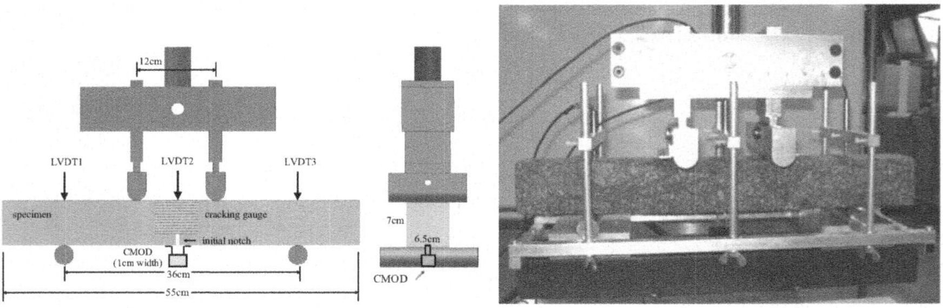


**Fig. 2.10** Four point bending test performed at University of Lyon/ENTPE

**Table 2.3** Size of pre-notch and ligament for the tested specimens

| Specimen | Initial notch $a_0$ (cm) | Ligament W-$a_0$ (cm) |
|---|---|---|
| ENTPE2 | 3.10 | 4.7 |
| ENTPE4 | 1.90 | 5.3 |
| ENTPE7 | 2.27 | 5.0 |
| ENTPE8 | 2.63 | 4.9 |

(crack opening or other). Monotonic tests as well as loading-unloading cyclic tests can be applied. A 10 kN load cell measures the applied load (*P*).

A thermal chamber is used to control the temperature. The temperature inside the chamber may be regulated from −50 to 80°C. A thermal gauge is fixed on the specimen to measure its surface temperature.

Three linear variable differential transducers (LVDT) are used to measure displacements on top of the beam: deflection in the centre of the beam, *LVDT2*, and vertical displacements at the two lower supports, *LVDT1* and *LVDT3* (see Fig. 2.10). The range of the used LVDTs is 5 cm. Measurements by *LVDT1* and *LVDT3* give displacements due to punching effect of the two lower supports on the beam. As the same load is applied by each upper and lower support displacement due to punching effect can be taken into account and corrected. The "true" deflection of the beam, called *f*, is then obtained by:

$$f = \mathrm{LVDT2} - \frac{\mathrm{LVDT1} + \mathrm{LVDT3}}{2} \tag{2.3}$$

An extensometer is used to measure the crack mouth opening displacement (CMOD) at the top of notch. It is placed under the beam (see Fig. 2.10).

An important measurement is the crack length evolution during the test. In order to follow the crack propagation the cracking gauges are used. These gauges are constituted of 21 parallel wires separated each other by 2.5 mm (Fig. 2.11 left). The wires are designed to break sequentially as the crack propagates in the specimen, which indicates the crack tip position (Fig. 2.11 right). The gauges used are 8 cm long and 5 cm high. One gauge is glued on each lateral side of the specimen.

The four point bending test is conducted with constant imposed displacement rate of the piston. The displacement rate fixed for each test is chosen depending on the test conditions as described further. The realization of the test takes place in two stages.

First, two cycles of loading/unloading are carried out to allow setting of specimen inside the loading frame. The maximum loading value during these two cycles is $P_{1max} = 0.8$ kN, which is relatively small (less than a third of the failure value). The minimal value is $P_{2min} = 0.2$ kN. Then, after these loading/unloading cycles, monotonic loading is applied at constant displacement rate of the piston, until the final failure of the beam. Figure 2.12 shows the typical obtained response of load

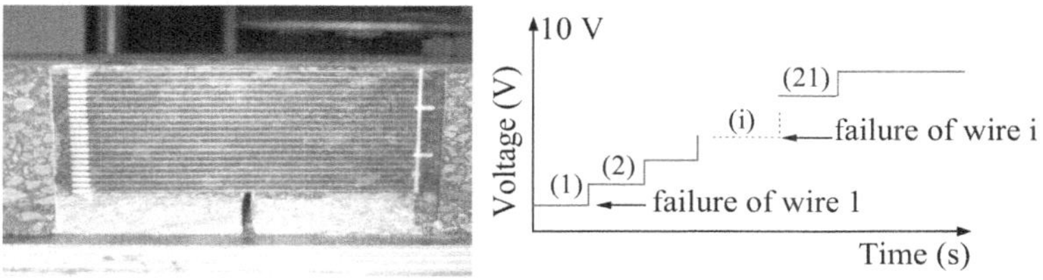


**Fig. 2.11** (left): Cracking gauge glued on the specimen; and (right): typical measured voltage–time curve when crack propagates

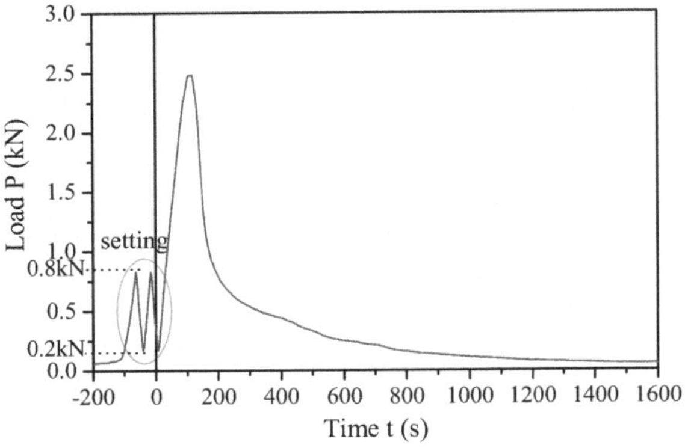


**Fig. 2.12** Loading path during the crack propagation test (ENTPE7: v = 5.9 mm/min; T = −1.5 °C)

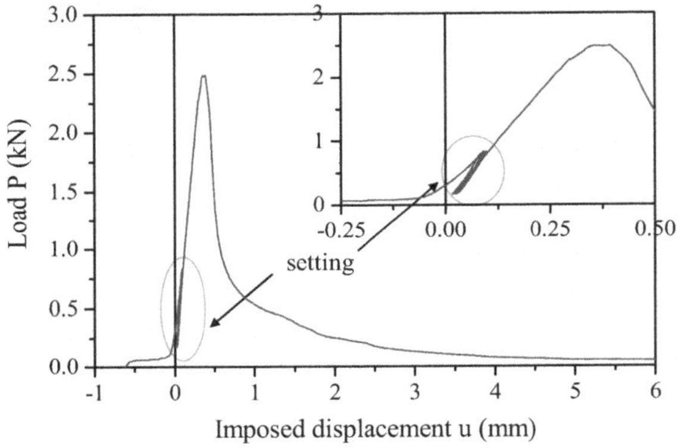


**Fig. 2.13** Load–displacement curve obtained for test ENTPE7: v = 5.9 mm/min; T = −1.5 °C (same test as Fig. 2.12)

versus time. Figure 2.13 presents result in the axes load versus displacement of piston, for the same test as Fig. 2.12 (test ENTPE7).

The determination of the crack length follows a type of measurement that has been applied in many previous studies (Lemaistre [14]; Jiang et al. [15]; Wendling et al. [16]; Nguyen et al. [13]).

As one gauge is used on each side of the specimen, two different values of crack length are obtained, noted $a_{gauge1}$ and $a_{gauge2}$. Results given by the two cracking gauges are plotted in Fig. 2.16, for test ENTPE7 already chosen in Figs. 2.12 and 2.13. As shown in Fig. 2.16, some differences appear between the two values $a_{gauge1}$ and $a_{gauge2}$. The measured crack length is slightly different on each side. The crack may also have a different evolution inside the beam.

A drawback of this method is that detection of crack is delayed due to the relatively higher ductility of the gauge wires as compared to asphalt concrete. Only macro-cracks above a certain minimal width can be detected. Figure 2.16 shows clearly that, the detected macro-crack using cracking gauges, appears after the peak of the load. This result probably does not reflect the reality. To address this shortcoming, a novel method was developed by the authors. The DRCL (Displacement Ratio method for predicting Crack Length) method is a new approach which makes it possible to calculate the crack length during the four point bending crack propagation test [13]. This method is based on the relation between two measured displacements: the crack mouth opening displacement "*CMOD*" and the deflection of the beam "*f*". The ratio between *CMOD* and *f* is called $r_d$:

$$r_d = \frac{CMOD}{f} \tag{2.4}$$

The DRCL method was developed at DGCB laboratory of University of Lyon/ ENTPE and was introduced for the first time for cyclic tests then extended to monotonic loading cases. During analysis of cyclic tests, it was observed that linear extrapolation of the linear part of the reloading branch passes through the origin in the axes CMOD-f. Linear viscoelastic behaviour, without crack propagation can be reasonably assumed for the material, during this reloading period.

The second part of this method relies on FEM calculation considering the hypothesis of isotropic linear elastic behaviour. The calculation is made in order to determine the displacements field in the beam during the test. When considering the boundary conditions imposed on the beam (Fig. 2.11 left), it can be shown that the displacement and the strain fields are the same for elastic or viscoelastic materials [17] if the crack does not propagate. The fields are independent of the modulus value of the material. They only depend of the fixed crack length.

The Finite Element Method (FEM) software COMSOL is used for the calculation, which is described in detail in Nguyen [18] and Tapsoba [19]. The calculation is repeated for different values of crack length *a*, in order to obtain the displacement ratio $r_d$ (Eq. 2.4) as a function of the crack length *a* (Fig. 2.14b).

The last part of the method consists in linking the measured experimental ratio $r_d$ with the crack length *a* considering relation of Fig. 2.14b.

The DRCL method can also be applied to monotonic (non-cyclic) test data, as considered in this study. In that case, a crack length can be obtained from any point of the monotonic curve of CMOD versus f, as explained in Fig. 2.15. Then the crack length evolution is obtained during the entire monotonic loading test.

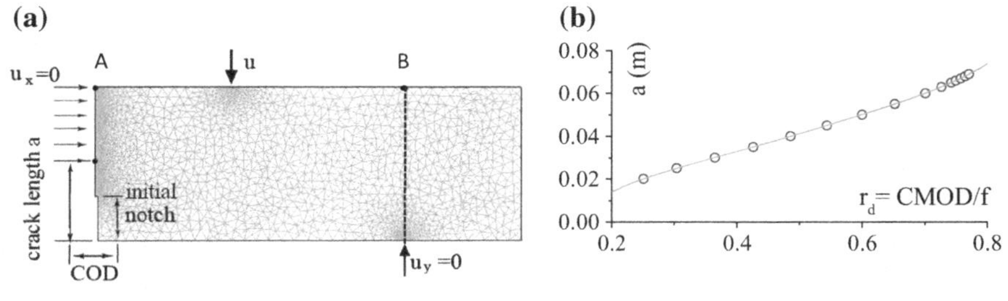


**Fig. 2.14** **a** boundary conditions and mesh used in FEM calculation: vertical displacement at A and B points and CMOD displacement computed; **b** displacement ratio $r_d$ values calculated by FEM as a function of crack length *a*

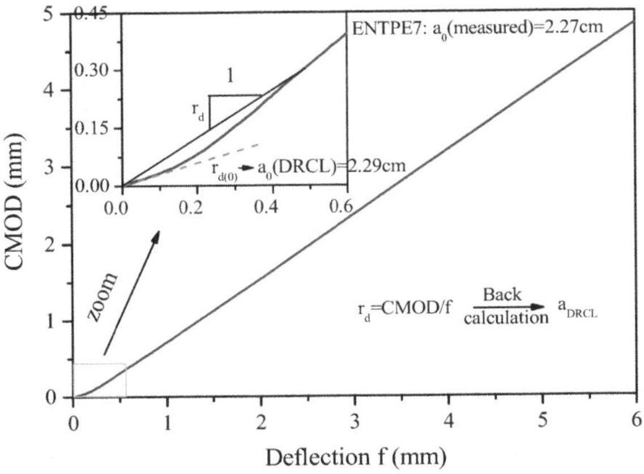


Fig. 2.15 Calculation of the crack lengths "$a_{DRCL}$" thanks to the displacement ratios $r_d$ determined from the CMOD-f curve during monotonic loading test (test ENTPE7)

### 2.2.3.2 Results

Figure 2.16 presents a comparison of the crack length obtained from cracking gages ($a_{gauge}$) and by DRCL method ($a_{DRCL}$) for ENTPE7 test (same test as presented in Figs. 2.12, 2.13 and 2.15). The evolution of crack length calculated by the DRCL method ($a_{DRCL}$), also plotted in Fig. 2.16, is slightly different from the evolution of crack length obtained by the cracking gauges ($a_{gauge1}$ and $a_{gauge2}$). A significant difference can be noted at the beginning of the test.

At the beginning of the curve: the deflection f increases, but the calculated crack $a_{DRCL}$ remains constant. The accuracy of the method is confirmed by the value of this calculated initial pre-notch, whose value is quite close to the measured initial notch (for example: 2.29 cm to be compared to 2.27 cm for test ENTPE7 in Fig. 2.15). The calculated crack clearly starts to propagate (point A in Fig. 2.16) before the load reaches its peak value (point B in Fig. 2.16). Whereas, cracking

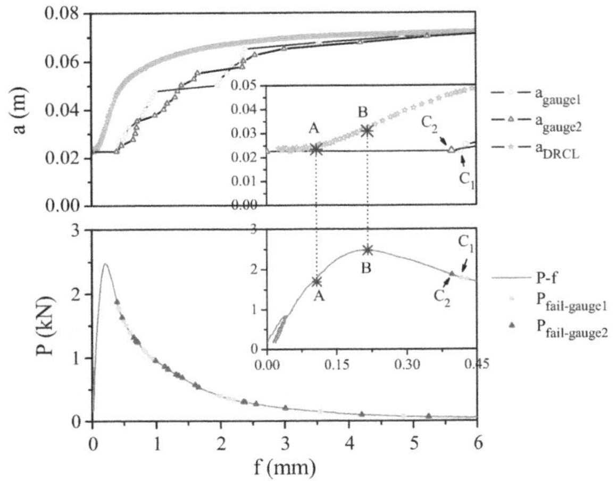


Fig. 2.16 ENTPE7 test: v = 5.9 mm/min; T = −1.5 °C (above): crack length in function of deflection of the beam. $a_{gauge1}$, $a_{gauge2}$: crack length measured by cracking gauges 1 and 2 respectively; $a_{DRCL}$: crack length obtained by the DRCL method; (below): load–deflection curve P–f. $P_{fail\text{-}gauge1}$, $P_{fail\text{-}gauge2}$: points located on the P–f curve where a wire fails for gauge1 and gauge2 respectively

gauges detect initiation of the crack (points $C_1$ and $C_2$ in Fig. 2.16) after the peak of the load. It is well admitted that, before macro-crack propagation, an initiation phase exists where damage occurs [20]. Only micro-cracks develop. Some studies showed that large damage occurs before the peak of the load—deflection curve, which can be considered as the start of macro-crack propagation [16, 21, 22]. The crack length calculated by the DRCL method could be considered as the sum of the macro-crack and a fictitious crack representing the damage zone at the crack tip. More details of the crack propagation mechanism evaluated by the DRCL method are presented in Nguyen et al. [13] and Nguyen [23].

#### 2.2.3.3 Summary

The DC(T) and C(T) tests are viable monotonic fracture tests available to characterize thermal and reflective cracking. Each test produces repeatable fracture energy results with coefficients of variation generally less than 15%. The DC(T) test has shown the ability to classify a variety of mixtures from RAP to WMA and to relate to transverse cracking in the field. Similarly, the C(T) test displayed the potential to differentiate among various TBO's and tack coat types. Few limitations are present with these tests. The most prominent limitations are the lack of mode mixity available with the tests and the lack of knowledge regarding deformation and stress states across the entire specimen during loading. Future extensions with these tests include the employment of digital image correlation to evaluate full field fracture behavior.

The Fenix test is an easy procedure that allows to compare the cracking resistance of different mixtures at different temperatures in terms of their dissipated energy or fracture energy and their tensile stiffness index (IRT). The procedure presents a good repeatability and it is sensitive to changes in temperature, binder type and binder content of the mixture. Furthermore, the mechanization of the specimen and their geometry makes possible to test not only lab specimens but also field cores. The four-point bending notch fracture test is another promising method to characterize Mode I fracturing of asphalt concrete. The novel DCRL method is a convenient and accurate method to characterize crack growth in this test and related fracture tests, including cyclic and monotonic types.

## 2.3 Lab Tests: Cyclic Cracking and Damage

### *2.3.1 Complex Modulus Testing Using Uniaxial Cylindrical Test Setup, Procedures and Analysis*

Complex modulus can be measured in different loading modes. Here the measurement of complex modulus using the test configuration designed by the DGCB laboratory of University of Lyon/ENTPE [21–27] is presented. The test consists in applying on cylindrical specimen, sinusoidal axial strain at different temperatures and frequencies. Axial stress ($\sigma$) is measured from a load cell. Axial strain ($\varepsilon_{ax}$) and radial strain ($\varepsilon_{rad}$) are obtained from non-contact displacement transducers (Fig. 2.17).

Small axial strain amplitudes (less than $10^{-4}$ m/m) were applied ensuring the behaviour remained inside the linear domain. The following Eq. (2.5) gives the sinusoidal evolution of the stress and strain measured with time (t).

$$\begin{cases} \sigma(t) = \sigma_0 \sin(\omega t + \phi_E) \\ \varepsilon_{ax}(t) = \varepsilon_{0ax} \sin(\omega t) \\ \varepsilon_{rad}(t) = \varepsilon_{0rad} \sin(\omega t + \pi + \phi_v) = -\varepsilon_{0rad} \sin(\omega t + \phi_v) \end{cases} \quad (2.5)$$

where $\phi_E$ is the classical phase angle between the axial strain and the axial stress, $\phi_v$ is the phase angle between the axial strain and the opposite of radial strain, $\omega$ is the pulsation ($\omega = 2\pi * frequency$).

Considering the complex notations where $j$ is the complex number ($j^2 = -1$), the measured values are written in Eq. (2.6).

$$\begin{cases} \sigma^*(t) = \sigma_0 e^{j(\omega t + \phi_E)} \\ \varepsilon^*_{ax}(t) = \varepsilon_{0ax} e^{j(\omega t)} \\ \varepsilon^*_{rad}(t) = -\varepsilon_{0rad} e^{j(\omega t + \phi_v)} \end{cases} \quad (2.6)$$

The complex modulus $E^*$ and the complex Poisson's ratio $v^*$ are obtained from Eq. (2.7).

$$\begin{cases} E^*(\omega) = \dfrac{\sigma^*(t)}{\varepsilon^*_{ax}(t)} = |E^*(\omega)| e^{j\varphi_E} = \dfrac{\sigma_0}{\varepsilon_{0ax}} e^{j\varphi_E} \\ v^*(\omega) = -\dfrac{\varepsilon^*_{rad}(t)}{\varepsilon^*_{ax}(t)} = |v^*(\omega)| e^{j\varphi_v} = \dfrac{\varepsilon_{0rad}}{\varepsilon_{0ax}} e^{j\varphi_v} \end{cases} \quad (2.7)$$

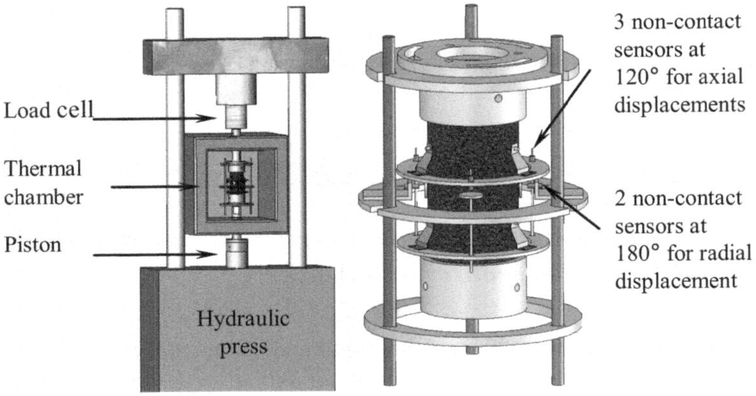


**Fig. 2.17** Experimental device used for complex modulus test: general view (left) and specimen view (right)

The specimen is loaded at 6 different frequencies (from 0.03 to 10 Hz) and 8 different temperatures (from −20 to 35 °C).

#### 2.3.1.1 Test Results

The norm of complex modulus $|E^*|$ and its phase angle $\phi_E$ (and norm of complex Poisson's ratio $|\nu^*|$ and its phase angle $\phi_\nu$, respectively) are presented in Fig. 2.18 (Fig. 2.19, respectively) for the tested material. It can be seen (Fig. 2.19) that the norm of the complex Poisson's ratio $|\nu^*|$ is not a constant as generally admitted for pavement design. For the considered mixture, $|\nu^*|$ varies from about 0.17, for high frequencies and/or low temperatures, to 0.35, for low frequencies and/or high temperatures. It decreases when temperature decreases and when frequency increases.

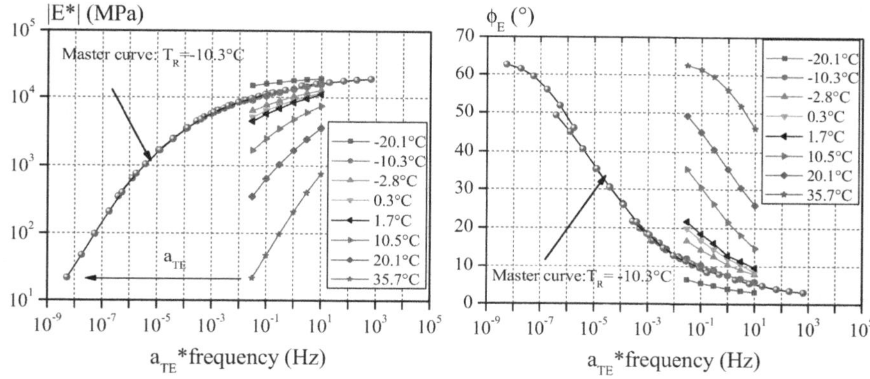


**Fig. 2.18** Experimental values (test MCE1) of complex modulus $E^*$ and master curves, plotted at reference temperature of −10.3 °C (left): Norm of complex modulus $|E^*|$; (right): phase angle $\phi_E$

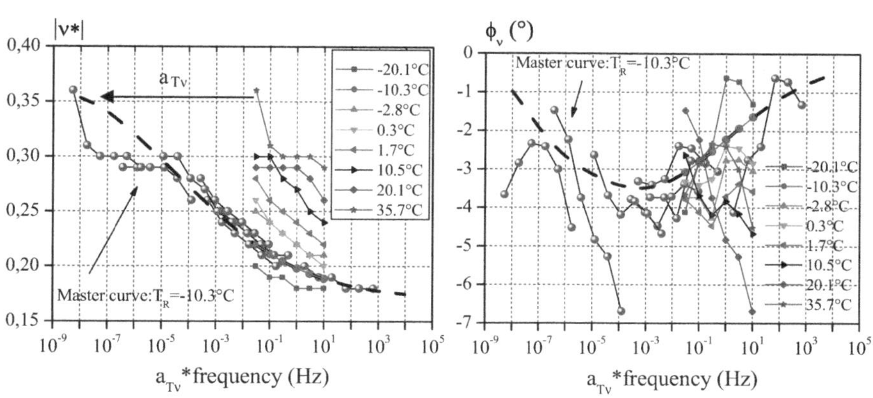


**Fig. 2.19** Experimental values (test MCE1) of complex Poisson's ratio $\nu^*$ and master curves, plotted at a reference temperature of −10.3 °C with $a_{T\nu}$ values equal to $a_{TE}$ values previously determined (Fig. 2.19) (left): norm of complex Poisson's ratio $|\nu^*|$; (right): phase angle $\phi_\nu$

Figure 2.19 also confirms results already pointed out by Nguyen et al. [13] and Pouget et al. [28], the phase angle of the complex Poisson's ratio $\phi_v$ is very small and negative but not nil as it is generally supposed in the literature. Negative values of $\phi_v$ mean that the radial strain is slightly delayed compared with the axial strain, which appears as physically acceptable.

As mentioned in the introduction, many researchers considering only unidirectional experiments have shown for this monodimensional limited case that bituminous mixtures respect the time—temperature superposition principle (TTSP) in the small strain domain, where the behaviour can be considered as linear viscoelastic (LVE). Application of this principle enables to plot a master curve at a given reference temperature ($T_R$). This complex modulus master curve is constructed by shifting along the frequency axis the complex modulus values obtained at different temperatures. The master curve of $E^*$ at the reference temperature −10.3 °C obtained for tests on specimen MCE1 is plotted in Fig. 2.18, which also shows the master curve of the phase angle $\phi_E$ at the same reference temperature.

An interesting output of this test is that the unidirectional result can be extended to the three dimensional case. Figure 2.19 shows that a master curve can also be obtained for the complex Poisson's ratio $v^*$ and its phase angle $\phi_v$. Some scatter appears in this Fig. 2.19. It can be explained by the practical difficulty for measuring Poisson's ratio, as involved strain levels remain very low. Furthermore, the shift factors used for complex Poisson's ratio ($a_{Tv}$) are identical to the shift factors obtained for complex modulus ($a_{TE}$). This experimental result confirms the same tendency observed in recent studies [25, 29, 30]. An important output is that a unique shift factor can be considered in Eq. (2.8).

$$a_T = a_{TE} = a_{Tv} \tag{2.8}$$

The classical WLF law (Williams, Landel and Ferry) [31] is used to fit the shift factors $a_T$ in Eq. (2.9). The results are plotted in Fig. 2.20.

$$\log(a_T) = \frac{-C_1(T - T_R)}{(T - T_R) + C_2} \tag{2.9}$$

## 2.3.2 *Complex Modulus and Fatigue Testing Using Cantilever Trapazoidal Beam Test*

### 2.3.2.1 Complex Modulus Testing: Specimen Geometry and Equipment

During the round robin testing of asphalt mixtures for the RILEM TC CAP due to smaller thickness of each layer, the sawing procedure (NF P 98-250-3:1992) had been adapted for Illinois materials to extract the specimen horizontally in each layer. For each slab, 8 specimen are sawn, 4 for each layer (refer to Figs. 2.21 and 2.22).

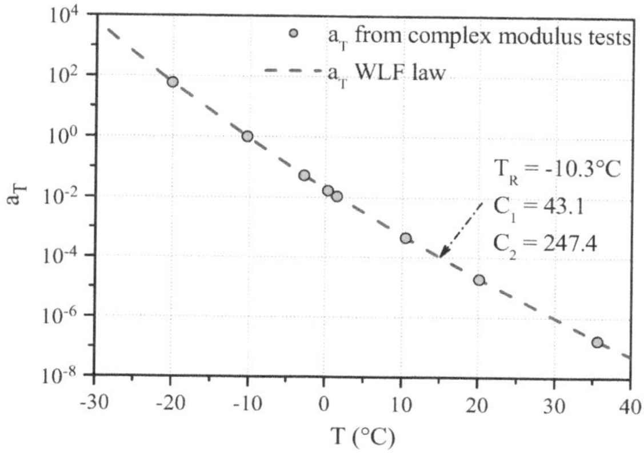


**Fig. 2.20** Shift factors $a_T$ of complex modulus and complex Poisson's ratio. Comparison with WLF law prediction for the test MCE1

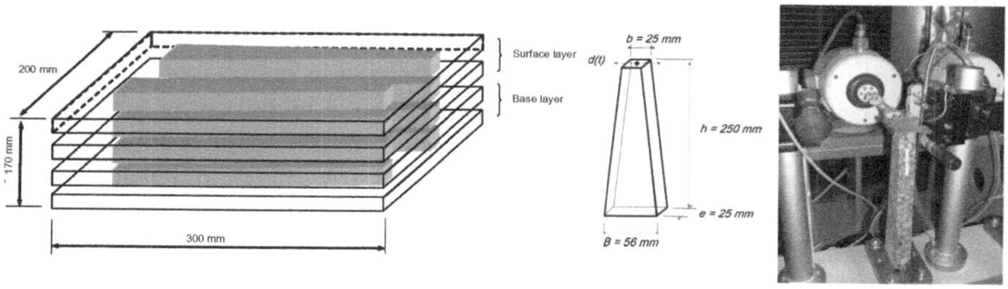


**Fig. 2.21** Specimen geometry and equipment

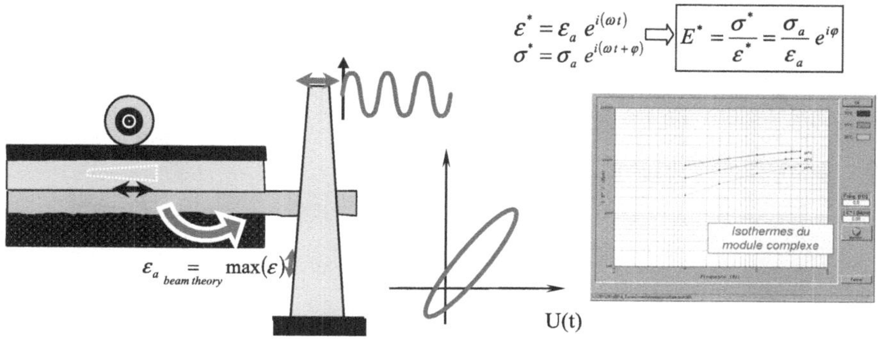


**Fig. 2.22** Experimental setup [32]

#### 2.3.2.2 Complex Modulus: Brief Test Description

Evaluation of the complex modulus of the materials has been performed according the current European standard EN 12697-26 on cantilever trapezoidal specimens (Fig. 2.23). Materials extracted from the two layers are lightly different. The top layer material's is "stiffer" than the top layer ones. This difference is logically correlated to the difference of void content of the two layers evaluated on the two sets of specimen (around 2%).

This test is a French standard test for the behaviour characterization of bituminous material and is systematically carried out to provide the information required for product standards (modulus at 15 °C, 10 Hz) for the French Design method [33]. It allows for the determination of linear viscoelastic performance of bituminous materials depending on their frequency and operating temperature ranges. Four specimens can be tested simultaneously on the 3MC device within the following maximum temperature and frequency ranges: from −10 to 60 °C and from 1 to 40 Hz so as master curve can be drawn. The results are used to set some rheological models before bituminous material modelling.

Rheology of bituminous materials is strongly dependent of loading time and temperature. Interrelationship between frequency and temperature for these materials makes possible to get the same mechanical behaviour in different experimental conditions. Hence, mechanical properties, determined at high loading time (or low frequency) and at low temperature, can be found at low loading time (or high frequency) and at high temperature. In some cases (thermo rheologically simple bituminous materials) the equivalency between time and temperature allows to build master curve from linear viscoelastic data by shifting measurement at different temperature in order to obtain a continuous curve at a reference temperature [34]. The mechanical behaviour of asphalt materials is assumed to be linear thermo-viscoelastic and represented by the five viscoelastic coefficients $E_0^i$, $E_\infty^i$, $k_i$, $h_i$, $\delta_i$ and the three thermal coefficients $A_0^i$, $A_1^i$, $A_2^i$ of the complex

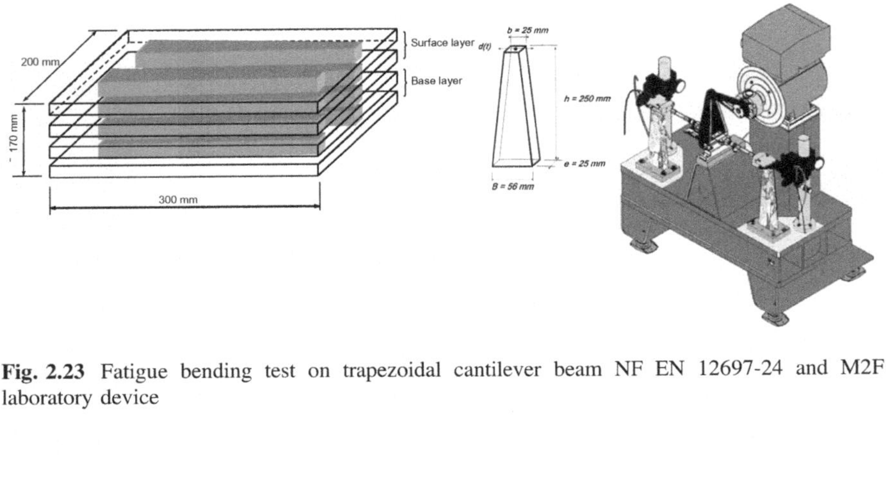


**Fig. 2.23** Fatigue bending test on trapezoidal cantilever beam NF EN 12697-24 and M2F laboratory device

modulus of the Huet-Sayegh model [35, 36] for the layer i. The Huet-Sayegh model consists in two parallel branches. The first branch is made up of a spring and two parabolic dampers that give the instantaneous and the retarded elasticity of asphalt, respectively. The second one is made up of a spring and it represents the static or the long-term elasticity of asphalt. Parameter $E_\infty$ is the instantaneous elastic modulus, $E_0$ is the static elastic modulus, k and h are the exponents of the parabolic dampers ($1 > h > k > 0$), and $\delta$ is a positive dimensional coefficient balancing the contribution of the first damper in the global behaviour. By means of parabolic creep laws associated to the two dampers, this rheological model predicts very accurately the complex modulus test obtained for asphalt mixes at different temperatures and frequencies.

The coefficients of the Huet-Sayegh model can be determined from experimental tests and by using the free software Viscoanalyse [34]. For 3D pavement analysis, a semi-analytical multi-layered solution using Fast Fourier Transforms and the linear behaviour of the Huet-Sayegh model for asphalt materials has been written in the free software ViscoRoute. ViscoRoute 2.0 enables users to consider multiple moving loads and elliptical-shaped loads [37].

#### 2.3.2.3 Fatigue Testing: Specimen Geometry and Laboratory Device

Due to the thickness of each layer, the sawing procedure (NF P 98-250-3:1992) has been adapted to extract the specimen horizontally in each layer. For each slab, 8 specimen are sawn, 4 for each layer. The experimental setup is shown in Fig. 2.24.

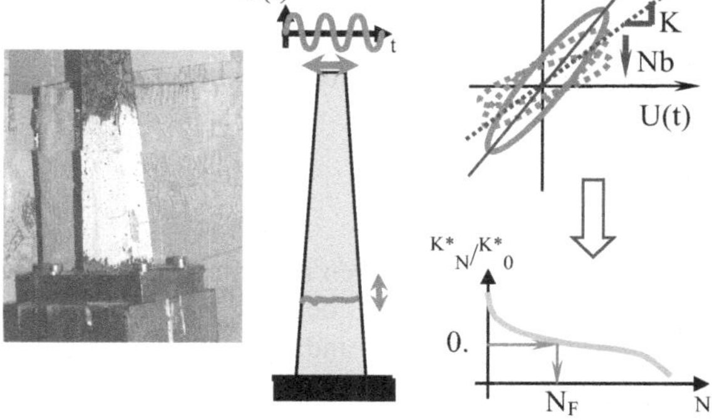


**Fig. 2.24** Experimental setup [32]

*Brief Test Description*

The fatigue resistance of the materials is evaluated in bending mode on trapezoidal cantilever beams according the standard [38] on the M2F device (Fig. 2.23). Results of this test are usually used for pavement thickness designing [33].

The trapezoidal specimen is clamped on its large base and loaded at the top small base. A constant amplitude displacement is applied. Tests are more often performed at 10 °C and at a frequency of 25 Hz. A variant of the test consist in applying a sinusoidal force at constant amplitude. A minimum of three loading levels sets need to be done. Each of them corresponds to a different loading level expressed by the maximal strain value in the specimen at the beginning of the test. To access to a mean value for each tests 6 specimen are tested at each loading level. For each test the force (and displacement) at the loaded boundary is measured as a function of the loading cycles.

The classical test analysis is based on a fatigue life criterion (Fig. 2.24). Classically, the specimen is supposed broken ($N_f$: number of cycles) when its stiffness reaches 50% of its initial value. A Wohler diagram is plotted presenting the fatigue life versus the loading levels ($\varepsilon_a$: maximal strain value around the specimen). Then, a linear regression is made in log-log coordinates leading to the relation between fatigue life and strain (Eq. 2.10).

For the French pavement design method, the strain level leading to specimen failure for one million cycles, $\varepsilon_6$ is used to calculate the admissible strain in each asphalt layers.

$$N_f = 10^6 \left(\frac{\varepsilon_a}{\varepsilon_6}\right)^p \tag{2.10}$$

where p is the fatigue line slope in the log-log diagram [33].

To study the effects of the size of the specimen on fatigue process, three size of homothetic specimen can be tested [39]. Then a fatigue damage model [40] can be used to assess intrinsic fatigue cracking parameters.

#### 2.3.2.4 Test Results

Fatigue tests were undertaken in controlled strain mode. Testing conditions were as follows: temperature 20 °C degrees; frequency 15 and 25 Hz; sinusoidal loading; strain levels between 120 and 190 με. [41, 42]. For each test stiffness modulus, phase angle and dissipated energy were calculated. The material chosen for the fatigue test is 10 mm Dense Bitumen Macadam (DBM) and 100 Pen Binder.

Figure 2.25 shows the typical trend of the stiffness modulus. It is calculated considering the recorded stress divided by the applied strain. As it can be possible to see, a three stage evolution process is recorded during a fatigue test. After a rapid reduction of stiffness (phase I), due to the internal heating phenomenon, the stiffness decrease seems more regular (phase II). Fracture occurs in the final stage (phase III)

Fig. 2.25 Stiffness evolution during a fatigue test (160 με)

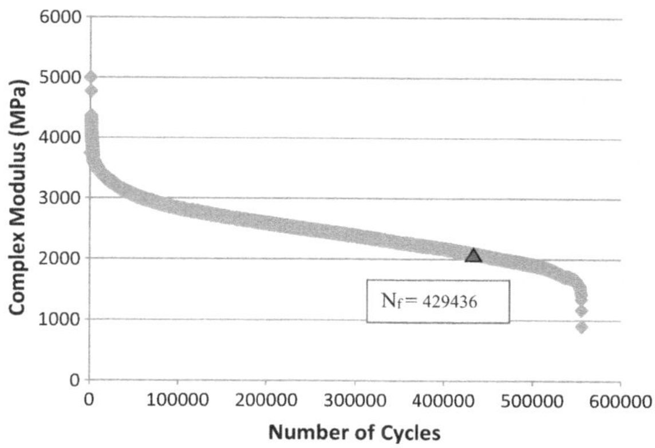


Fig. 2.26 Dissipated energy versus number of cycles during a fatigue test (160 με) fitted with power law ($Ax^k$)

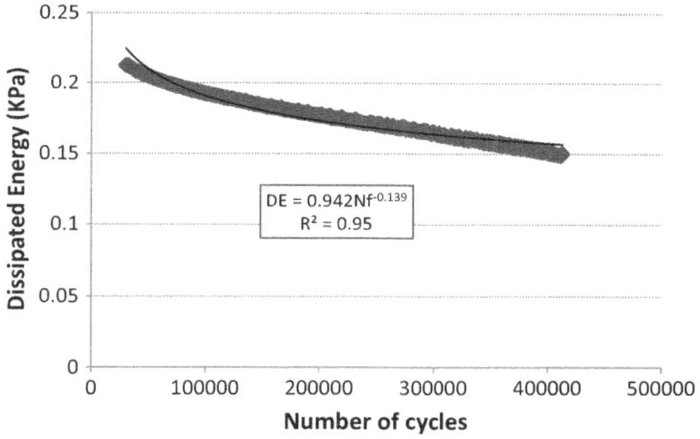


and it is characterised by an acceleration of stiffness drop [40–46]. Figure 2.26 shows the dissipated energy versus number of cycles.

## 2.3.3 *Skrinkage-Bending Test from LRPC of Autun*

### 2.3.3.1 Specimen Geometry and Equipment

Among all different lab tests used to get the behaviour of asphalt concrete beam with one or two layer(s) and an initial crack notch, the one coming from the regional laboratory of Autun (France) offers the possibility to simulate in laboratory, with thermal control possibilities crack propagations due to different shrinkage and cyclic bending load conditions. Readers may find more details about this test in Vecoven et al. [47] and Dumas et al. [48] (Fig. 2.27). By mean of a notch, this equipment has been used for testing the crack propagation in an asphalt material coming from the previous Rilem TC-CAP (Fig. 2.27).

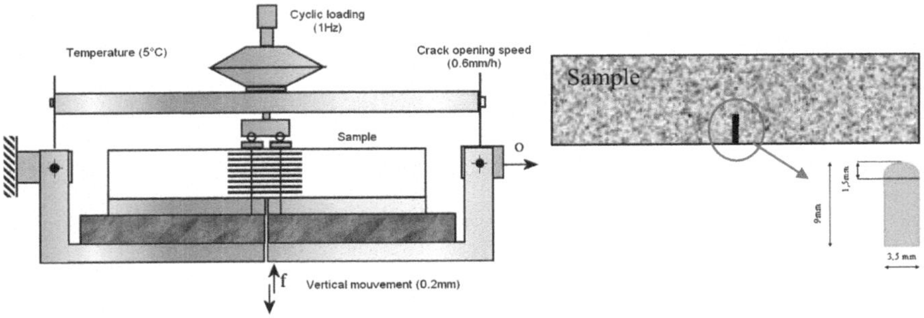


**Fig. 2.27** Equipment [47] and the "special mono-layer" specimen geometry tested for the Rilem TC-CAP

### 2.3.3.2 Brief Test Description

This test simulates at once the pavement thermal contraction and the heavy traffic solicitations at a constant temperature (+5 °C) in the following way (Fig. 2.27):

- The thermal shrinkage of precrack semi-rigid pavement is simulated by the opening of the movable plate (average speed 0.6 mm/h)
- The action of a track axle is simulated by a cyclic loading at a frequency of 1 Hz monitored by the deflexion set to 0.2 mm.

During the test the measure consists in following the crack (initiation, propagation in the first centimetres and breaking time) by mean of special cracks sensors, strain sensors, force and displacement sensors (Fig. 2.28).

Although the link between these lab experimental results and those obtain with some accelerating pavement testing has to be done with lots of care, this device is still used. It serves in France as a reference to test anti-crack systems for the

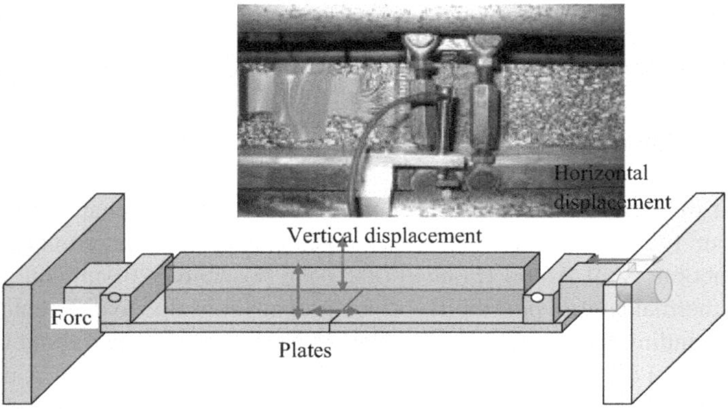


**Fig. 2.28** Experimental setup

knowledge of long term behaviour of 40% of French roads made of composite pavement (see SETRA Note No.57 for the typical French procedure used for testing anti-crack systems).

## 2.4 Constitutive Models for Crack Initiation and Propagation

The cracking phenomenon in asphalt concrete is quite complex and requires employment of fundamental fracture mechanics principles that are coupled with time and temperature dependent constitutive models to accurately represent the initiation and propagation of cracks. Due to the level of complexity involved with modelling of asphalt concrete as a heterogenous composite it is common for most modelling approaches to homogenize the asphalt mixture as an isotropic viscoelastic composite. This assumption holds true at lower temperatures, as temperature increases the presence of viscoplasticity becomes more prevalent and should be incorporated in the model formulations.

This section of the chapter is divided into three sub-sections. The first subsection presents an energy based approach to model crack growth initiation in viscoelastic materials. The formulation is based on approach proposed by Dubois et al. [49]. This is followed by cohesive zone approach, which is also an energy based formulation for representation of fracture quasi-brittle materials. Finally, a model proposed by University of Florida is presented that separates the elastic strain energy from the total strain energy to characterize fracture process in asphalt mixtures.

### *2.4.1 Crack Growth Initiation Model for Viscoelastic Materials*

#### 2.4.1.1 Model Introduction

For purposes of modeling, the bituminous concrete can be assumed as an isotropic viscoelastic material. It can be hypothesized that the crack growth process is generally concentrated within the bitumen or mastic phase of the bituminous concrete with crack bifurcations driven by the aggregates and their thin film interfaces. The study of the crack initiation and its propagation requires integration of time dependent behaviour associated with an energy based approach near the crack tip, thus making energy release rate concept a feasible approach. The crack growth initiation model described here focuses on a numerical approach that couples an incremental formulation for viscoelastic response and energy release rate evolution versus time allowing the simulation of the crack initiation and its propagation.

The model is implemented in a finite element software using spectrum decomposition scheme for Prony series in the complex domain. This algorithm allows the definition of the strain and stress responses by integrating the Boltzmann's formulation in the time domain. In this context, time frequency transposition is proposed using a spectrum decomposition technique. The fracture analysis for representation of crack initiation and growth in bitumen follows Dubois's approach [49]. In this approach the energy release rate is computed by performing a separation technique providing the exact the viscous dissipation isolation. Based on the non-dependence past integrals, the Gθ technique is adapted for creep crack growth [50].

#### 2.4.1.2 Model Formulation

The viscoelastic behaviour is generally written in terms of a Boltzmann's integral which defined the time relationship between the stress tensor $\underline{\underline{\sigma}}(t)$ and the strain tensor $\underline{\underline{\varepsilon}}(t)$. Assuming a linear response, the hereditary behaviour is traduced by the following form:

$$\underline{\underline{\varepsilon}}(t) = \int_{0}^{t} \mathbf{J}(t-\tau) \cdot \frac{\partial \underline{\underline{\sigma}}}{\partial \tau} d\tau \tag{2.11}$$

In which $\mathbf{J}$ is the four-order creep tensor. The bitumen can be characterized as an isotropic material. It is generally admitted that the Poisson's coefficient $v$ can be assumed as a constant property. With this consideration, Eq. (2.11) can be rewritten as follows:

$$\underline{\underline{\varepsilon}}(t) = \mathbf{A} \cdot \int_{0}^{t} C(t-\tau) \cdot \frac{\partial \underline{\underline{\sigma}}}{\partial \tau} d\tau \tag{2.12}$$

$C(t)$ designates a compliance function which can be defined as the inverse function of the tangent rigidity modulus noted $E(t)$. $\mathbf{A}$ is a constant tensor composed with the Poisson's ratio. The linear viscoelastic behavior can be expressed with a discrete spectrum representation which allows us the storage overcoming of the complete past history of stresses and strains. This method is based on a time discretization of the compliance function according to the following Prony's series:

$$C(t) = C_o + \sum_{m=1}^{M} C_m \cdot [1 - \exp(-t/\tau_m)] \tag{2.13}$$

$C_o$ and $C_m$ are positive compliance constants. $\tau_m$ are relaxation times. Then, the governing equations can be obtained using the finite difference integration and a step-by-step algorithm. Incremental constitutive equations, using a linear approximation of stresses in each time step $\Delta t_n = t_n - t_{n-1}$, are given by:

$$\underline{\underline{\Delta\varepsilon_n}} = \mathbf{A} \cdot M_n \cdot \underline{\underline{\Delta\sigma_n}} + \underline{\underline{\tilde{\varepsilon}}}(t_{n-1}) \tag{2.14}$$

$\underline{\underline{\Delta\varepsilon_n}}$ and $\underline{\underline{\Delta\sigma_n}}$ are the increments of $\underline{\underline{\varepsilon}}(t)$ and $\underline{\underline{\sigma}}(t)$, respectively, during the time increment $\Delta t_n$. $\underline{\underline{\tilde{\varepsilon}}}(t_{n-1})$ is a pseudo-strain tensor at time $t_{n-1}$ illustrating the influence of the past history of stress. $M_n$ designates the viscoelastic compliance function defined, for the specific time step $\Delta t_n$, by:

$$M_n = C_o + \sum_{m=1}^{M} C_m \cdot \left[1 - \frac{\tau_m \cdot (1 - \exp(-\Delta t_n/\tau_m))}{\Delta t_n}\right] \tag{2.15}$$

The finite element resolution of the incremental formulation in Eq. (2.14) is performed by employ the Ghazlan's method derived from the virtual displacement principle. According to the nodal displacement vector notation $\{\Delta u\}_n$, the balance equation, in the discredited domain $\Omega$, can be written as:

$$K_T \cdot \{\Delta u\}_n = \{\Delta F_{ext}\}_n + \{\tilde{F}\}(t_{n-1}) \tag{2.16}$$

$K_T$ is an equivalent stiffness matrix. $\{\Delta F_{ext}\}_n$ denotes the increment of nodal force vector. $\{\tilde{F}\}(t_{n-1})$ is the supplementary viscous loading vector representing the complete mechanical history. It overcomes the difficulty to perform, for each time step, the Boltzmann's integral without storage of stress and strain histories. It is given by:

$$\{\tilde{F}\}(t_{n-1}) = \int_{\Omega} B^T \cdot M_n \cdot \mathbf{A} \cdot \{\tilde{\varepsilon}\}_{n-1} d\Omega \tag{2.17}$$

where $\{\tilde{\varepsilon}\}_{n-1}$ is the strain vector derived from the strain history tensor $\underline{\underline{\tilde{\varepsilon}}}(t_{n-1})$.

The crack growth process in a viscoelastic media induces different sources of energy dissipation achieved through viscous dissipation, the crack surface separation and the kinetic energy caused by the crack tip advance and dynamic effects. According to Dubois's formulations [50], the crack growth initiation can be driven by considering the viscoelastic energy release rate definition calculated by a classical path independent integral. In this context, this work uses the Gθ integral defined, for axisymmetric description, as follow:

$$G\theta = \frac{1}{P} \cdot \int_{C} \left(-W_{\varepsilon} \cdot \theta_{k,k} + \sigma_{ij} \cdot u_{i,k} \cdot \theta_{k,j}\right) dC \tag{2.18}$$

$C$ is path surrounding the crack tip. $\vec{\theta}$ is a derivable vector field defined by it boundary conditions on the integration path according to a crack growth in the x direction or the r direction for axisymmetric cases, respectively. $P$ is a geometric parameter allowing to adapt the formalism for axisymmetric configurations (circular crack tip length). For viscoelastic behaviours, Staverman has demonstrated the $W_{\varepsilon}$ form by integrating creep properties. It form is given by:

$$W_{\varepsilon} = \frac{1}{2} \cdot \int_{0}^{t} \int_{0}^{t} A_{ijkl} \cdot [2 \cdot C(t - \xi) - C(2 \cdot t - \xi - \beta)] \cdot \frac{\partial \sigma_{ij}}{\partial \xi} \cdot \frac{\partial \sigma_{kl}}{\partial \beta} d\xi d\beta \tag{2.19}$$

$A_{ijkl}, \sigma_{ij}$ and $\sigma_{kl}$ are components of the tensor **A** and the stress tensor, respectively, see Eq. (2.12).

#### 2.4.1.3 Example Application of the Crack Growth Initiation Model: Numerical Modelling of the Bitumen Fracture

The researchers at IFSTTAR have developed a test in order to evaluate the bitumen contribution in the fracture phenomena in bituminous concrete [51, 52]. This test allows to simulate the crack growth process in bitumen around aggregates. The specimen is composed by a thin film fixed between two steel convex protuberances onto which the testing machine imposes a displacement with different loading rates and temperature (c.f. Fig. 2.29). The specimen geometry allows assuming an axisymmetric approach. The crack initiation is localized in the specimen center. The vertical symmetry enables the assumption of fracture in mode—I with a concentric crack front along the privileged crack line. The discretized geometry through the finite element mesh is also shown in Fig. 2.29. The contact between steel convex protuberances and bitumen is modeled with a line interface on to which a vertical displacement boundary condition is imposed. It can be noted that the film thickness is a parameter called *e*. The crack propagation line is a symmetric line in which vertical displacements are blocked. The crack tip advance is operated by freeing vertical boundary conditions with respect to the actual crack length [53].

For the materials used in this example, the viscoelastic properties, in the frequency domain, have been measured on a pure bitumen of PEN50/70 grade by employing a viscoanalyser Metralib at different temperatures. The Prony series parameters for linear viscoelastic characterization are determined at a reference temperature of 0 °C. The decomposition of Prony series is based on a discrete

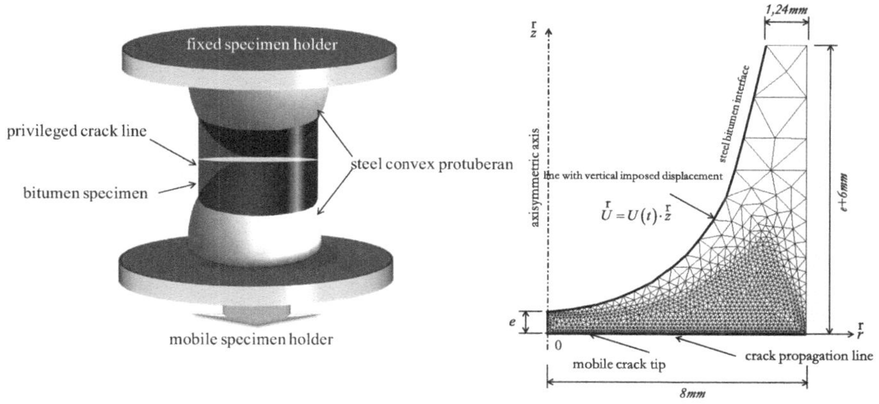


**Fig. 2.29** Experimental model and axisymmetric finite element mesh

**Table 2.4** Prony series parameters at reference temperature, T = 0 °C

| Compliances ($Mpa^{-1}$) | | Relaxation times (s) | |
|---|---|---|---|
| $C_o$ | $1.02 \times 10^{-3}$ | – | – |
| $C_1$ | $6.37 \times 10^{-4}$ | $\tau_1$ | $2.39 \times 10^{-3}$ |
| $C_2$ | $6.72 \times 10^{-4}$ | $\tau_2$ | $1.53 \times 10^{-2}$ |
| $C_3$ | $4.16 \times 10^{-3}$ | $\tau_3$ | $2.07 \times 10^{-1}$ |

spectrum decomposition allowing the limited number of parameters. The compliance components, $C_m$ and relaxation times $\tau_m$ are given in Table 2.4.

The tenacity test results showed the intrinsic separation work rate $G_s$ to be 4 J/m$^2$ and the tension toughness $f_t$ of 2 MPa at the reference temperature of 0 °C.

It is necessary to adjust the Poisson's ratio throughout the tensor A (Eq. (2.12)) in order to make adjustments for the assumed initial crack length. In this context, it considers that, for very small crack lengths, the reaction force can be assimilated as a function of Poisson's ratio, $v$. A first test is analyzed with a film thickness ($e$) of 320 μm, a displacement speed of 11 m/s. The crack initiation is given for a critical displacement value of 6.7 μm and an equivalent reaction force of 150 N. The correspondence between displacement time and force leads to resolving the Poisson's ratio value at 0.35. This first simulation deals with the crack growth initiation by imposing different speeds for displacements. Results are given in terms of displacement-force curves for 11 (reference result), 5.5, 2.0 and 1.1 m/s, see Fig. 2.30a. These results show that the relaxation effects are evident during the crack growth initiation phase. At high speeds, the crack growth initiation is driven by the elastic response inducing high force value and lower displacements. In the other hand, at low speeds the long term properties with a softening behaviour is apparent through lower induced force levels.

Figure 2.30b shows comparisons between the experimental and finite element simulation results. A good prediction for the crack growth initiation phase can be

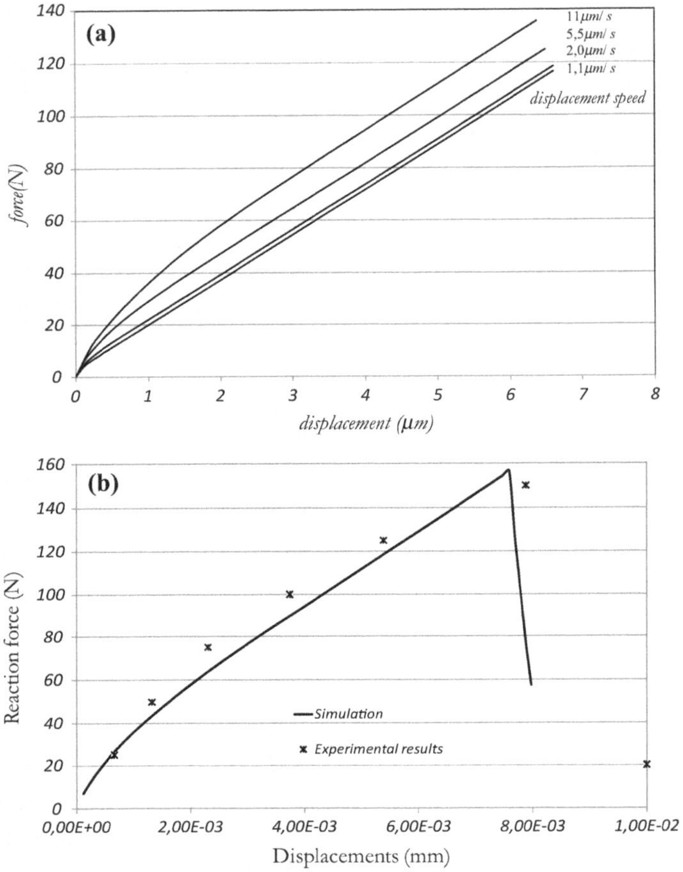


**Fig. 2.30** Force-displacement, **a** experimental results; **b** experimental and simulation results

observed. The differences can be partly due to difficulties in defining the time discretization of creep function. However, the crack growth instability traduces more differences with experimental results. In fact, the time domain resolution and the crack growth algorithm can trace the progressive crack lip decohesion but does not take into account the damage state around the crack tip characterized by the process zone. The cohesive zone approach discussed next focusses on capturing the damage state within the process zone.

## 2.4.2 *Cohesive Zone Fracture Model*

### 2.4.2.1 Model Introduction

The geometric and non-linear material response in the vicinity of the crack tip and the importance of accurate simulation of crack initiation and propagation make a standard "strength of materials" type analysis insufficient for the realistic simulation of cracking in asphalt concrete. Wagoner et al. [1, 6], Li et al. [54] and other researchers involved in fracture testing of asphalt concrete at low and intermediate temperatures have reported the presence of a fracture process zone ahead of an advancing crack tip. The cohesive zone fracture modelling approach allows for better representation of the traction-separation behaviour of quasi-brittle materials by linking intrinsic material properties such as fracture energy and cohesive strength with the length scale associated with the size of the aforementioned fracture process zone (FPZ). For simulation of crack initiation and propagation, a cohesive zone model is recommended for its accuracy in representing the FPZ and efficiency in numerical simulation framework, such as finite element analysis. Several researchers including Song et al. [55] have demonstrated the capabilities of the cohesive zone model (CZM) for accurate simulation of cracking in asphalt concrete materials. The bi-linear CZM described by Song et al. [55] has been applied for simulation of thermal and reflective cracking in asphalt pavements and overlays, for example by Dave et al. [56]. Use of temperature dependent exponential and power law softening cohesive zone models have been utilized by Dave and Behnia [57] to design high performance asphalt overlays.

### 2.4.2.2 Model Formulation

Here a temperature dependent bi-linear CZM is presented. This approach allows the fracture energy of asphalt concretes to be simulated as function of temperature. At present, the tensile strength of the material is modelled to be independent of temperature. This simplification was made on the basis of the laboratory test results, which indicated very limited dependence of strength with temperature within the simulation thermal boundaries, i.e. for temperatures below 10 °C. Figure 2.31 illustrates the temperature dependent bi-linear CZM for an asphalt mixture from in mode-I (pure opening) fracture. The CZM describes the relationship between traction and the displacement jump along the potential crack path. Notice that the intrinsic model form of the present CZM requires use of an initial, linearly elastic response regime with very high stiffness selected based upon the parametric study of Song et al. [55], as described below. The location 'A' in the figure illustrates the peak traction corresponding to the material strength as the threshold for material damage. The displacement at this location can be written as $\delta_A$. The region between points 'A' and 'B' represents the progression of damage in the material and could represent the coalescence of the micro-cracks within the fracture process zone. The

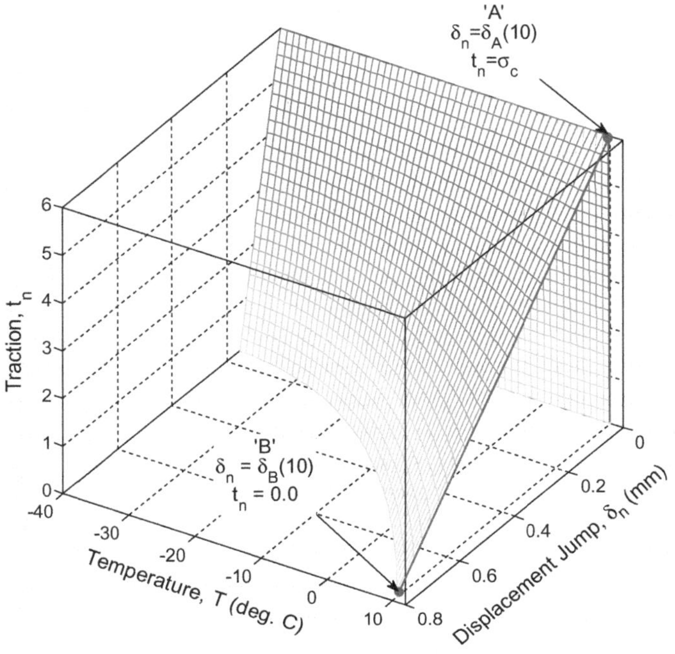


**Fig. 2.31** Temperature dependent bi-linear cohesive zone model for MnROAD section 34 mixture (cohesive strength = 5.94 MPa)

region past of location 'B' represents a fully formed macro-crack, the cohesive displacement at point 'B' is commonly referred as the critical displacement and could be given by $\delta_B$. At any given temperature, the area under the traction-displacement curve represents the fracture energy of the material. For the given example the traction-displacement surface illustrates its evolution with temperature. The traction-displacement relationship for the temperature dependent bi-linear CZM in mode-I could be written as (2.20):

$$\begin{aligned} \text{For } \delta_n < \delta_A, t_n(T) &= \sigma_c \frac{\delta_B(T)}{\delta_A(T)} \left( \frac{\delta_n}{\delta_B(T)} \right) \\ \text{For } \delta_n > \delta_A, t_n(T) &= \sigma_c \left( \frac{\delta_B(T) - \delta_n}{\delta_B(T) - \delta_A(T)} \right) \end{aligned} \tag{2.20}$$

where, $\sigma_c$ represents the material cohesive strength and $T$ is the temperature. The displacements at any given instance is the element is $\delta_n$ and the corresponding traction is $t_n$. The displacement at peak traction and at the complete separation are $\delta_A$ and $\delta_B$ respectively. The fracture energy of material, $G_f$, links the strength, $\sigma_c$, with the displacement jump at complete separation, $\delta_B$ as (2.21):

$$G_f(T) = \frac{1}{2}\sigma_c \delta_B(T) \tag{2.21}$$

The displacement jump at peak traction, $\delta_A$ is chosen to minimize undesired or artificial compliance in the cohesive zone interface elements prior to material softening (Fig. 2.31). It is typically recommended for this to be less than or equal to 1% of $\delta_B$. This recommendation is made in order to minimize the artificial compliance induced due to model in the simulation results.

#### 2.4.2.3 Model Verification, Calibration and Validation

Verification, calibration, and validation steps provide a realistic evaluation of the accuracy and robustness of a numerical model. As part of the RILEM CAP TG-03 work, the researchers from University of Illinois completed these steps to evaluate the capabilities of cohesive zone models. The verification step in this case employed the double cantilever beam (DCB) geometry due to the presence of a known analytical solution. As shown in Fig. 2.32a, the numerical results matched well with the closed-form solution. Next, the calibration step included the use of the DC(T) geometry with experimentally evaluated strength and fracture energy parameters embedded in the CZM. The numerical solution in the calibration stage was adjusted by a constant factor of 0.84 to produce a global load-CMOD response similar to the experimental results. In general, calibration factors are employed in considering global fracture energy because CZM models tend to over-predict the fracture resistance of the material at the local level. Finally, validation was completed using the semi-circular bend (SCB) geometry. In this study, the cohesive zone model performed quite well as shown in Fig. 2.32b. Therefore, the verification, calibration, and validation steps displayed the abilities of the CZM approach in asphalt materials.

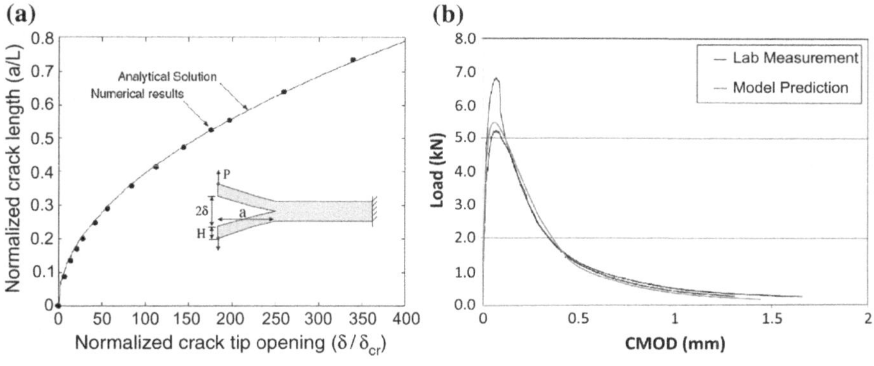


**Fig. 2.32** **a** Verification using DCB **b** validation using SCB

## 2.4.3 *Hot Mix Asphalt Fracture Mechanics Model*

### 2.4.3.1 Model Introduction

Laboratory and field studies at the University of Florida have led to the development of a viscoelastic fracture mechanics-based cracking model, termed HMA-Fracture mechanics (HMA-FM) model. The model is focused on describing both initiation and propagation of cracks in asphalt mixtures for any combination of loading and temperature conditions. Previous research work has shown that conventional fracture mechanics approach [58, 59] may provide a rational way to consider the effects of stress redistribution induced by flaws or micro-cracks in a material. The HMA-FM model assumes that crack growth in asphalt pavements occurs in discontinuous, step-wise manner on basis of the anecdotal evidence from field observations as reported in previous research [60, 61].

Viscoelasticity and continuum damage approach [61, 62] analyzes micro-cracking behavior of asphalt mixtures under realistic loading conditions and healing effect. However, continuum damage mechanics is not applicable after crack initiates (the system is no longer a continuum) and, thus, a continuous crack growth law is assumed for crack propagation in the HMA-FM approach.

The HMA-FM model provides a viscoelastic fracture mechanics-based cracking model able to predict both crack initiation and step-wise propagation, under generalized loading and healing conditions, based on an energy-based threshold concept.

### 2.4.3.2 Model Formulation

The HMA-FM model utilizes a concept of threshold or limit that is based on the observation that micro-damage in asphalt mixtures (i.e., damage that may potentially lead to crack initiation or crack propagation) appears to be fully healable, whereas macro-damage (i.e., damage associated with crack initiation or propagation) does not appear to be healable [59–66]. This indicates that a damage threshold exists below which damage is fully healable. Once the threshold is exceeded, the developed macro-damage is no longer healable. As with other two approaches discussed earlier in the section, the threshold limit is defined in terms of energy. The energy threshold defines the development of macro-cracks (macro-damage), at any time during either crack initiation or propagation, at any point in the mixture. If loading and healing conditions are such that the induced energy does not exceed the mixture threshold, then the mixture may never crack, regardless of the number of load repetitions applied.

As discussed by Zhang [63] and Roque et al. [65], fracture (crack initiation or crack propagation) can develop in asphalt mixtures in two distinct ways, defined by two distinct thresholds or failure limits. It has been determined that the dissipated

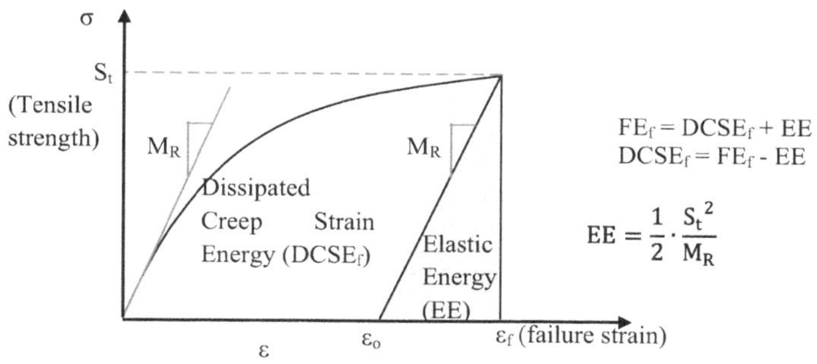


**Fig. 2.33** Determination of failure limits: FE and DCSE [59–65]

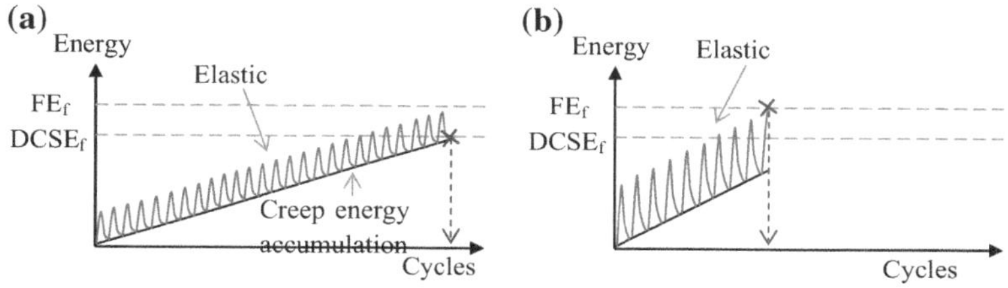


**Fig. 2.34** Potential failure conditions under continuous loading **a** repeated loading **b** critical load

creep strain energy (DCSE) density and the fracture energy (FE) density of asphalt mixtures suitably define the lower and the upper failure limits, respectively.

In the HMA-FM approach the fracture energy density and dissipated creep strain energy density is determined from the stress-strain response of a tensile strength test conducted on asphalt mixture, as shown in Fig. 2.33. The fracture energy density limit ($FE_f$) is determined as the area under the stress-strain curve, while the dissipated creep strain energy density limit ($DCSE_f$) is the fracture energy minus the elastic energy (EE) at the time of fracture. Resilient modulus ($M_R$) and tensile strength ($S_t$) are used to define EE. These two energy failure limits ($DCSE_f$ and $FE_f$) have been identified as fundamental material properties of asphalt mixtures, independent of mode of loading, rate of loading and specimen geometry [67].

The lower threshold ($DCSE_f$) is selected as the failure criterion under repeated loading condition (Fig. 2.34a). When repeated stresses significantly below the tensile strength occur, cracking will eventually take place if the rate of damage accumulation exceeds the rate of healing during the loading period. In contrast, the upper energy threshold ($FE_f$) corresponds to the failure criterion under critical loading condition (Fig. 2.34b). In this case, fracture would occur if any single load applied during the loading cycle exceeds the threshold required to fracture the mixture.

One unique feature of the HMA-FM model is the consideration of both failure limit and rate of damage to predict fracture. The rate of damage of an asphalt mixture under repeated loading conditions is defined as the dissipated creep strain energy density accumulated per load cycle (DCSE/cycle). Therefore, there is an inherent assumption that damage can be quantified in terms of the viscous response (creep) of the asphalt mixture.

For a haversine load consisting of a 0.1 s loading period followed by a 0.9 s rest period, DCSE/cycle is defined as the integral of the stress $\sigma(t)$ multiplied by the creep strain rate $\dot{\varepsilon}_{cr}(t)$:

where $\sigma_{ave}$ represents the average stress in the zone of interest, and $\dot{\varepsilon}_{cr,max}$ is the maximum creep strain rate.

$$\frac{DCSE}{cycle} = \int_0^{0.1} \sigma(t) \cdot \dot{\varepsilon}_{cr}(t) \cdot dt = \int_0^{0.1} \underbrace{\sigma_{ave} \cdot \sin(10\pi t)}_{\sigma(t)} \cdot \underbrace{\dot{\varepsilon}_{cr,max} \cdot \sin(10\pi t)}_{\dot{\varepsilon}_{cr}(t)} \cdot dt \quad (2.22)$$

The maximum creep strain rate can be estimated from a creep test, based on the relationship between creep strain $\varepsilon_{cr}(t)$ and creep compliance D(t):

$$\varepsilon_{cr}(t) = \sigma_{ave} \cdot D(t) \quad (2.23)$$

One model used for creep compliance representation is power law function:

$$D(t) = D_0 + D_1 \cdot (t_{cr})^m \quad (2.24)$$

where $D_0$, $D_1$, and m are fitting parameters, and $t_{cr}$ is the duration of the creep test. Typically, a 1000-second creep test has been considered to be enough to capture the viscous response exclusively for intermediate temperatures associated with fatigue (0–25 °C). Then, the maximum creep strain rate is determined as:

$$\dot{\varepsilon}_{cr,max} = \sigma_{ave} \cdot \frac{dD(t)}{dt} = \sigma_{ave} \cdot m \cdot D_1 \cdot (t_{cr})^{m-1} \quad (2.25)$$

Substituting Eq. (2.25) into Eq. (2.22), the rate of damage (DCSE/cycle) is obtained:

$$CSE/cycle = \frac{1}{20} \cdot (\sigma_{ave})^2 \cdot m \cdot D_1 \cdot (t_{cr})^{m-1} \quad (2.26)$$

It should be pointed out that healing, which can be described in terms of recovered DCSE per cycle [68], must be determined and used to adjust the rate of damage.

The parameters utilized in this approach can be used not only to predict damage and crack growth in mixtures subjected to generalized loading conditions, but they may also be used in mixture design and optimization. For example, the fracture tolerance of mixture improves as the $DCSE_f$ limit increases. Similarly, a lower creep rate, which is a function of $D_1$ and m-value, will result in a lower rate of

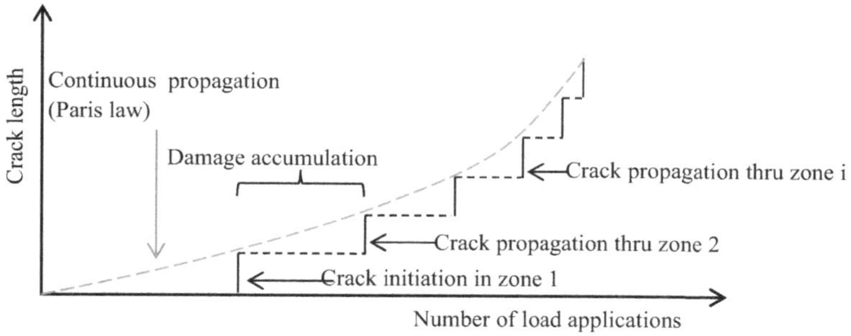


**Fig. 2.35** Step-wise crack growth law [59–65]

damage accumulation. However, lower rate of damage does not necessarily assure improved cracking performance, since failure limit and healing rate are also involved in cracking mechanism.

The HMA-FM model assumes a step-wise crack growth law, as shown in Fig. 2.35. When the DCSE accumulated in the zone with the highest average tensile stress equals the $DCSE_f$ limit of the mixture, a crack initiates. Then, crack propagates through the next zone every time that the DCSE accumulated in that zone equals the $DCSE_f$ limit of the mixture. Note that the model keeps track of the DCSE induced in all zones where crack growth may be of interest (at every step).

The basic analysis framework of the model is as follows:

1. Divide potential crack path(s) into zones. Previous analysis showed that zone size had little effect on the predicted rate of crack propagation [69], so zones of fixed size are considered for the analysis. Typically, a zone size of 5 mm is used, which is assumed to capture the effect of stress concentration near the contact points between aggregates.
2. Calculate or estimate the stress distribution and compute the average tensile stress in each zone during condition 1, i.e. $(\sigma_{ave})_{i,\ 1}$, where i represents the zone number.
3. Determine the rate of damage in each zone for condition 1, i.e. $(DCSE/cycle)_{i,1}$ according to Eq. (2.26) for the average tensile stress calculated in step 2, i.e. $(\sigma_{ave})_{i,1}$.
4. Determine the number of cycles for crack initiation in zone 1 ($N_{f,1}$):

$$N_{f,1}(initiation) = \frac{DCSE_f}{(DCSE/cycle)1,1} \tag{2.27}$$

in which $DCSE_f$ is the failure limit of the asphalt mixture, and $(DCSE/cycle)_{1,1}$ is the rate of damage for zone 1 during condition 1.

5. Determine the new stress distribution as a result of crack growth. Once the crack initiates, the stress distribution ahead of the crack tip $\sigma(r)$ can be estimated as:

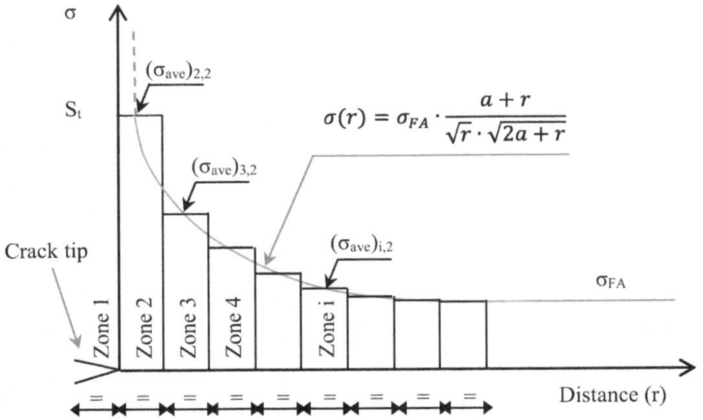


**Fig. 2.36** Stress distribution ahead of the crack tip after crack initiation

$$\sigma(r) = \sigma_{FA} \frac{a+r}{\sqrt{r \cdot \sqrt{2a+r}}} \tag{2.28}$$

where r is the distance measured from the crack tip, a is half the crack length, and $\sigma_{FA}$ represents the far away stress.

As shown in Fig. 2.36, tensile stresses in the material have an upper limit equal to the tensile strength of the asphalt mixture ($S_t$) and a lower limit equal to the far way stress ($\sigma_{FA}$).

6. Compute the average tensile stress in each zone during condition 2, i.e. $(\sigma_{ave})_{i,2}$.
7. Calculate the rate of damage in each zone for condition 2, i.e. $(DCSE/cycle)_{i,2}$ according to Eq. (2.26) for $(\sigma_{ave})_{i,2}$ obtained in step 6.
8. Determine the number of cycles for crack propagation through zone 2 ($N_{f,2}$):

$$N_{f,2}(propagation) = \frac{DCSE_f - \overbrace{\left(\frac{DCSE}{cycle}\right)_{2,1} \cdot N_{f,1}}^{Accumulated\ from\ previous\ condition}}{\left(\frac{DCSE}{cycle}\right)_{2,2}} \tag{2.29}$$

where $(DCSE/cycle)_{2,1}$ is the rate of damage for zone 2 during condition 1, $(DCSE/cycle)_{2,2}$ is the rate of damage for zone 2 during condition 2, and $N_{f,1}$ is the number of cycles to fail zone 1.

9. Repeat steps 5 through 8 to determine crack propagation through new zones. Note that the fracture energy density limit ($FE_f$) also has to be checked for every condition.

It should be pointed out that crack propagation predicted by the model (crack length versus number of load repetitions) agreed closely with crack growth measured in laboratory tests [59–65]. These results support that the HMA-FM model is rational and reliable to predict cracking performance of asphalt mixtures.

#### 2.4.3.3 Use of Model for Evaluation of Cracking Performance

A detailed analysis and evaluation of 22 field test sections throughout the state of Florida resulted in the qualification, calibration and validation of the energy-based criteria for cracking evaluation in asphalt mixtures [70]. The work clearly indicated that there is no single mixture property or characteristic (volumetrics, viscosity of recovered binder, resilient modulus, tensile strength, creep compliance, etc.) that can reliably predict cracking performance of asphalt mixtures. A parameter termed energy ratio (ER), which was derived using the HMA-FM model, was determined to accurately distinguish between pavements that exhibited cracking and those that did not, except for mixtures with excessively low or unusually high dissipated creep strain energy thresholds.

Using data from the field sections and laboratory testing the parameters such as ER and A can be calibrated to be linked to laboratory measured properties. The energy ratio (ER) is defined as the dissipated creep strain energy density limit ($DCSE_f$) of the mixture divided by the minimum dissipated creep strain energy required can be represented in terms of material properties as:

$$ER = \frac{DCSE_f}{DCSE_{min}} = \frac{DCSE_f}{\frac{m^{2.98} \cdot D_1}{A}} \tag{2.30}$$

where, m and $D_1$ are the creep compliance power law fitting parameters.

A is a parameter that depends on the tensile stress in the pavement section ($\sigma_t$) and the tensile strength of the mixture ($S_t$):

$$A = \frac{(6.36 - S_t)}{33.44 \cdot \sigma_t^{3.1}} + 2.46 \times 10^{-8} \tag{2.31}$$

The energy ratio accounts for the effects of pavement structural characteristics ($\sigma_t$) and material properties ($DCSE_f$, $S_t$, m, $D_1$) on cracking performance: the higher the value of the energy ratio, the better the expected cracking performance of the section. Therefore, ER can be used to integrate asphalt mixture properties in the pavement design process as well as to predict the performance of in-service pavement sections.

**Table 2.5** Summary of discrete crack models for asphalt mixtures

| Model (contributor) | Theoretical basis for model | Implementation in pavement structural model (framework) | Laboratory testing requirements | Model qualification (Q), verification (Vr), calibration (C) and validation (Vd) status | | | | | Field performance prediction status |
|---|---|---|---|---|---|---|---|---|---|
| | | | | Q | Vr | C | Vd (lab) | Vd (field) | |
| Crack growth initiation method (U. Limoges) | Viscoelasticity, energy release rate concept (Gθ approach) | No | – Viscoelastic characterization<br>– Crack growth test at different temperatures and loading rates | X | X | X | X | | Not applied at present but application is planned |
| Cohesive zone model (U. Illinois) | Quasi-brittle fracture process zone characterization using fracture energy and cohesive strength. Potential based approach | Yes (finite element analysis) | – Fracture energy testing (Typ.: Disk-shaped compact tension test, ASTM D7313)<br>– Cohesive strength (Typ.: Indirect tensile strength test)<br>– Viscoelastic characterization (Typ.: 1000 s indirect tensile creep test at 3 temperatures) | X | X | X | X | X | Applied for reflective and thermal cracking |
| HMA-fracture mechanics model (U. Florida) | LEFM extended to step-wise crack growth model and dissipated creep strain energy concept | Yes (layered elastic analysis/ boundary element analysis) | – Resilient modulus (Typ.: Indirect tensile resilient modulus @ 10 °C)<br>– Viscoelastic characterization (Typ.: 1000 s indirect tensile creep test at @ 10°C) | X | X | X | | X | Applied for fatigue (top-down and bottom-up) cracking |

(continued)

**Table 2.5** (continued)

| Model (contributor) | Theoretical basis for model | Implementation in pavement structural model (framework) | Laboratory testing requirements | Model qualification (Q), verification (Vr), calibration (C) and validation (Vd) status | | | | | Field performance prediction status |
|---|---|---|---|---|---|---|---|---|---|
| | | | | Q | Vr | C | Vd (lab) | Vd (field) | |
| | | | – Tensile strength (Typ.: Indirect tensile strength @ 10 °C)<br>– Healing test (Typ.: @ 20 °C) | | | | | | |

#### 2.4.3.4 Summary of Models for Discrete Cracking in Asphalt Mixtures

This section discussed three models that are developed and implemented for modelling of cracks in asphalt mixtures. All three of the approaches described herein have their advantages and disadvantages in terms of their capabilities and limitations. Each of these models have provided insight into the mechanisms of cracking in pavements. Table 2.5 provides a brief summary of the three approaches in terms of the basis (theory/approach) behind the models, required laboratory tests/material parameters, status of the model in terms of model qualification, verification, calibration and validation, and the status of application of the models to simulate cracking performance of pavements and improve understanding of cracking mechanisms for asphalt and composite pavements. The validation status is divided into two categories, namely laboratory and field. Laboratory validation refers to use of model to predict a different test using calibration and material parameters from a different test. Field validation refers to comparisons between model predictions of pavement cracking performance against the actual field cracking performance.

## 2.5 Constitutive Models for Cyclic Degradation

### *2.5.1 Dissipated Energy Concept*

#### 2.5.1.1 Model Introduction

Asphalt is a viscoelastic material, thus it dissipates energy under mechanical work (loading and relaxation). Usually, in an elastic material the energy is stored in the system when the load is applied, all the energy is recovered when the load is removed; in this case the unloading and the loading curves coincide. Viscoelastic materials are characterised by a hysteresis loop because the unloaded material traces a different path to that when loaded (phase lag is recorded between the applied stress and the measured strain); in this case the energy is dissipated in the form of mechanical work, heat generation, or damage [71, 72].

The area of the hysteresis loop represents the dissipated energy in a load cycle and the following equation can be used to calculate its value in a linear viscoelastic material subjected to sinusoidal loading:

$$W_i = \pi \sigma_i \varepsilon_i \sin \varphi_i \tag{2.32}$$

where,

$W_i$ *Dissipated energy in cycle i,*
$\sigma_i$ *stress level in cycle i,*
$\varepsilon_i$ *strain level in cycle i, and*
$\varphi_i$ *phase angle in cycle i.*

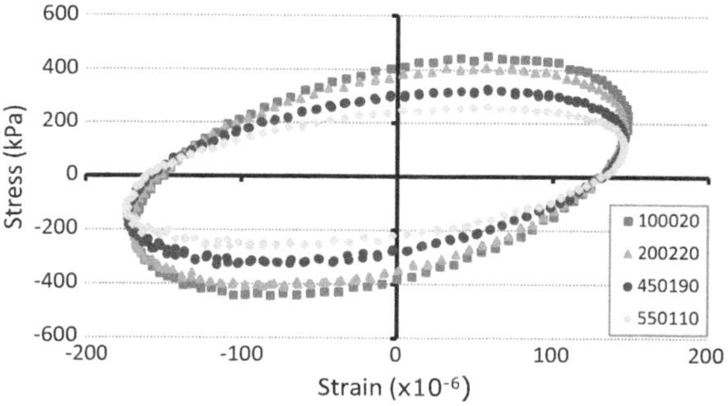


**Fig. 2.37** Evolution of hysteresis loop during a fatigue test

During a fatigue test, the stiffness reduces, the fatigue process starts and microcracks are induced in the material; therefore the dissipated energy, $W_i$, varies per loading cycle and it, usually, increases for controlled stress tests and decreases for controlled strain tests.

An example of the evolution of hysteresis loop (dissipated energy) during a fatigue test is shown in Fig. 2.37.

#### 2.5.1.2 Model Formulation

According to the classical fatigue analysis, fatigue life is conventionally defined as a number of cycles at which the stiffness modulus has decreased to the half of the initial value (see Fig. 2.39). The number of cycles to reach 50% of initial stiffness is indicated by $N_f^{50}$. However, dissipated energy methods have been considered in order to determine the number of cycles to failure. Thus, a comparison between different fatigue failure criteria, based on dissipated energy concepts, and the conventional analysis is presented here.

The Energy Ratio ($R\varepsilon$) method is considered first. It was introduced by Hopman et al. in 1989 [45]. The Energy ratio is the cycle number where cracks are considered to initiate ($N_f^{R\varepsilon}$). $N_1$ is defined as the point at which the slope of the energy ratio versus the number of cycles deviates from a straight line. In a controlled strain test, energy ratio is defined as follows:

$$R_\varepsilon = \frac{nW_0}{W_i} = \frac{n(\pi\sigma_0\varepsilon_0 \sin\varphi_0)}{\pi\sigma_i\varepsilon_i \sin\varphi_i} \tag{2.33}$$

where $W_0$ is the energy dissipated in the first cycle, $W_i$ is the energy dissipated at $i$th cycle. If the stress is replaced by the product of strain and modulus, and considering that the strain level remains constant for a strain controlled test, then the Eq. (2.33) can be simplified and written as follow [71]:

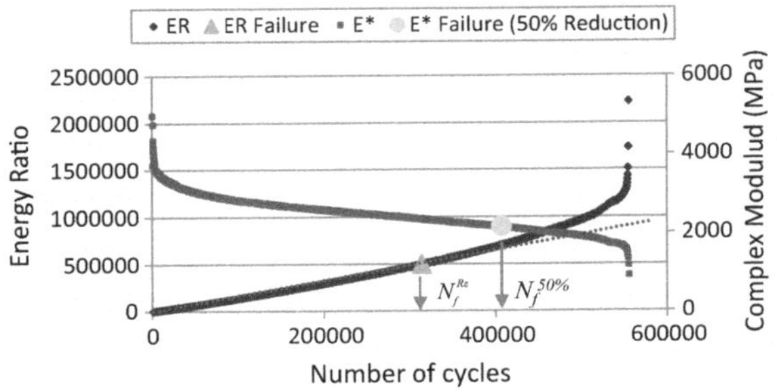


**Fig. 2.38** Energy ratio versus number of load cycles at 20 °C and at 15 Hz loading frequency

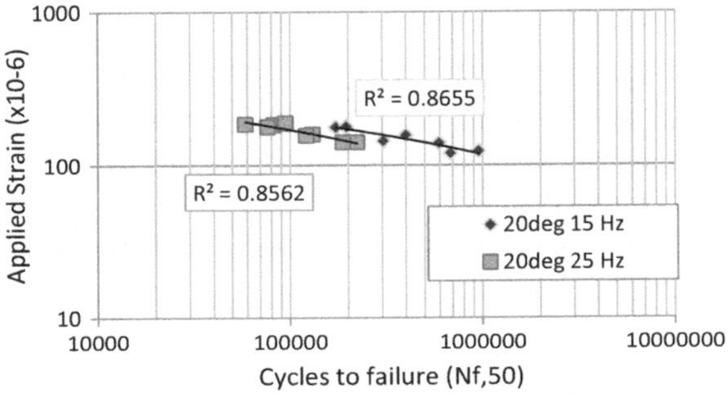


**Fig. 2.39** Classical fatigue life curves obtained at 20 °C at 15 and 25 Hz loading frequencies

$$R_{\varepsilon} = \frac{n}{E_i^*} \tag{2.34}$$

Figure 2.38 shows the energy ratio and complex modulus plotted as function of load cycles for an asphalt mixture. It can be seen that the number of cycles obtained by means of the classical analysis is generally greater than the energy ratio method.

Figure 2.39 shows the classical Whöler curves that represents the life duration versus applied strain amplitude. In the conventional approach, the relationship

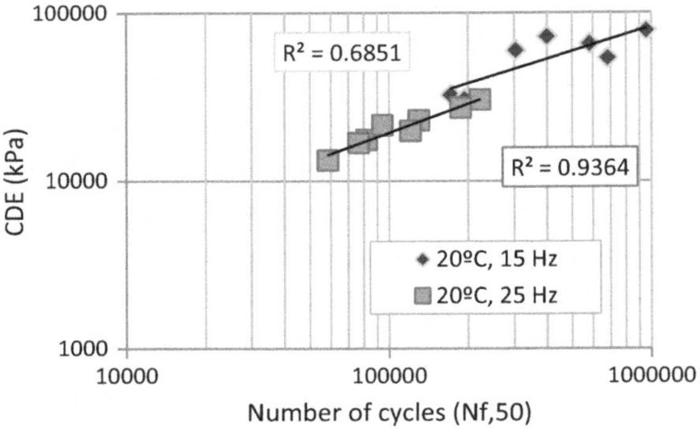


**Fig. 2.40** Cumulative dissipated energy versus load cycle curve obtained at 20 °C at 15 and 25 Hz loading frequencies

between applied strain $\varepsilon_0$ and load cycle to failure $N_f$ is often times presented in a power function [24]:

$$N_f = K_1(\varepsilon_0)^{-k2} \tag{2.35}$$

where $K_1$ and $K_2$ are the intercept and the slope, respectively. These are determined experimentally and they seem to be highly correlated. Researchers [69–75] believe that mode of loading, testing temperature, frequency and asphalt content has a more significant effect on the $K_1$–$K_2$ relation than asphalt type, air voids levels and aggregate gradation. Figure 2.40 shows the cumulative dissipated energy versus load cycle curve.

Van Dijk and Visser [76] performed some of the early research to consider dissipated energy as a fatigue parameter. This work determined an equation that relates the cumulative dissipated energy ($W_f$) to the number of cycles to failure as shown in Eq. (2.36):

$$W_f = A\left(N_f^{W_f}\right)^z \tag{2.36}$$

where $W_f$ is the cumulative dissipated energy to failure, i.e. the total energy dissipated by the material during the fatigue test (sum of all areas within the stress-strain hysteresis loop for every cycle until failure); $N_f^{W_f}$ is the number of load cycle to failure, and $A$, $z$ are the mixture dependent parameters that are determined experimentally.

The cumulative dissipated energy has not shown very good promise to be a parameter to describe the fatigue performance in asphalt materials, as it does not distinguish the amount of dissipated energy due to damage as opposed to viscous dissipation. Also, Eq. (2.36) is not unique it was found that it changes depending on the mode of loading, loading frequency and temperature. The same result was found by SHRP study for stress controlled tests [77].

Some researchers [78, 79] have suggested the Ratio of Dissipated Energy Change (*RDEC*) as a parameter to describe fatigue in asphalt materials. The same researchers believe that the *RDEC* is a true indicator of damage because it is able to eliminate the other forms of dissipated energy due to mechanical work or heat generation. Therefore, it can be considered a good parameter to describe the fatigue process in asphalt, and is calculated with the following expression:

$$RDEC = \frac{DE_{n+1} - DE_n}{DE_n} \tag{2.37}$$

where *RDEC* is ratio of the dissipated energy change per load cycle, $DE_n$ is dissipated energy produced in load cycle $n$, and $DE_{n+1}$ is dissipated energy produced in load cycle $n + 1$.

Figure 2.41 shows the variation of the *RDEC* and the complex modulus ratio ($E^*/E_0$) plotted against the number of load cycles. Complex modulus ratio is ratio of

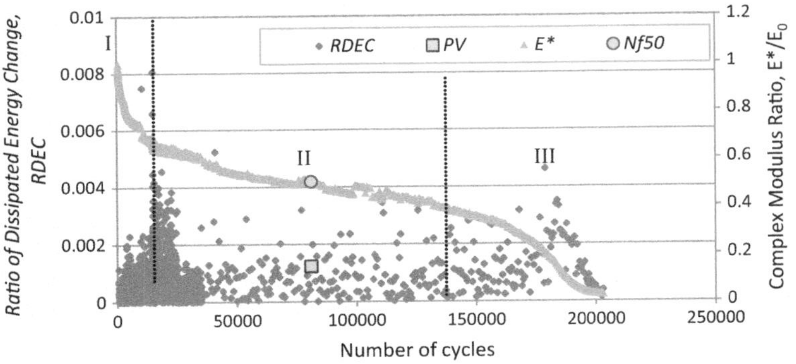


**Fig. 2.41** RDEC versus load cycle curve obtained at 20 °C at 15 Hz

the complex modulus in the current cycle normalized against complex modulus from first load cycle.

In case of asphalt mixtures, three main phases during a fatigue test have been suggested. The *RDEC*, after a rapid decrease (I stage), reaches a plateau stage in which a plateau value (*PV*) can be obtained, corresponding to the *RDEC* value when initial stiffness modulus has been reduced of half. This represents an energy plateau where an almost constant rate of energy input is being turned into damage. Researchers have validated that the *PV* is uniquely related to fatigue life. After the *RDEC* increases rapidly until true fatigue failure (III stage).

One challenge associated with the use of *RDEC* plateau value based approach is to find the value of *PV*. Often it is not easy to obtain the *PV* from experimental data and also they contain a high amount of dispersion. A hypothesis that has been proposed is that if the dissipated energy curve follows a power law relationship ($DE = AN_f^k$), the *RDEC* can be simplified, considering the exponential slope of the power law *k*, as follows:

$$\mathrm{RDEC} = \frac{1 - \left(1 + \frac{100}{a}\right)^k}{100} \tag{2.38}$$

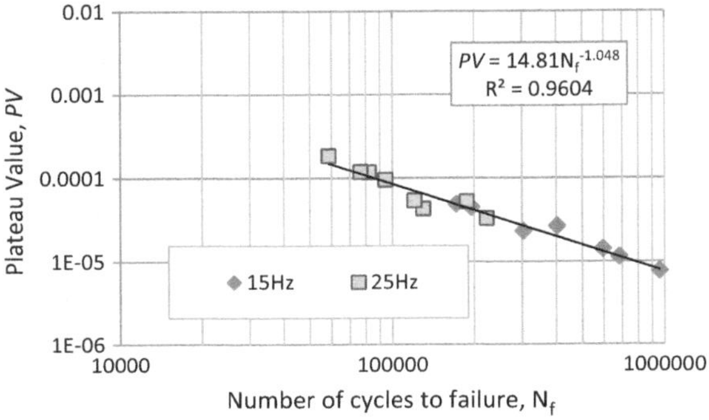


**Fig. 2.42** PV versus number of cycles obtained at 20 °C at 15 and 25 Hz

Also in this case, it is not so easy to determine the PV depending on the evolution of the dissipated energy during the fatigue test. However, the plateau value is correlated with a number of fatigue cycles to failure by means of a statistical approach, using the following equation [78, 80].

$$PV = cN_f^d \tag{2.39}$$

The value of the constants $c$ and $d$ are determined experimentally. The coefficient $d$ typically varies from −0.80 to −1.60 [80]. It can be seen from the experimental curve the fit the experimental data from both frequencies (15 and 25 Hz) is characterised by a coefficient $d$ equal to −1.048 (see Fig. 2.42).

### 2.5.2 *Non-local Modeling of Fatigue Microcracking with Application to Specimen Size Effects*

#### 2.5.2.1 Model Description

For a fatigue test with a sinusoidal loading and an offset equal to zero, the initial complex modulus of an undamaged material at the beginning of the test can be expressed as follows:

$$E_O^*(i\omega) = \frac{\sigma_0 e^{i\omega t}}{\varepsilon_0 e^{(i\omega t - \phi_0)}} = \frac{\sigma_0}{\varepsilon_0} e^{i\phi_0} = |E_O^*| e^{i\phi_0} \tag{2.40}$$

where, $E^*$ is complex modulus, $\omega$ is the frequency, $\phi$ is phase angle for initial cycle, $\sigma$ is uniaxial stress and $\varepsilon$ is uniaxial strain. The subscript for these variables represents the number of load cycle. After N cycles, the complex modulus is $|E_N^*| e^{i\phi_N}$ and the fatigue damage $D$ induced by the micro cracks development within the material can be interpreted as a function of the ratio of $|E_N^*| e^{i\phi_N}$ to the initial complex modulus:

$$D = 1 - \frac{|E_N^*| e^{i\phi_N}}{|E_O^*| e^{i\phi_0}} \tag{2.41}$$

Assuming the phase angle of the complex viscoelastic modulus of the material is not locally affected by the damage growth, $\phi_N = \phi_0$ and the viscoelastic behaviour of the material can be deduced from an elastic model. The damage is then expressed as a function of the ratio of the corresponding Young's modulus $E_O$ and $E_N$:

$$D = 1 - \frac{|E_N^*|}{|E_O^*|} \approx 1 - \frac{E_N}{E_O} \tag{2.42}$$

with $D$ growing from 0 for the undamaged material to 1 when the material does not transfer stress anymore.

This damage $D$ is introduced in the constitutive equations as a scalar affecting the material isotropic elastic tensor $C_{ijkl}$, function of the Young modulus $E$ and the Poisson ratio $v$:

$$\sigma_{ij} = (1 - D)C_{ijkl}\varepsilon_{kl} \tag{2.43}$$

where $\sigma_{i_j}$ and $\varepsilon_{kl}$ are respectively the elastic stress and strain tensor components.

To be representative of a mode I cracking (opening), only the strains induced by the positive principal stresses (traction) $\sigma_i$ are assumed to contribute to the cracking damage in the form of a scalar equivalent strain:

$$\tilde{\varepsilon} = \sqrt{\sum_{i=1,\ldots,3} \frac{\langle\sigma_i\rangle^2}{E_o(1-D)}} \quad \text{where} \quad \langle\sigma_i\rangle = \frac{1}{2}(|\sigma_i| + \sigma_i) \tag{2.44}$$

To avoid localisation problems in the structural computation, a non local integration of the equivalent strain can be applied as a mesh regularization method. This non local integration is realized through a convolution of the strain field with a normalized 3D Gaussian kernel: the local equivalent strain value $\tilde{\varepsilon}$ is replaced by a non-local equivalent strain $\bar{\varepsilon}$ equal to the Gaussian weighted average of $\tilde{\varepsilon}$ taken over a region about the point of interest.

$$\bar{\varepsilon} = \frac{1}{V_r(x)} \int_\Omega \Psi(x-s)\tilde{\varepsilon}(s)ds \quad \text{where} \quad V_r(x) = \int_\Omega \Psi(x-s)ds \tag{2.45}$$

and $\Psi(x-s) = \mathrm{e}^{\frac{-4|x-s|^2}{l_c^2}}$ a Gaussian function

For a sinusoidal loading, if the Miner's rule of linear additive damage is assumed, the Wöhler curves are respected for a damage evolution law of the form [81]:

$$\dot{D} = f(D)\varepsilon^\beta \langle\dot{\varepsilon}\rangle \tag{2.46}$$

The damage growth relationship component $f(D)$ is defined by the following mathematical expression:

$$f(D) = \frac{\alpha 2}{\alpha 1 \alpha 3}\left(\frac{D}{\alpha 2}\right)^{1-\alpha 3} e^{\left(\frac{D}{\alpha 2}\right)^{\alpha 3}} \tag{2.47}$$

The model is implemented in a finite element code with parameters $\beta, \alpha 1, \alpha 2, \alpha 3$ to be material properties.

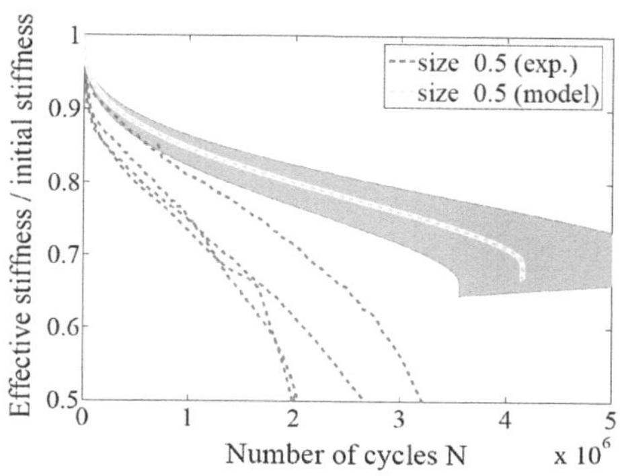


**Fig. 2.43** Comparison between model predictions and experimental results (size 0.5)

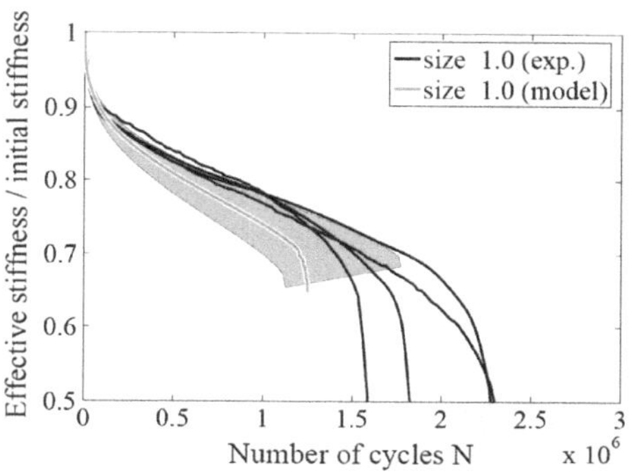


**Fig. 2.44** Comparison between model predictions and experimental results (size 1)

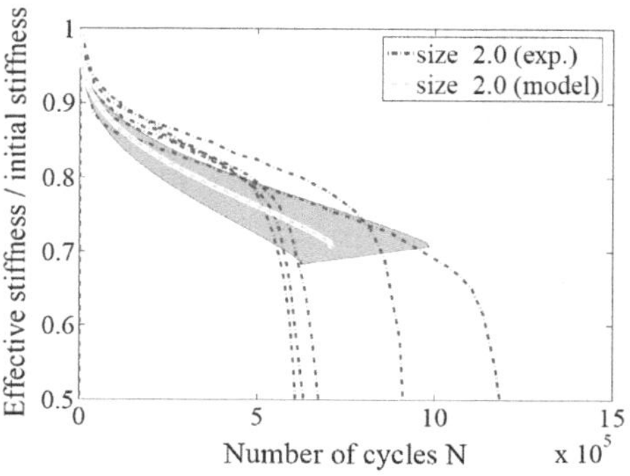


**Fig. 2.45** Comparison between model predictions and experimental results (size 2)

#### 2.5.2.2 Results of the Model Prediction for Study of Specimen Size Effects

Numerical simulations were conducted for the three specimen sizes with a set of material parameters $\beta, \alpha 1, \alpha 2, \alpha 3$ evaluated on the bituminous mix used in previous research.

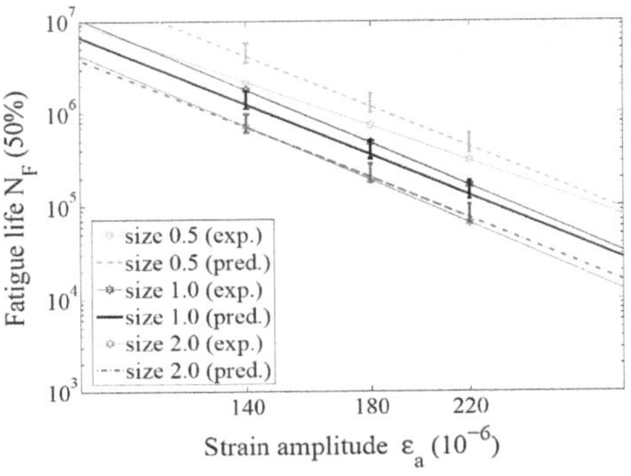


**Fig. 2.46** Comparison between model predicted and experimental Wöhler curves

Comparison between the model predictions and the experimental results are presented for the three sizes (Figs. 2.43, 2.44 and 2.45) for the strain level $140 \times 10^{-6}$:

The experimental trend for the number of cycle before rupture to increase as the size of the specimen decreases is reproduced by the simulation but this effect is over estimated for the smaller size. Model predicted and experimental Wöhler curves are compared in Fig. 2.46:

The model chosen implies parallel predicted Wöhler curves with a material dependent log-log slope equal to $-(\beta + 1)$. This is not the case for the experimental Whöler curves with a significantly different slope for the experimental curve obtained for the smaller samples (size 0.5).

The model prediction/experimental differences could thus be explained by viscoelastic thermal effects, affecting the $\beta$ parameter induced by the material self-heating neglected in the modelling.

## 2.5.3 *Viscoelstoplastic Continuum Damage Model*

### 2.5.3.1 Model Introduction

The viscoelastoplastic continuum damage (VEPCD) model in tension is based on (1) the elastic-viscoelastic correspondence principle, (2) continuum damage mechanics to account for the effect of microcracking on the constitutive behavior, (3) a time- and stress-dependent viscoplastic model to account for the plastic and viscoplastic behavior, and (4) the time-temperature superposition (TTS) principle with growing damage to describe the effect of temperature on the constitutive behavior. The resulting models are integrated by the strain decomposition approach to form the VEPCD model. These principles are briefly described below.

#### 2.5.3.2 Time-Temperature Superposition (TTS) with Growing Damage

If the time-temperature superposition principle can be extended to outside of the linear viscoelastic (LVE) range for asphalt mixtures, its impact is significant in terms of testing requirements and efficiency in modeling.

The TTS with growing damage can be verified using a simple technique where stress and time are determined at a strain level from the constant crosshead rate monotonic tests at different rates and temperatures. The corresponding time is converted to reduced time with the time-temperature shift factors from LVE characterization (e.g., frequency and temperature sweep dynamic modulus tests) and plotted with the respective stress. If the resulting graph, for a wide range of strain levels, appears continuous, then TTS with growing damage is said to be verified. Details on the theoretical background of this technique can be found in Chehab et al. [82].

Chehab et al. [82] demonstrated that the thermorheologically simple (TRS) behavior extends well beyond the LVE limits to highly damaged levels for asphalt concrete in tension. Underwood et al. [83] proved that TTS with growing damage is valid for various modified asphalt mixtures. Other researchers (Zhao [84]; Gibson et al. [85]; Kim et al. [86]) also found that asphalt concrete in compression is TRS at high strain levels with damage.

The importance of this finding lies in the reduction of testing conditions required for modeling purposes. Once the behavior at a given temperature is known, the behavior at any other temperature can be predicted using the LVE shift factors. In terms of the VEPCD model, this principle is considered by replacing physical times with reduced times, generally calculated from Eq. (2.48), or more specifically by Eq. (2.49) if temperature does not change with time:

$$\xi = \int_0^t \frac{dt}{a_T} \tag{2.48}$$

$$\xi = \frac{t}{a_T} \tag{2.49}$$

For the remainder of the discussion on this model, the formulations are presented using the reduced time ($\xi$) instead of the physical time ($t$), based on the TTS principle with growing damage.

#### 2.5.3.3 Viscoelastic Continuum Damage (VECD) Model

The viscoelastic strain in the VEPCD model covers both linear viscoelasticity and the damage due to microcracking. The VECD model forms the basis for the viscoelastic strain. This model is based on two principles: the elastic-viscoelastic correspondence principle based on pseudo strain and the continuum damage mechanics-based work potential theory.

The stress-strain relationships for many viscoelastic materials can be represented by elastic-like equations through the use of so-called *pseudo* variables. This simplifying feature enables a class of extended correspondence principles to be established and applied to linear as well as some nonlinear analyses of viscoelastic deformation and fracture behavior [87]. Using these correspondence principles, one can obtain viscoelastic solutions from their elastic counterparts through a simple conversion procedure. The usual Laplace transform-based correspondence principle is limited to LVE behavior with time-varying boundary conditions, whereas the correspondence principles based on pseudo variables are applicable to both linear and nonlinear behavior of a class of viscoelastic materials with stationary or time-dependent boundary conditions. Also, the latter does not require a transform inversion step to obtain the viscoelastic solutions but rather requires a convolution integral which is much easier to handle than the inversion step.

Consider a stress-strain equation for linear viscoelastic materials,

$$\sigma_{ij} = \int_{o}^{\xi} E_{ijkl}(\xi - \tau)\frac{\partial \varepsilon_{kl}}{\partial \tau} d\tau \tag{2.50}$$

where

| | |
|---|---|
| $\sigma_{ij}$, $\varepsilon_{kl}$ | stress and strain tensors; |
| $E_{ijkl}(t)$ | the relaxation modulus matrix; |
| $\xi = t/a_T$ | reduced time; |
| $t$ | physical time; |
| $a_T$ | the time-temperature shift factor; and |
| $\tau$ | the integration variable. |

Equation (2.50) can be written as

$$\sigma_{ij} = E_R\, \varepsilon_{kl}^R \quad \text{or} \quad \varepsilon_{kl}^R = \frac{\sigma_{ij}}{E_R} \tag{2.51}$$

if we define

$$\varepsilon_{kl}^R = \frac{1}{E_R}\int_{o}^{\xi} E_{ijkl}(\xi - \tau)\frac{\partial \varepsilon_{kl}}{\partial \tau} d\tau \tag{2.52}$$

where $E_R$ is termed the *reference modulus*, which is a constant and has the same dimension as the relaxation modulus, $E_{ijkl}(t)$. The usefulness of Eq. (2.51) is that a correspondence can be found between Eq. (2.51) and the linear elastic stress-strain relationship. That is, the equations in Eq. (2.51) take the form of *elastic* stress-strain equations even though they are actually *viscoelastic* stress-strain equations. The $\varepsilon_{kl}^R$

is called the *pseudo strain*. The pseudo strain accounts for all the hereditary effects of the material through the convolution integral. The reference modulus, $E_R$, is introduced here because it is a useful parameter in discussing special material behaviors and introducing dimensionless variables. For example, if we take $E_{ijkl}(t) = E_R$ in Eq. (2.52), we obtain $\varepsilon_{kl}^R = \varepsilon_{kl}$, and Eq. (2.51) reduces to the linear elastic equation $\sigma_{ij} = E_{ijkl}\varepsilon_{kl}$ or $\varepsilon_{kl} = \sigma_{ij}/E_{ijkl}$. If we take $E_R = 1$ in Eq. (2.52), pseudo strains are simply the linear viscoelastic stress response to a particular strain input. For the remainder of this chapter, $E_R$ is set to one. The results from experimental verification of the correspondence principle have been documented by Kim and Little [88], Kim et al. [89], and Lee and Kim [90] using uniaxial monotonic and cyclic data of asphaltic materials under a wide range of test conditions.

The material property needed to calculate the pseudo strain in Eq. (2.52) is the relaxation modulus. Typically, the relaxation modulus test is not easy to perform due to the large amount of stress that develops at the beginning of the test from the step input of displacement. Therefore, the relaxation modulus is determined from the complex modulus using the theoretical inversion process. Alternatively, the relaxation modulus can also be determined from creep compliance mastercurve through inversion process.

Because the complex modulus tests are performed on a separate set of representative specimens for the mixture in question, the relaxation modulus determined from the complex modulus tests may not be the same as the one for specimens used in damage testing due to sample-to-sample variability. In order to minimize the effect of the sample-to-sample variability, the dynamic modulus ratio, *DMR*, is introduced. The *DMR* is the ratio between the dynamic modulus of a test specimen determined from a fingerprint dynamic modulus test performed at 10 Hz right before the fatigue test and the dynamic modulus determined at the fingerprint test temperature and 10 Hz from the mastercurve generated from separate specimens using the time-temperature superposition.

Thus, the governing constitutive equation in uniaxial mode becomes

$$\sigma = DMR\ \varepsilon^R \tag{2.53}$$

where

$$\varepsilon^R = \frac{1}{E_R}\int_o^{\xi} E(\xi-\tau)\frac{\partial\varepsilon}{\partial\tau}d\tau \tag{2.54}$$

In most cases, the *DMR* value remains between 0.9 and 1.1. When the *DMR* value is significantly out of this range, a re-examination of the data (both the relaxation modulus and test results) is necessary.

A state variable based approach is recommended for pseudo-strain calculation due to its efficiency over piece-wise linear approximation which is fundamentally sound but computationally inefficient.

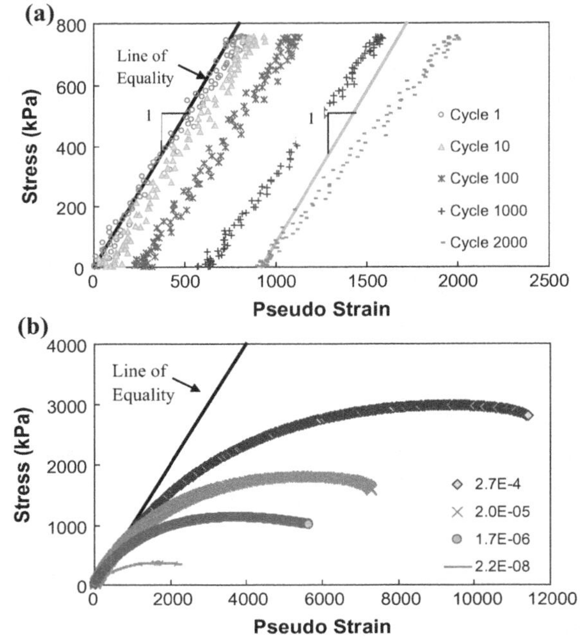


**Fig. 2.47** Stress-pseudo strain behavior of asphalt concrete in: **a** Controlled-stress cyclic test; **(b)** Constant-strain-rate monotonic test (different symbols represent different strain rates)

#### 2.5.3.4 Work Potential Theory

Figure 2.47a shows typical stress versus pseudo strain hysteresis loops for an asphalt mixture at different numbers of cycles in the controlled-stress cyclic test. Relatively high stress amplitude is used to induce significant damage in the specimen. Unlike the negligible damage case, change in the slope of each $\sigma - \varepsilon^R$ cycle (i.e., reduction in the pseudo stiffness of the material) can be observed from this figure due to the damage incurred in the specimens.

The effect of damage on pseudo stiffness can also be seen in the monotonic data shown in Fig. 2.47b. In this figure, the stress-pseudo strain curves deviate from the line of equality as damage grows. Also, the onset of this deviation occurs at different times as the rate of loading changes, indicating the presence of the rate-dependent damage mechanism. On basis of these observations, the following uniaxial versions of constitutive equations are presented for linear elastic and linear viscoelastic bodies with and without damage. They also show how models of different complexity may evolve from simpler ones.

$$\textit{Elastic Body without Damage}: \sigma = E\varepsilon \tag{2.55}$$

$$\textit{Elastic Body with Damage}: \sigma = C(S_m)E\varepsilon \tag{2.56}$$

$$\textit{Viscoelastic Body without Damage} : \sigma = E_R \varepsilon^R \tag{2.57}$$

$$\textit{Viscoelastic Body with Damage} : \sigma = C(S_m) E_R \varepsilon^R \tag{2.58}$$

where $C(S_m)$ indicates that $C$ is a function of damage parameters, $S_m$. The function, $C(S_m)$, represents the changing *stiffness* of the material due to growing damage. Equation (2.58) results from Eqs. (2.55), (2.56), and (2.57). The form of Eq. (2.58) is also supported by the observations made in Fig. 2.47; i.e., the pseudo stiffness changes as the damage grows. To determine an analytical representation of the damage function, $C$, the work potential theory, the continuum damage mechanics principle developed by Schapery [91], is adopted.

In continuum damage mechanics, the damaged body can be viewed as a homogeneous continuum on a macroscopic scale, and the influence of damage is typically reflected in terms of reduction in stiffness or strength of the material. The state of damage can be quantified by a set of parameters often referred to as *internal state variables* or *damage parameters* in the context of thermodynamics of irreversible processes. The growth of damage is governed by an appropriate damage (or internal state) evolution law. The *stiffness* of the material, which varies with the extent of damage, is determined as a function of the internal state variables by fitting the theoretical model to available experimental data.

When damage occurs due to external loading, the work done on the body is not entirely stored as strain energy; part of it is consumed in causing damage to the body. The amount of energy required to produce a given extent of damage is expressed as a function of *internal state variables*. The total work input to the body during the processes in which damage occurs depends, in general, on the path of loading. However, it has been observed that, for certain processes in which damage occurs, the work input is independent of the path of loading [92, 93].

Schapery [91] applied the method of thermodynamics of irreversible processes and the observed phenomenon of path independence of work in damage-inducing processes to develop the work potential theory so that it may be applicable to describing the mechanical behavior of elastic media with growing damage and other structural changes. The theory is general enough to allow for strong nonlinearities and coupling between the internal state variables and to describe a variety of mechanisms including micro- and macrocrack growth in monolithic and composite materials. Sicking [94] applied the theory to model the damage-related material nonlinearity in graphite-epoxy laminates, and Lamborn and Schapery [95] showed the existence of a work potential for suitably limited deformation paths using experimental data from axial and torsional deformation tests on angle-ply fiber-reinforced plastic laminates. The elements of work potential theory in terms of a strain energy formulation may be represented as follows:

$$\textit{Strain Energy Density Function} : W = W(\varepsilon_{ij}, S_m) \tag{2.59}$$

$$Stress - Strain\,Relationships : \sigma_{ij} = \frac{\partial W}{\partial \varepsilon_{ij}} \tag{2.60}$$

$$Damage\,Evolution\,Laws : -\frac{\partial W}{\partial S_m} = \frac{\partial W_S}{\partial S_m} \tag{2.61}$$

where $\sigma_{ij}$ are stresses, $\varepsilon_{ij}$ are strains, $S_m$ are internal state variables (or damage parameters), and $W_S = W_S(S_m)$ is the dissipated energy due to damage growth. The internal state variables, $S_m$ ($m$ = 1,2,…, M), account for the effects of damage, and the number of internal state variables (i.e., M) is typically determined by the number of different mechanisms governing the damage growth. Equation (2.61) is similar to a crack growth equation (e.g., $\boldsymbol{G} = \boldsymbol{G}_c$, where $\boldsymbol{G}$ is the energy release rate and $\boldsymbol{G}_c$ is the fracture toughness) and, in fact, Eq. (2.61) is used to find $S_m$ as functions of $\varepsilon_{ij}$. The left-hand side of Eq. (2.61) is the *available* thermodynamic force, while the right-hand side is the *required* force for damage growth.

Based on the elastic-viscoelastic correspondence principle, the strains, $\varepsilon_{ij}$, that appear in the elastic damage model, Eqs. (2.59–2.61), are replaced with corresponding pseudo strains, $\varepsilon_{ij}^R$, defined by Eq. (2.52). Then, according to the correspondence principle, the set of equations written in terms of pseudo strains now governs the corresponding viscoelastic damage problem. It was found from experimental studies (e.g., [96]), however, that the damage evolution laws for elastic materials cannot be translated directly into evolution laws for viscoelastic materials through the correspondence principle. Not only is the available force for growth in $S_m$ rate-dependent, but the resistance against the growth of $S_m$ is rate-dependent for most viscoelastic materials. The following evolution laws, which are similar in form to the well-known power-law crack growth laws for viscoelastic materials [97], are adopted in this study as they can reasonably represent the actual damage evolution processes in many viscoelastic materials:

$$\dot{S}_m = \left(-\frac{\partial W^R}{\partial S_m}\right)^{\alpha_m} \tag{2.62}$$

where $W^R = W^R(\varepsilon_{ij}^R, S_m)$ is the pseudo strain energy density function, $\dot{S}_m$ is the damage evolution rate, and $\alpha_m$ are material-dependent constants related to the viscoelasticity of the material. Equation (2.62) is similar to the crack propagation rate equation. The same form of evolution laws has been used successfully in describing the behavior of a filled elastomer with growing damage [96, 98] also adopted the work potential theory in modeling the rate-dependent behavior of asphalt-aggregate mixtures under constant strain rate monotonic loading.

Finally, the work potential theory applied to viscoelastic media with the rate type damage evolution law is represented by the following three components for the uniaxial loading condition:

$$Pseudo\,Strain\,Energy\,Density\,Function: W^R = W^R(\varepsilon^R, S_m) \tag{2.63}$$

$$Stress - Strain\,Relationship: \sigma = \frac{\partial W^R}{\partial \varepsilon^R} \tag{2.64}$$

$$Damage\,Evolution\,Law: \dot{S}_m = \left(-\frac{\partial W^R}{\partial S_m}\right)^{\alpha_m} \tag{2.65}$$

#### 2.5.3.5 Determination of Material Parameters

The work potential theory specifies an internal state variable, $S_m$, to quantify damage, which is defined as any microstructural changes that result in an observed stiffness reduction. The method used to solve the damage evolution law in Eq. (2.65) is a matter of preference. Lee and Kim [90] proposed a solution that utilizes the chain rule and makes no assumption regarding $\alpha$. The solution of these researchers is presented in Eq. (2.66). It is noted that both methods have been successfully applied in asphalt concrete research [94–100].

$$S_{i+1} = S_i + \left[-\frac{1}{2}(C_i - C_{i-1})\left(\varepsilon_i^R\right)^2\right]^{\frac{\alpha}{1+\alpha}} \Delta t^{\frac{1}{1+\alpha}} \tag{2.66}$$

This method assumes that the rate of change in damage is constant over some discrete time step. This rate of change is determined at a point near the current value of damage ($S_i + \delta S$) where the extrapolation error is minimized. The method begins with an initial calculation of $S$ by either of the approximate methods, both of which require results from constant crosshead rate tests for the stress-pseudo strain relationship. The initial $S$ values are plotted with the pseudo stiffness values, $C$, calculated from the following relationship, which is obtained from Eq. (2.58):

$$C = \frac{\sigma}{I \times \varepsilon^R} \tag{2.67}$$

The $C$ and initial $S$ values are then fit to some mathematical form, such as the one presented in Eq. (2.68), where $a$ and $b$ are fitting parameters:

$$C = e^{aS^b} \tag{2.68}$$

Returning to the damage evolution law, and noting that the increments of time are generally small, one can write the rate of change in damage as

$$\frac{dS}{dt} = \frac{\Delta S}{\Delta t} \tag{2.69}$$

Substituting this expression into Eq. (2.65), and rearranging and writing in the discrete form, one finds the following equation:

$$S_{i+1} = S_i + \Delta t\left(-\frac{\left(\delta W_d^R\right)_i}{\delta S}\right)^{\alpha} \tag{2.70}$$

It must be observed that for the uniaxial case, the work function ($W^R$) is given by

$$W^R = \frac{1}{2}C(S)\varepsilon^R \tag{2.71}$$

Substituting Eq. (2.71) into (2.70) and simplifying, one arrives at

$$S_{i+1} = S_i + \Delta t\left(-\frac{1}{2}\left(\varepsilon^R\right)^2\frac{(\delta C)_i}{\delta S}\right)^{\alpha} \tag{2.72}$$

In Eq. (2.72), it is assumed that before loading occurs, $S$ and $C$ are zero and one, respectively. Further, $\delta S$ must be specified and should be significantly less than the change in damage over a time step (typically, 0.1 is used). After calculating the value of damage ($S_i$) and the incremental damage ($S_i + \delta S$) at a given time step, the corresponding values of $C$ are found by Eq. (2.68). The difference between these values ($\delta C$) is then used to calculate damage at the next time step. The process is repeated until all data points are processed.

After completing this first iteration, the new values of $S$ are plotted against the original pseudo stiffness values, and a new analytical relationship is found. The entire process is repeated until the change in successive iterations is small. In this research, eight such iterations were performed, but it was noted that very little improvement was made after the third or fourth iteration.

Daniel and Kim [99] studied the relationship between damage parameter ($S$) and the normalized pseudo secant modulus ($C$) under varying loading conditions. The most significant finding from their study is that a unique damage characteristic relationship exists between $C$ and $S$, regardless of loading type (monotonic versus cyclic), loading rate, and stress/strain amplitude. In addition, the application of the TTS principle with growing damage to the $C$ versus $S$ relationships at varying temperatures yields the same damage characteristic curve in the reduced time scale. The only condition that must be met in order to produce the damage characteristic relationship is that the test temperature and loading rate combination must be such that only the elastic and viscoelastic behaviors prevail with negligible, if any, viscoplasticity. When the test temperature is too high or the loading rate is too slow, it was found that the $C$ versus $S$ curve deviates from the characteristic curve.

To ensure that the test temperature is low enough and the loading rate is fast enough not to induce any significant viscoplastic strains, the tests are performed at a low temperature (typically 5°C) with varying loading rates. If the $C$ versus $S$ curves at different rates overlap to form a unique relationship, the combinations of the temperature and loading rate are sufficiently satisfactory to develop the damage characteristic relationship.

Finally, the VECD model is

$$\sigma = C(S)\varepsilon^R \tag{2.73}$$

or

$$\varepsilon_{ve} = E_R \int_0^{\xi} D(\xi - \tau) \frac{d\left(\frac{\sigma}{C(S)}\right)}{d\tau} d\tau \tag{2.74}$$

by converting Eq. (2.73) to predict the viscoelastic strain. Note that $E_R$ in Eq. (2.74) is set to one and that the *DMR* is not used in Eq. (2.73). *DMR* is only necessary in calibrating the model using the experimental data from several replicate specimens.

The major advantage of the damage characteristic relationship is that it allows a reduction in testing requirements. Since the same relationship exists in monotonic and cyclic tests, the material behavior under cyclic loading can be predicted from the damage characteristic curve characterized from the much simpler monotonic tests. Daniel and Kim [99] have verified that this approach can predict the fatigue life of asphalt concrete within the sample-to-sample variation.

#### 2.5.3.6 Determination of VECD Model Parameters with Example from RILEM TC-CAP Material

The modeling techniques presented in the previous sections are applied to the asphalt concrete specimens fabricated by the research team at the University of Illinois at Urbana-Champaign. Two types of tests are required for the full VEPCD characterization: (1) temperature and frequency sweep dynamic modulus tests and (2) constant crosshead rate tests until failure. These tests were performed in direct tension using a MTS closed-loop servo-hydraulic loading frame. Table 2.6

**Table 2.6** VEPCD coefficients for the UIUC mixture

| Model | | | Coefficients |
|---|---|---|---|
| LVE | Shift factor function | a1 | 0.000223 |
| | | a2 | −0.13514 |
| | | a3 | 0.670097 |
| | Sigmoidal function | a | 1.999011 |
| | | b | 2.40541 |
| | | d | 1.646788 |
| | | g | 0.729349 |
| | Alpha | | 3.44 |
| VECD | a | | −0.00051 |
| | b | | 0.589123 |

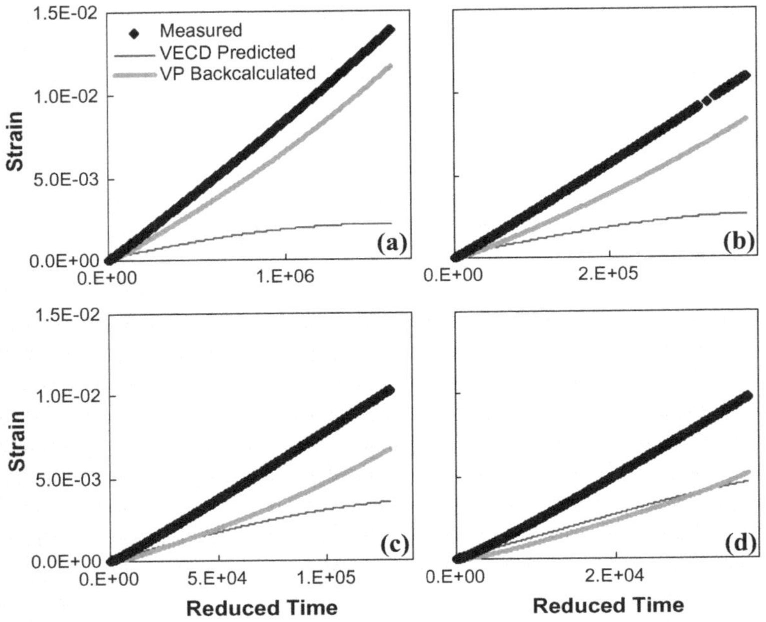


**Fig. 2.48** Backcalculated viscoplastic strains from 40 °C constant crosshead rate tests at: **a** 0.0003 ε/s, **b** 0.001 ε/s, **c** 0.003 ε/s and **d** 0.01 ε/s

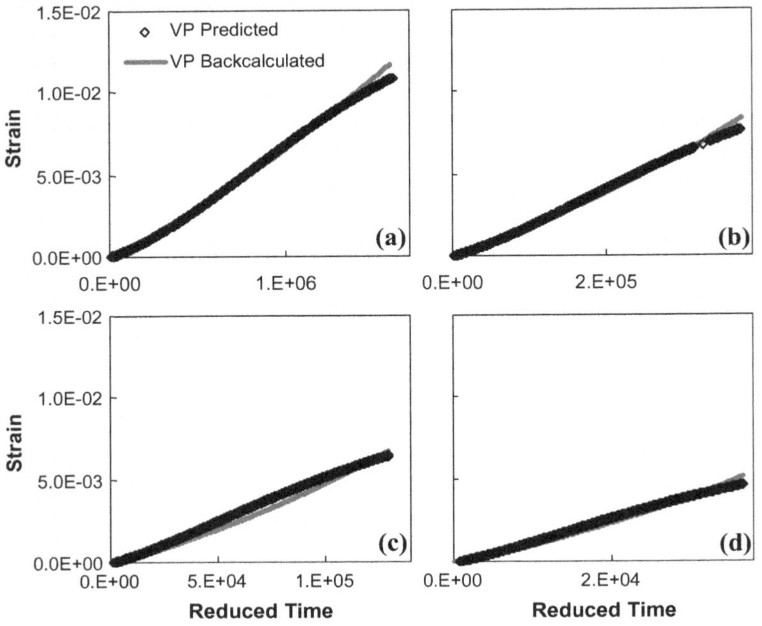


**Fig. 2.49** Characterization of viscoplastic models from backcalculated viscoplastic strains from 40 °C constant crosshead rate tests at: **a** 0.0003 ε/s, **b** 0.001 ε/s, **c** 0.003 ε/s and **d** 0.01 ε/s

summarizes the values of the VEPCD coefficients determined from the testing of the UIUC mixture. Figures 2.48 and 2.49 show backcalculated viscoplastic strains.

It is noted that the VEPCD modeling techniques currently investigated by the NCSU research team are slightly different from the methodologies presented in this document. The basic principles of the VECD model are the same; however, the most recent form of the VECD model (so-called simplified VECD model) uses the data from a constant crosshead amplitude cyclic test instead of the monotonic test. More details on the simplified VECD model and its application can be found in Underwood et al. [101] and Hou et al. [102]. Also the current form of the viscoplastic model is a new type of rate model for explaining the behavior of asphalt concrete in compression using repeated creep and recovery tests. This model places an emphasis on capturing load history effects directly using viscoelastic-like convolution integrals. This approach leads naturally to a back stress concept that can be used to explain the effect of rest periods on the permanent deformation evolution.

#### 2.5.3.7 Summary of Models for Cyclic Degradation

The discussion of the models for characterization of asphalt mixtures in terms of cyclic degradations or fatigue damage was presented in three sections. First section presented several different approaches that are used to represent asphalt mixture's cyclic degradation characteristics. These are typically used for purposes of comparing various mixtures and have been used in an empirical manner to predict field performances. The other two approaches present formulations that are implemented within structural models (such as, finite element analysis) and have been applied to better understand the cracking mechanisms in pavements. A summary of the approaches presented herein on the topic of cyclic degration models is described in Table 2.7. Similar to the table presented previously for comparisons of discrete cracking models the basis (theory/approach) behind the models, required laboratory tests/material parameters, status of the model in terms of model qualification, verification, calibration and validation, and the status of application of the models to simulate cracking performance of pavements are also presented.

## 2.6 Summary

### *2.6.1 Summary and Observations on the State of the Art for Laboratory Tests and Constitutive Models for Characterization of Fracture in Bulk Asphalt Material*

This chapter provided a state of the art review of laboratory characterization and modelling tools that are currently available for understanding the fracture

**Table 2.7** Summary of discrete crack models for asphalt mixtures

| Model (contributor) | Basis for model | Implementation in pavement structural model (framework) | Laboratory testing requirements | Model qualification(Q), verification(Vr), calibration (C) and validation(Vd) status | | | | | Field performance prediction status |
|---|---|---|---|---|---|---|---|---|---|
| | | | | Q | Vr | C | Vd (lab) | Vd (field) | |
| Dissipated energy concepts for fatigue characterization (different approaches: energy ratio, DCSE, RDEC, PV) (U. Nottingham and IFSTTAR) | Empirical, energy dissipation rate, viscoelasticity | No | – Cyclic (fatigue) tests (uniaxial and flexural) | X | | X | X | | Applied for mixture design and preliminary pavement life predictions |
| Non-local Model of fatigue microcracking (IFSTTAR) | Viscoelasticity, non-local damage distribution | Yes (Finite element analysis) | – Viscoelastic characterization (Typ.: Unixial complex modulus test)<br>– Cyclic (fatigue) tests | X | X | | X | | Not applied at present but application is planned |
| Continuum damage approach (North Carolina State U.) | Continuum damage, work potential theory, viscoelasticity | Yes (Layered viscoelastic analysis) | – Viscoelastic characterization (Typ.: Uniaxial complex modulus test)<br>– Cyclic tests (Typ.: Constant cross-head pull-pull tensile tests) | X | X | X | | X | Applied for fatigue (traffic and thermal effects) cracking |

mechanisms of asphalt mixtures. This review demonstrated that significant progress has been made on this topic over past decade. The following table provides a summary of the laboratory tests presented in this review in terms of temperature conditions, loading rates and specimen requirements.

Below are some observations and remarks on the basis of the review presented herein:

- The use of fracture energy based testing procedures through use of notched specimens has shown great potential and such methods are becoming routine testing practice. The fracture energy has shown better correlation to extent of field cracking as opposed to strength based measures.
- The laboratory methods for evaluation of cyclic response of asphalt mixtures has progressed significantly as well. Major progress has been made on the use of advanced constitutive models to analyse the test results, such as, incorporation of material's viscoelastic response as well as use of dissipated energy concepts. Furthermore, the cyclic tests have become more robust through use of multi-axial strain measurements.
- The constitutive models for fracture in asphalt mixtures have become exceedingly robust and more representative of the physical processes associated with material's failure. The energy dissipation based approaches have been adopted for modelling of discrete cracks in the asphalt material, these approaches have shown promising results at capturing the critical cracking events for pavement thermal and reflective cracks. The use of Paris law based linear-elastic fracture mechanics models have been superseded by the energy based models.
- The constitutive models capable of capturing cracking in asphalt pavements under cyclic loading conditions have progressed in the phenomenological direction and have moved away from older empirical approaches that were based on data-fitted regression equations. Through use of thermo-viscoelastic modelling framework the recent modeling approaches are able to capture the effects of loading rate and temperature leading to higher prediction reliability.

### 2.6.2 *Scientific Lack and Future Research Suggested*

A unified laboratory test and model that can characterize and capture response of asphalt mixtures under all the modes that affect its properties (temperatures, loading rates, aging effect etc.) is desired. The present report demonstrates that the current state of the art in characterization and modelling of asphalt bulk fracture is capable of capturing material response under wide-range of conditions, however the present approaches are still incapable of capturing all possible range of variables that affect fracture behaviour of asphalt mixtures.

The major scientific areas that is lacking in the field of fracture in bulk asphalt materials identified through this review include:

- Limited ability of models to capture both static and cyclic crack growths in an interchangeable manner.
- Models and tests are lacking fundamental understanding of healing processes.
- Most laboratory tests and analysis procedures focus on use of global specimen responses; more research is needed on incorporation of local material response from laboratory tests (see Chap. 4).
- Development of universal criteria from model qualification, verification and validation is needed.

## References


1. Wagoner MP, Buttlar WG, Paulino GH (2005) Disk-shaped compact tension test for asphalt concrete fracture. Exp Mech 45:270–277
2. Ahmed S, Dave EV, Buttlar WG, Exline MK (2010) Fracture properties of gap and dense graded thin bonded overlays. J Assoc Asphalt Paving Technol 79:443–446
3. Ahmed S, Dave E, Behnia B, Buttlar WG, Exline MK (2010) Fracture characterization of gap-graded asphalt mixtures and thin bonded wearing courses. Int J Pavement Res Technol 3:128–134
4. Ahmed S, Dave EV, Buttlar WG, Behnia B (2012) Compact tension test for fracture characterization of thin bonded asphalt overlay systems. Mat Struct 45(8): 1207–1220. https://dx.doi.org/10.1617/s11527-012-9827-0.
5. Behnia B, Dave EV, Ahmed S, Buttlar WG, Reis H (2011) Investigation of effects of the recycled asphalt pavement (RAP) amounts on low temperature cracking performance of asphalt mixtures using acoustic emissions (AE). J Transp Res Board 2208:64–71
6. Wagoner MP, Buttlar WG (2007) Influence of specimen size on fracture energy of asphalt concrete. J Assoc of Asphalt Paving Technol 76:391–426
7. Braham AF, Buttlar MP, Clyne TR, Marasteanu MO, Turos MI (2009) The effect of long-term laboratory aging on asphalt concrete fracture energy. J Assoc Asphalt Paving Technol 78:417–454
8. Hill BC, Behnia B, Hakimzadeh S, Buttlar WG, Reis H (2012) Evaluation of the low temperature cracking performance of WMA mixtures. J Transp Res Board
9. Buttlar WG, Ahmed S, Dave EV, Braham AF (2010) Comprehensive database of asphalt concrete fracture energy and links to field performance. Paper presented at the 89th annual meeting of the transportation research board. Washington, D.C.
10. Dave EV, Hoplin C, Helmer B, Dailey J, Van Deusen D, Geib J, Dai S, Johanneck L (2016) Effects of mix design and fracture energy on transverse cracking performance of asphalt pavements in Minnesota. Transp Res Rec 2576:40–50. http://dx.doi.org/10.3141/2576-05.
11. Dave EV, Hoplin C (2015) Flexible pavement thermal cracking performance sensitivity to fracture energy variation of asphalt mixtures. Road Mater Pavement Des 16(1):423–441. http://dx.doi.org/10.1080/14680629.2015.1029697
12. Ahmed S (2010) Fracture characterization of thin bonded asphalt concrete overlay systems. University of Illinois, Urbana-Champaign
13. Nguyen ML, Di Benedetto H, Sauzéat C, Wendling L (2008) Investigation of cracking in bituminous mixtures with a four point bending test. In: 6th RILEM International Conference on Cracking in Pavements. Chicago, Illinois. 16–18
14. Lemaistre H (1998) Etude des propriétés thermomécaniques de divers réfractaires. INSA, Lyon

15. Jiang F, Rohatgi A, Vecchio KS, Adharapurapu RR (2004) Crack length calculation for bend specimens under static and dynamic loading. Eng Fract Mech 71:1971–1985
16. Wendling L, Xoline E, Gimenez D, Reynaud P, De La Roche C, Chevalier J, Fantozzi G (2004) Characterisation of crack propagation in butuminous mixtures. In: Fifth international RILEM conference on cracking in pavements. Limoges, France. 639–646
17. Mandel J (1955) Sur les corps visco-élastiques à comportment linéaire. Comptes rendus del l'Académie des Sciences 241:1910–1912
18. Nguyen ML (2005) Etude de la propagation de fissure dans les enrobés bitumineux. Master of Civil Engineering, University of Lyon/ENTPE, Lyon [in French]
19. Tapsoba N (2008) Fissuration des enrobés bitumineux. University of Lyon/ENTPE, Lyon
20. Di Benedetto H, Corté JF (2005) Matériaux routiers Tome 2, Vol. 2, Hermes [in French] (2005)
21. Birgisson B, Montepara A, Romeo E, Roncella R, Tebaldi G, Roque R (2008) The use of digital image correlation for accurate determination of fracture energy density in hot mix asphalt (HMA). In: 6th RILEM international conference on cracking in pavements. Chicago Illinois, 2008: 811–820
22. Simonin F (2000) Comportement thermomécanique de béton réfractaires alumineux contenant du spinelle de magnésium. Ph.D. Thesis, INSA Lyon [in French]
23. Nguyen ML (2009) Etude de la fissuration et de la fatigue des enrobés bitumineux. Ph.D. Thesis, University of Lyon/ENTPE, Lyon [in French]
24. Di Benedetto H, De La Roche C, Baaj H, Pronk A, Lundstrom R (2004) Fatigue of bituminous mixtures. Mater Struct 37(267):202–216
25. Di Benedetto H, Delaporte B, Sauzéat C (2007) Three-dimensional linear behavior of bituminous materials: experiments and modeling. Int J Geomech 7(2):149–157
26. Delaporte B, Di Benedetto H, Chaverot P, Gauthier G (2007) Linear viscoelastic properties of bituminous materials: from binders to mastics. In: Proceeding AAPT. San Antonio, TX, pp 455–494
27. Delaporte B, Di Benedetto H, Chaverot P, Gauthier G (2009) Linear viscoelastic properties of bituminous materials including new products made with ultrafine particles. Road Mater Pavement Des 10(1):7–38
28. Pouget S, Sauzeat C, Di Benedetto H, Olard F (2010) From the behavior of constituent materials to the calculation and design of orthotropic bridge structures. Road Mater Pavement Des 11(SI EATA):111–144
29. Di Benedetto H, Nguyen HM, Pouget S, Sauzéat C (2008) Time-temperature superposition principle for bituminous mixtures: three dimensional approach and extension in the non-linear domain. In: First international conference on transport infrastructure. Beijing
30. Nguyen HM, Pouget S, Di Benedetto H, Sauzeat C (2008) Generalization of the time-temperature superposition principle for bituminous mixtures: experimentation and modeling. Orgagec, Paris
31. Williams ML, Landel RF, Ferry JD (1955) The temperature dependence of relaxation mechanisms in amorphous polymers and other glass-forming liquids
32. Bodin D (2002) Modèle d'endommagement par fatigue: application aux enrobés bitumineux. École Centrale de Nantes, Nantes
33. Corte JF, Goux MT (1996) Design of pavement structures: the French technical guide. Transport Research Report 1539:116–124
34. Chailleux E, Ramond G, Such C, de la Roche CA (2006) Mathematical-based master-curve construction method applied to complex modulus of bituminous materials. Roads Mater Pavement Des 7(SI EATA):75–92
35. Huet C (1999) Coupled size and boundary-condition effects in viscoelastic heterogeneous and composite bodies. Mech Mater 31:787–829
36. Sayegh G (1967) Viscoelastic properties of bituminous mixes. In: Proceedings of the 2nd international conference on the structural design of asphalt pavements proceedings, University of Michigan, Ann Arbor, Michigan, USA, pp 743–755

37. Chabot A, Chupin O, Deloffre L, Duhamel D (2010) Viscoroute 2.0: a tool for the simulation of moving load effects on asphalt pavement. RMPD Spec Issue Recent Adv Numer Simul Pavements 11(2):227–250
38. EN 12697-7:2002 Bituminous mixtures. Test methods for hot mix asphalt. Determination of the bulk density of bituminous specimens by gamma rays
39. Bodin D, de la Roche C, Pijaundier-Cabot G (2006) Size effect regarding fatigue evaluation of asphalt mixtures. Roads Mater Pavement Des 7(SI EATA):181–200
40. Bodin D, Pijaundier-Cabot G, De La Roche C, Piau JM, Chabot A (2004) Continuum damage approach to asphalt concrete fatigue modelling. ASCE J Eng Mech 130(6):700–708
41. EN-12697-24-2007. Bituminous mixtures—test methods for hot mix asphalt—part 24: resistance to fatigue
42. EN-12697-26-2004. Bituminous mixtures—test methods for hot mix asphalt—part 26: Stiffness
43. Baburamani P (1999) Asphalt fatigue life prediction models: a literature review. ARRB Transp Res
44. Bankowski W, Sybilski D (2007) Energetic method as an alterative for conventional method in fatigue life analysis of bituminous mixtures. In: 4th international SIIV congress. Palermo, Italy, pp 12–14
45. Hopman PC, Kunst P, Pronk AC (1989) A renewed interpretation method for fatigue measurement verification of Miners rule. In: 4th Eurobitume symposium. Madrid, pp 557–561
46. Pell PS (1967) Fatigue of asphalt pavement mixes. In: Proceedings of the second international conference on the structural design of asphalt pavements. Ann Arbor, Michigan
47. Vecoven JH (1989) Méthode détude de systèmes limitant la remontée de fissures dans les chausses. In: Rigo JM, Degeimbre R (ed) Proceedings of the 2th international RILEM conference on reflective cracking in pavements—assessment and control, March 8 to 10, Liège. pp 57–62
48. Dumas P, Vecoven J (1993) Processes reducing reflective cracking; synthesis of laboratory tests. In: Rigo JM, Degeimbre R, et Francken L (ed) Proceedings of the 2th international RILEM conference on reflective cracking in pavements—state of the art and design recommendations. March 10 to 12, Liège. pp 246–253
49. Dubois F, Petit C (2005) Modelling of the crack growth initiation in viscoelastic media by the Gθ-integral. Eng Fract Mech 72:2821–2836
50. Dubois F, Chazal C, Petit C (1999) A finite element analysis of creep-crack growth in viscoelastic media. Mech Time Depend Mater 2:269–286
51. Nguyen HN (2008) Etude numérique de la fissuration dun milieu viscoélastique: Analyse de l'essai de rupture sur bitume, Ph.D. thesis, LCPC
52. Maillard S, de La Roche C, Hammoum R, Gaillet L, Such C (2004) Experimental investigation of fracture and healing of bitumen at pseudo-contact of two aggregates. Eurobitume 1291–1304
53. Dubois F, Moutou Pitti R, Picoux B, Petit C (2012) Finite element model for crack growth process in concrete bituminous. Adv Eng Softw 44:35–43
54. Li X, Braham A, Williams A, Marasteanu M, Buttlar W (2008) Effect of factors affecting fracture energy of asphalt concrete at low temperature. Road Mater Pavement Des 9(SI):397–416
55. Song SH, Paulino GH, Buttlar WG (2006) A bilinear cohesive zone model tailored for fracture of asphalt concrete considering rate effects in bulk materials. Eng Fract Mech 73(18):2829–2848
56. Dave E, Buttlar W (2010) Low temperature cracking prediction with consideration of temperature dependent bulk and fracture properties. Road Mater Pavement Des 11(SI):33–59
57. Dave EV, Behnia B (2017) Cohesive zone fracture modeling of asphalt pavements with applications to design of high performance asphalt overlays. Int J Pavement Eng. http://dx.doi.org/10.1080/10298436.2017.1353393

58. Majidzadeh K, Kaufmann EM, Ramsamooj DV (1971) Application of fracture mechanics in the analysis of pavement fatigue. J Assoc Asphalt Paving Technol 40:227–246
59. Ramsamooj DV (1991) Prediction of fatigue life of asphalt concrete beams from fracture test. J Test Eval 19(3):231–239
60. Jacobs MJ, Hopman PC, Molenaar AA (1996) Application of fracture mechanics principles to analyze cracking in asphalt concrete. J Assoc Asphalt Paving Technol 65:1–39
61. Kim YR, Lee HJ, Little DN (1997) Fatigue characterization of asphalt concrete using viscoelasticity and continuum damage theory. J Assoc Asphalt Paving Technol 66:520–569
62. Kim YR, Little DL, Benson F (1990) Chemical and mechanical evaluation on healing mechanism of asphalt concrete. J Assoc Asphalt Paving Technol 59:240–276
63. Zhang Z (2000) Identification of suitable crack growth law for asphalt mixtures using the superpave indirect tensile test (IDT), Ph.D. Dissertation, University of Florida, Gainesville
64. Zhang Z, Roque R, Birgisson B, Sangpetngam B (2001) Identification and verification of a suitable crack growth law. J Assoc Asphalt Paving Technol 70:206–241
65. Roque R, Birgisson B, Sangpetngam B, Zhang Z (2002) Hot mix asphalt fracture mechanics: a fundamental crack growth law for asphalt mixtures. J Assoc Asphalt Paving Technol 71:816–827
66. Zhang Z, Roque R, Birgisson B (2001) Evaluation of laboratory measured crack growth rate for asphalt mixtures. Transp Res Rec J Transp Res Board 1767:67–75
67. Birgisson B, Montepara A, Romeo E, Roque R, Roncella R, Tebaldi G (2007) Determination of fundamental tensile failure limits of mixtures. J Assoc Asphalt Paving Technol 76:303–344
68. Kim B, Roque R (2006) Evaluation of healing property of asphalt mixture. Transp Res Rec J Transp Res Board 1970:84–91
69. Sangpetngam B (2003) Development and evaluation of a viscoelastic boundary element method to predict asphalt pavement cracking, Ph.D. Dissertation, University of Florida, Gainesville
70. Roque R, Birgisson B, Drakos C, Dietrich B (2004) Development and field evaluation of energy-based criteria for top-down cracking performance of hot mix asphalt. J Assoc Asphalt Paving Technol 73:229–260
71. Rowe GM (1993) Performance of asphalt mixtures in the trapezoidal fatigue test. In: Proceedings of the association of asphalt paving technologists, vol 62
72. Rowe GM (1996) Application of the dissipated energy concepts to fatigue cracking in asphalt pavements. Ph.D. thesis at University of Nottingham, UK
73. Ghuzlan KA, Carpenter SH (2003) Traditional fatigue analysis of asphalt concrete mixtures. TRB, Transportation Reasearch Board
74. Monismith CL, Deacon JA (1969) Fatigue of asphalt paving mixtures. J Transp Eng ASCE, vol 95
75. Pell PS, Cooper KE (1975) The fatigue of testing and mix variable on the fatigue performance of bituminous materials. Association of asphalt paving technologist 44 proc., Phoenix, AZ
76. Van Dijk W, Visser W (1977) The energy approach to fatigue for pavement design. In: Proceedings of the association of asphalt paving technologists (AAPT), vol 46
77. SHRP-A-404 (1994) Fatigue response of asphalt-aggregate mixes. Washington, D.C.: Report for the asphalt research program institute of transportation studies University of California, Berkeley. Strategic highway research program national research council
78. Ghuzlan K, Carpenter S (2000) Energy-derived, damage-based failure criterion for fatigue testing. Transp Res Record—TRR 1723
79. Shen S (2007) Dissipated energy concepts for HMA performance: fatigue and healing. Ph.D. thesis, Department of Civil and Environmental Engineering. University of Illinois at UrbanaChampaign: Urbana, Illinois
80. Ghuzlan K, Carpenter S (2001) Fatigue damage analysis in asphalt concrete mixtures using the dissipated energy approach. Can J Civ Eng
81. Stéfani C (1990) Unicité des états spaciaux dendommagement. Internal note LCPC

82. Chehab GR, Kim YR, Schapery RA, Witczak MW, Bonaquist R (2002) Time-temperature superposition principle for asphalt concrete mixtures with growing damage in tension state. J Assoc Asphalt Paving Technol 71:559–593
83. Underwood BS, Kim YR, Chehab GR (2006) A viscoelastoplastic continuum damage model of asphalt concrete in tension. In: Proceedings of the 10th international conference of asphalt pavements
84. Zhao Y (2002) Permanent deformation characterization of asphalt concrete using a viscoelastoplastic model. North Carolina State University
85. Gibson NH, Schwartz CW, Schapery RA, Witczak MW (2003) Viscoelastic, viscoplastic, and damage modeling of asphalt concrete in unconfined compression. Transp Res Rec 1860:3–15
86. Kim YR, Guddati BS, Underwood TY, Yun Subramanian V, Heidari AH (2005) Characterization of ALF mixtures using the viscoelastoplastic continuum damage model. Final report to the Federal Highway Administration
87. Schapery RA (1984) Correspondence principles and a generalized J-integral for large deformation and fracture analysis of viscoelastic media. Int J Fract 25:195–223
88. Kim YR, Little DN (1990) One-dimensional constitutive modeling of asphalt concrete. ASCE J Eng Mech 116(4):751–772
89. Kim YR, Lee YC, Lee HJ (1995) Cooresponcence principle for characterization of asphalt concrete. J Mater Civ Eng ASCE 7(1):59–68
90. Lee HJ, Kim YR (1998) A uniaxial viscoelastic constitutive model for asphalt concrete under cyclic loading. ASCE J Eng Mech 124(1):32–40
91. Schapery RA (1990) A theory of mechanical behavior of elastic media with growing damage and other changes in structure. J Mech Phys Solids 38:215–253
92. Schapery RA (1987) Deformation and fracture characterization of inelastic composite materials using potentials. Polym Eng 27:63–76
93. Lamborn MJ, Schapery RA (1989) An investigation of deformation path-independence of mechanical work in fiber-reinforced plastics, sicking, Washington, D.C.; USA, pp. 991–1000
94. Sicking DL (1992) Mechanical characterization of nonlinear laminated composites with tranverse crack growth. College station: Texas A&M University
95. Lamborn MJ, Schapery RA (1993) An investigation of the existence of a work potential for fiber reinforced plastic. J Compos Mater 27(4):352–382
96. Park SW (1994) Development of a nonlinear thermo-viscoelastic constitutive equation for particulate composites with growing damage. Texas A&M University
97. Schapery RA (1975) A theory of crack initiation and growth in viscoelastic media, part I: theoretical development, part II: approximate methods of analysis, part III: analysis of continuous growth. Int J Fract 11: 141–159, 369–388, 549–562
98. Park SW, Kim YR, Schapery RA (1996) A viscoelastic continuum damage model and its application to uniaxial behavior of asphalt concrete. Mech Mater 24(4):241–255
99. Daniel JS, Kim YR (2002) Development of a simplified fatigue test and analysis procedure using a viscoelastic continuum damage model. J Assoc Asphalt Paving Technol 71:619–650
100. Chehab GR, Kim YR, Schapery RA, Witczak MW, Bonaquist R (2003) Characterization of asphalt concrete in uniaxial tension using a viscoelastoplastic model. J Assoc Asphalt Paving Technol 315–355
101. Underwood BS, Kim YR, Guddati MN (2010) Improved calculation method of damage parameter in viscoelastic continuum damage model. Int J Pavement Eng 11(6):459–476
102. Hou T, Underwood BS, Kim YR (2010) Fatigue performance prediction of North Carolina mixtures using the simplified viscoelastic continuum damage model. J Assoc Asphalt Paving Technol 79:35–80

# Chapter 3
# Interface Debonding Behavior

**Christophe Petit, Armelle Chabot, Alexandra Destrée and Christiane Raab**

## 3.1 Introduction

During the 1990s, reflective cracking was the main damage observed in pavement structures [1]. This phenomenon was driven by fatigue or other forms of cracking in bottom layers reflecting damage up to the surface layers. Pavement design tools have been built using the Burmister model [2] that address the initial multi-layered structure without any discontinuities and assume either perfectly separated interface or perfectly bonded interfaces. For the material layers, these tools are still mainly based on damage criteria such as:

- Bottom-up cracking occurring in base layers (not for studying top-down cracking occurring in thin overlays).
- Rutting due to high vertical strain level.

Although bonding properties have often been neglected, particularly for pavements with a thin or ultra-thin surface course [3, 4], many researchers have stressed the importance of a good interlayer bond for pavement durability and performance [5–12]. In addition, pavement design tools [13–16] mainly assume that traffic loading may be reduced to a vertical constant normal pressure applied on the

C. Petit (✉)
Laboratory GC2D, Université de Limoges, Egletons, France
e-mail: christophe.petit@unilim.fr

A. Chabot
Department of Materials and Structures (MAST), IFSTTAR, Nantes, France
e-mail: armelle.chabot@ifsttar.fr

A. Destrée
Belgian Road Research Centre (BRRC), Brussels, Belgium

C. Raab
EMPA, Swiss Federal Laboratories for Materials Science and Technology, Dübendorf, Switzerland



W. G. Buttlar et al. (eds.), *Mechanisms of Cracking and Debonding in Asphalt and Composite Pavements*, RILEM State-of-the-Art Reports 28,
https://doi.org/10.1007/978-3-319-76849-6_3

surface of the pavement. However, increasing traffic load magnitudes and frequency are increasing pavement stresses, especially damaging the interface between surface and sub-layers. There is a critical need to take into account interface criteria in forthcoming pavement design tools. Apart from some advanced models [17], design tools often neglect surface layer damage due to horizontal tire forces and shear or mixed stresses leading to debonding between layers and surface damage. It is mathematically and numerically challenging to accurately model interface discontinuities in a pavement system, particularly for thin surface layers and thin reinforcement layers, where high stress gradients exist. Some of the common interface debonding distress manifestations are now summarized.

Surface damage and top-down cracking has been observed in connection with different developments such as new surface technologies using very thin surface layers or increasing tire-pavement contact stresses [18], and several researchers [19, 20] found insufficient interface bonding to be one of the main reasons for failure. Simultaneously, most of the road networks are becoming older, and new road construction techniques including complex systems are under development. These pavements are often subjected to premature distresses (corrugation, peeling, slippage or fatigue cracking) that may be due to poor interface bonding (see Fig. 3.1).

Most heavily travelled roads with thick pavement structures are now thirty or forty years old, and only the top layers have undergone one or more repair operations. Traditional rehabilitation works consist of overlaying the old wearing course or milling it before covering it with a new overlay. These new composite pavement configurations have given rise to new, rapid deterioration modes in highway structures, such as a series of potholes or top layer delaminations. These phenomena generally occur during spring thaw and are often associated with frost and heavy rainfall. In her paper, Vulcano-Greullet [21] investigated three types of degradation observed for three distinct rehabilitation strategies (see for example Fig. 3.2).

(a) (b)

**Fig. 3.1** Pavements damages in Belgium **a** Corrugation in asphalt pavement **b** Peeling in composite pavement

**Fig. 3.2** Pavement wearing course laid on a smooth surface **a** Potholes in sequence **b** Damaged binder course under wearing course [21]

The mechanism causing upper layer deterioration was found to be connected to freeze-thaw cycles in combination with humidity and traffic. There are other causes which can lead to pavement degradation long before any surface deterioration appears. The presence of water near the surface, for instance, can explain some of the damage observed. Fatigue failure at the interface between layers during hot weather periods can also be the reason for damage [19, 22]. Similar deterioration processes are also discussed and illustrated in a study conducted on the LCPC's circular test track facility. The study reveals fatigue phenomena at the level of the first interface in hot weather, which, combined with traffic and rainy, wintry conditions, can weaken top layers and lead to their rapid degradation [23].

These new technologies and systems with very thin layers and/or reinforcement interlayers have been found to even increase surface damage rates and loss of bond integrity. These early pavement failures have led researchers to seek improved interlayer bonding techniques and surface damage resistance through the development of new tests and through new insights gained with modeling. Some of the modern research themes in pavement bonding are:

- Debonding of interface layers due to higher interface shear stresses combined with interface tensile stresses;
- Top-down cracking due to surface shear stresses, inducing surface tensile stresses;
- Rutting combining higher levels of normal and shear stresses in the case of asphalt pavement (not considered in this book);
- Edge cracking of Ultrathin White-Topping (UTW) pavements with marginal bond levels, especially in relation to moisture and other environmental effects.

This chapter deals with cracking, debonding, fatigue and damage at or near the interface of multi-layered pavements. The focus is not only on asphalt pavements, but also on so-called composite pavements, consisting of Portland cement and asphalt concrete layers. Before significant advances can be made in improving interlayer bond, field observations combined with laboratory tests are necessary to characterize (qualify) debonding mechanisms. Next, advanced models of interface debonding and bulk pavement layer fracturing can be constructed. Based on this understanding, new material systems, lab tests, and specifications can then be developed to create pavement systems designed against early bond-related failures.

Regarding interlayer bond testing of pavement layers, the topic has gained more and more importance throughout Europe and in many other countries over the past ten years (USA, Switzerland, etc.) [24–26]. Several test methods are in use today to evaluate the interlayer bond of pavement layers, but they vary with respect to test devices, test temperature, loading solicitation as well as sample geometry and preparation [27–29]. The aim of TC 206-ATB was to perform inter-laboratory testing for interlayer bonding of asphalt pavements. Chapter 6 of a previous TC 206-ATB RILEM State of the Art Report [30] entitled "Mechanical testing of interlayer bonding in asphalt pavements", provides a comprehensive summary of different mechanisms behind pavement debonding. Due to the differences in experimental conditions, a direct comparison of the test results obtained from the varied test methods employed is often not possible. Indeed, despite many efforts, recommendations and information are still missing regarding the differences between the methods and the selection of a given method for the assessment of damage related to adhesion between layers and internal cohesion of these layers.

The aim of this chapter is to identify and to quantify the interface key parameters (type of adhesion material and its content, roughness, humidity and water presence, cleanliness, etc.) for debonding between layers in order to analyze debonding mechanisms and to improve modelling. The proposed classification was adopted in this chapter—Interface debonding behavior.

As stated in the RILEM STAR Book on cement concrete materials by TC193-RLS [31], various local physical and interfacial morphological parameters, such as roughness or texture, are thought to heavily influence global interface bonding behaviour and performance (Fig. 3.3).

Mesoscale parameters (roughness, adhesion) and environmental or construction parameters (cleanliness, water content, etc.) should be introduced in future design methods for hot or warm mixed asphalt (HMA, WMA) and Portland cement concrete (PCC) pavements. In summary, to investigate the mechanisms of debonding that often occur in pavement structures (Figs. 3.4 and 3.5), different parameters must be considered:

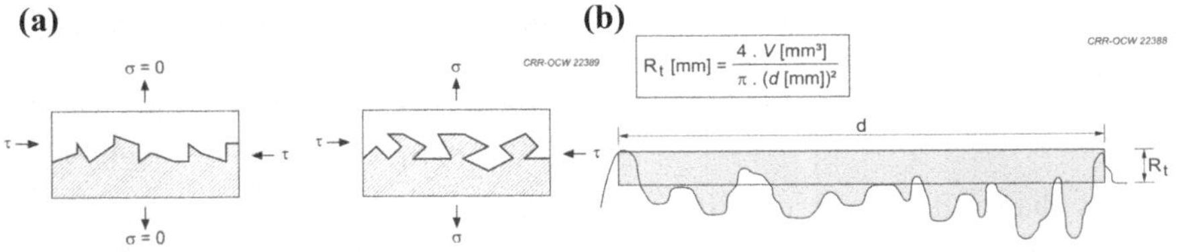


**Fig. 3.3** Interface roughness parameters [31] **a** Mechanical tensile and shear bond between substrate and overlay **b** Determination of roughness using the sand area method [32]

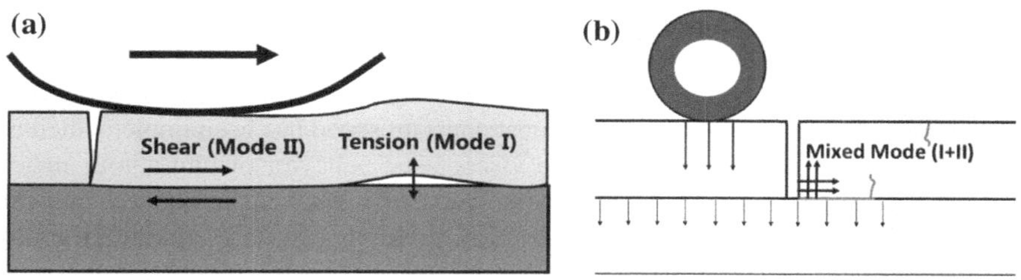


**Fig. 3.4** Pavement interface fracture debonding modes **a** far from a vertical crack [7] **b** at the edge of a vertical crack [36, 37]

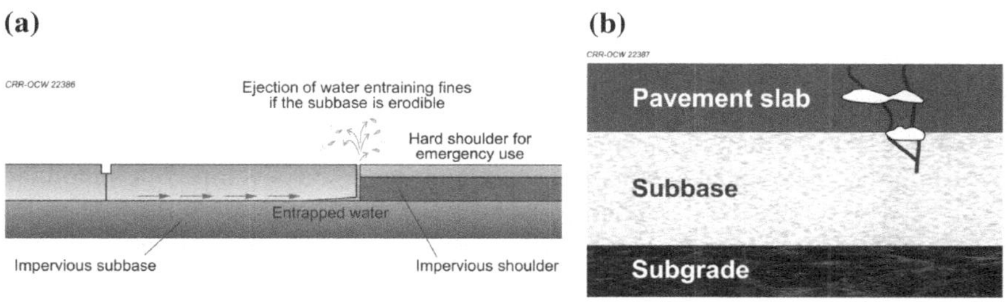


**Fig. 3.5** Water propagation in pavements from the top and its effect on interface bonding **a** [34] **b** [35]

- When the first layer is cracked, water may play an important role in the duration of interface bonding life. This has been clearly demonstrated for concrete pavements [33, 34] (Fig. 3.5a). As previously mentioned, this phenomenon is also very applicable to asphalt pavements [21, 35] (Fig. 3.5b).
- The mechanical behavior of the bulk materials such as the effect of the viscoelasticity of the asphalt layers linked to traffic speed and load should be rigorously considered in the development and interpretation of lab tests.
- Environmental effects such as the ratio of modulus dependence between two layers (linked to the temperature sensitivity of the asphalt material) and the moisture content of the materials should be considered.
- Interface physical condition such as cleanliness and roughness conditions, which may lead to different interlocking aspects, should be accounted for.
- Specimen sizes for laboratory testing should carefully consider representative volume element principles and fracture size effects.
- A sufficiently large number of specimens should be considered, especially considering the inherent variability associated with cyclic testing and fracture.

For the modelling of cracked pavement structures especially regarding bond conditions, some new developments have occurred or are still under way around the world, such as:

- *For studying the influence of a sliding interface in a flexible pavement (IFSTTAR ViscoRoute© software)*
  A semi-analytical method to compute the response of a layered viscoelastic medium under a uniform load moving at constant speed has been implemented in the ViscoRoute© software [38]. The viscoelastic behavior of bituminous materials is described using the Huet-Sayegh law. Emphasis was put on the treatment of interlayer relations so that fully-bonded or sliding interface conditions could be considered (and mixed) in the solution process. To compute the solution, a method based on a global matrix gathering all the interlayer and boundary conditions was developed, which proved to be efficient in solving the governing system of equations [39]. One can also use this type of software to take into account a highly viscoelastic tack coat interface. This method has been applied in a major APT test, which has validated the accuracy of the approach [40].
- *Simplified models for cracked pavements (IFSTTAR FissuRoute project)*
  The primary objective of this study was to build a 3D modelling approach for the study of cracking in pavements. For ease of use and short computation time, a simplified theoretical and numerical tool needed to be developed. A simplified cracked pavement model, denoted M4-5nB (for multiparticle model of multi-layered materials, with 5n (n: number of layers) equilibrium equations lying on a Boussinesq soil mass) was proposed as the kernel of this software tool [36, 37, 41]. To accelerate the solution process even further within the scope of multi-cycle computations, a method for extrapolating 2D results in order to estimate 3D results has been proposed. Describing the soil with a Winkler model and an additional "shear" layer instead of Boussinesq simplifies and accelerates the numerical solution [42–45].
- *The RCCM model for interface modelling (Université de Marseille—LMA)*

  1. This study involved the modelling of interfaces in masonry construction [46] by taking into account a mechanical phenomenon which governs their rupture such as adherence, sliding, friction, damage. For this issue, a test campaign was carried out on various masonry components: bricks; mortar and bricks assemblies composed of two and three bricks (full and hollow) bonded by mortar joints. The aim of the experimental study was to identify the mechanical characteristics of each masonry material and to describe the behavior of masonry structures that will be considered for the modelling study. For the numerical modelling, the LMGC90 software was used. The goal was to validate the RCCM (Raous-Cangémi-Cocou-Monerie) model of a material interface [47, 48], which takes into account nonlinearities such as damage. Firstly, the RCCM model was validated in the local size on samples tested experimentally. Secondly, the model was more broadly applied to other structures (walls of various sizes) and further validated.
  2. The modelling of interfaces taking into account friction with applications to bolted junctions in nuclear engineering [49], ground-tire in aeronautics [50], pellet-cladding interaction and metal forming in nuclear engineering [51] was pursued.

3. The theoretical modelling of interfaces using asymptotic techniques [52–55]. These techniques were applied to plane interfaces, roughness, viscoelastic behavior, plasticity, curved interfaces, etc., was also pursued.

- *New Finite Element methods* such as the extended finite element method (XFEM) [56, 57] or the generalized finite element method (GEM) [58, 59] may also be used to help the modelling of such fracture conditions in pavement structures or specimens in lab tests keeping the initial mesh during the crack propagation.

This chapter—Interface debonding behavior summarizes the different tests available around the globe to characterize, mostly in the laboratory, the bond between pavement layers. The aim of this document is to classify them and to analyze and discuss the types of applications for which these tests may be useful.

Existing interlayer adhesion test methods can be divided into four categories (Fig. 3.4):

- Mode I: Opening mode (see Sect. 3.2)
- Mode II: In plane shear mode (see Sect. 3.3)
- Mode III: Out of plane shear mode (see Sect. 3.4)
- Mixed Mode: Combination of modes I and II (see Sect. 3.5).

The four following sections discuss separately the different fracture mode tests linked to the existing possible fracture modes in pavements. Mostly, "pure" mode test methods (opening mode I or in plane shear mode II or out of plane mode III) are currently used worldwide for determining the interlayer bond of pavement layers. Most of the mixed mode test methods (mainly for the combination of Modes I and II) were developed by a few research teams and they are therefore not standardized tests. Many of the tests presented here can be performed on specimens prepared in laboratory or on cores (or slabs) extracted from the pavement.

The final section in this chapter identifies the main shortcomings of the reviewed tests and summarized the remaining challenges needed to improve knowledge related to the mechanisms of debonding in pavements and to develop testing and modelling solutions accordingly.

## 3.2 Mode I—Opening Mode

Interface Mode I opening mode is a debonding or an interface damage phenomenon created by means of a tension loading (Fig. 3.4a).

In this section, the objective is to highlight the main mechanisms that favor this type of fracture mode in asphalt and composite pavement systems. Herein, adhesion properties are viewed as the primary mechanism to investigate. Whatever the type of interface being used, cleanliness of the underlayers to be bonded to is important. In the case of using tack coat at the interface between two layers, adhesion properties are mainly driven by their properties (viscosity, breaking index, binder content, and the characteristics of the residual binder) and their conditions of application (application rate, curing time, and nature and preparation of the underlayer). Interface roughness probably plays a second role for Mode I opening mode, and it may be considered as a secondary parameter.

Two types of Mode I tension loading devices are described below. The paragraph 3.2.1 deals with Tensile Bond Test (TBT). It describes only tests that use specimens without any initial notch. On the contrary, the paragraph 3.2.2, called Tensile Notch Bond Test (TNBT), presents tests that are applied on specimens with an initial notch.

## 3.2.1 *Tensile Bond Test (TBT)*

The tensile bond test (TBT) can be applied to assess the capability of tack coats as well as the internal cohesion of the two road construction layers. Some care should be taken with the adhesive (glue); it should be strong enough to avoid failure in the adhesive (Fig. 3.6).

Usually two types of specimens are used: cylindrical and cubical specimen. Due to the geometry of this specimen, the difficulty of this test is to get uniform tensile stress distribution close to the interface (Figs. 3.7 and 3.8).

With the help of digital image correlation (DIC) analysis [61] (see Chap. 4), the Université de Limoges gives information about the thickness of interface named

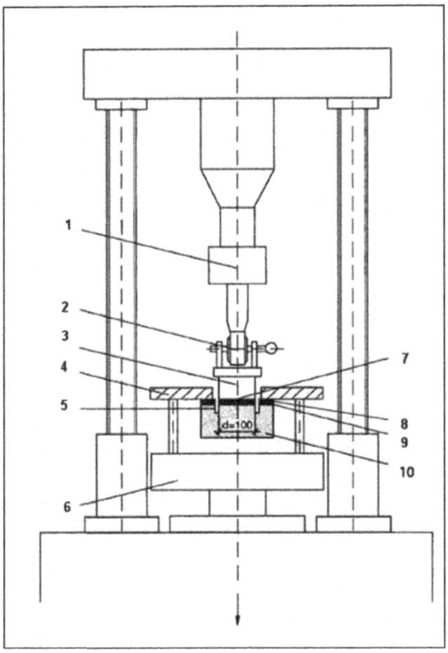


Key

1. load cell
2. globe joint
3. test-plunger
4. circular counter-plate
5. ring-groove
6. counter plate
7. adhesive glue
8. slurry surfacing or thin-layer hot mixed asphalt
9. interface between slurry surfacing or thin-layer hot mixed asphalt and bottom layer
10. bottom layer

**Fig. 3.6** Example of a test apparatus for tensile adhesion tests [60]

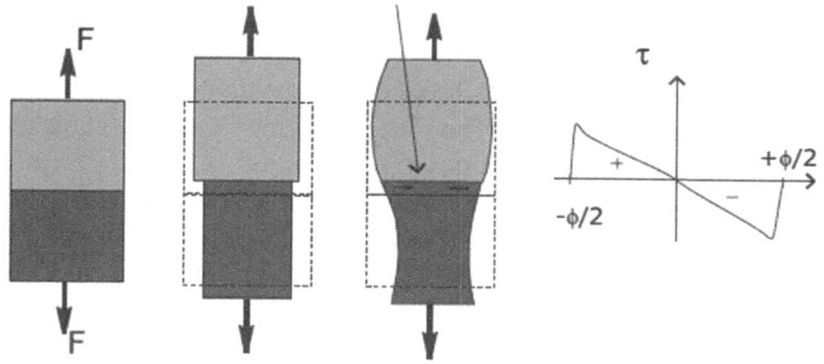


**Fig. 3.7** Schematic strain and stress in the sample [28]

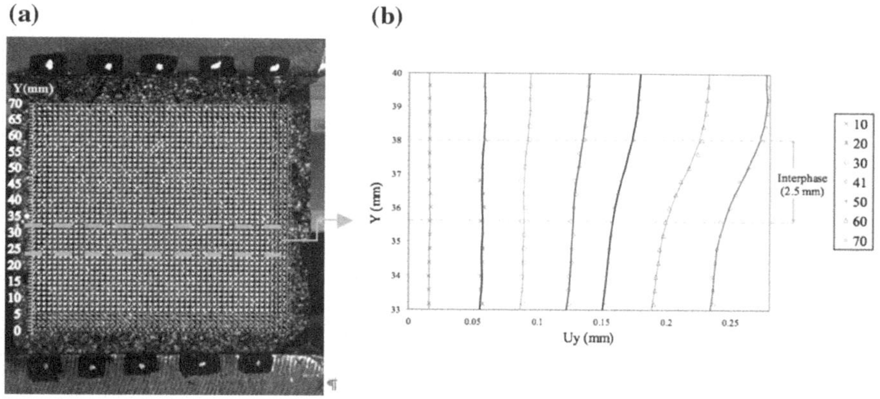


**Fig. 3.8** DIC measurements (see Sect. 3.3) **a** DIC grid **b** Displacement interface area [63]

interphase the area affected by interface behavior (Fig. 3.8a, b) [62, 63]. Such general description can be done in different field of material studies on the area affected by interface behavior.

Additional mechanical problems stem from the fact that a highly heterogeneous materials are used in the pavement field. Accordingly, scale and size effects have to be carefully considered.

#### 3.2.1.1 Cylindrical Specimen

The tensile adhesion test (TAT) described in the European pre-standard [60] is applicable on cores with a height of at least 60 mm, and it is intended to be carried out in the laboratory. The TAT can be applied to assess the capability of tack coats (if the failure occurs at the interface) as well as the internal cohesion of two pavement layers (if the failure occurs in the pavement layers). The specimens are cores with an outer diameter of 150 ± 2 mm (or 100 mm) and a height of at least 60 ± 5 mm. A concentric ring-shaped groove [inner diameter of 100 ± 2 mm (or 50 mm)] is drilled into the surfacing layer of the specimen, to a depth of approximately 10 mm below the interface. The plunger is glued with a suitable adhesive to the surface circumscribed by the groove, over its full area. The specimens are conditioned and tested at either (0.0 ± 0.5) °C or (10.0 ± 0.5) °C. The test is stress-controlled by applying a tensile force (200 N/s) until failure. The tensile strength is calculated from the maximum force and the surface area circumscribed by the groove (Eq. 3.1).

$$\sigma_{TAT} = \frac{F_{\max}}{\pi \times \left(\frac{D}{2}\right)^2} \tag{3.1}$$

where:

| | |
|---|---|
| σTAT | is the tensile bond strength in N/mm$^2$ to the nearest 0.1 N/mm$^2$; |
| Fmax | is the maximum force in N rounded to the nearest 1 N; |
| A | is the surface area inside the ring-groove in mm$^2$ to the nearest 0.1 mm$^2$; |
| D | is the diameter of the circular surface inside the ring-groove in mm to the nearest 0.1 mm. |

According to the pre-standard, at least two specimens shall be tested. The visual assessment of the surface after the failure is determined according to the classification depicted in Table 3.1. This kind of table can also be used for cubical and in situ test.

Very few countries specify tensile bond strength limits for tack coat. Austria [64, 65] uses a tensile test similar to the (pr EN 12697-48, 2015) [60] on 100 mm internal diameter cores at a test temperature of 0 °C. The minimum limits as a function of tack coat type are depicted in Table 3.2.

**Table 3.1** Failure modes for tensile bond test according to [60]

| Classification | Visual assessment | Mode of failure |
|---|---|---|
| A | Within the top layer | Cohesion |
| B | Partly at the interface, partly in the top layer | Mixed |
| C | At the interface | Adhesion |
| D | Partly in the bottom layer, partly at the interface | Mixed |
| E | In the bottom layer | Cohesion |
| F | Partly or completely in the adhesive | |

**Table 3.2** Tensile test specifications [64, 65]

| Tack coat type | Minimum average tensile strength (T = 0 °C) (MPa) |
|---|---|
| Unmodified | 1.00 |
| Polymer-modified | 1.50 |

A general problem of this test method lies in the fact that the specimen does not necessarily fail at the interface. Debonding happens along the weakest surface within the system, which in the worst case can include a poorly glued interface with the platens.

The following paragraph gives an overview of the normative tensile adhesion test (TAT) conditions and describes the main conclusions of a study in the laboratory and in the field where this interlayer bond test method was used.

*European method (TAT)* Specimens are conditioned in a climate chamber for a minimum of 12 h at 10 °C, and all tests are normally conducted at room temperature. From the tensile adhesion test the tensile force F (in N) as a function of the tensile extension (in mm) is obtained. From both laboratory and field studies described in three contributions of BRRC [66–69], their authors have shown that:

- In order to obtain a reliable estimation of tensile bond strength, three specimens are sufficient for thin wearing courses (thickness $\geq$ 30 mm) while six specimens are needed for ultra-thin wearing courses (thickness $\leq$ 20 mm).
- The direct tensile test executed according to the European method has a discriminating character to study the bonding of both thin and ultra-thin wearing courses to both asphalt and cement concrete base courses.
- The tensile strength can be obtained in general with the same precision for thin and ultra-thin wearing courses.
- A difference between tensile and shear loading (see Sect. 3.3) is that in the tensile test, the strength of the courses also affects the results in case the courses are weaker than the interface.
- The two loading modes (direct tension, direct shear) appear to be complementary, which means that one cannot replace the other to analyze the mechanical performance of the interface.

#### 3.2.1.2 Cubical Specimen

The tensile bond test described in the Belgian Road Research Centre working method is applicable on surface layers with a thickness $\geq$ 15 mm and can only be carried out in the laboratory [70]. The TBT can be applied to assess the capability of tack coats as well as the internal cohesion of the two road construction layers. The specimens have a square section of 80 by 80 mm and can be easily extracted from cores (for a core with a surface area of 400 $cm^2$, 3 specimens can be extracted) or from slabs manufactured in laboratory. Metal plates are glued with a suitable adhesive to both sides of the specimen (Fig. 3.9). These are conditioned and tested at (10 $\pm$ 0.5) °C. A tensile load is applied in displacement-controlled mode (0.5 mm/min) until the specimen fails. The tensile bond strength is calculated from the maximum force and the surface area of the cubical specimen. This latter is calculated at BRRC as the average of five specimens.

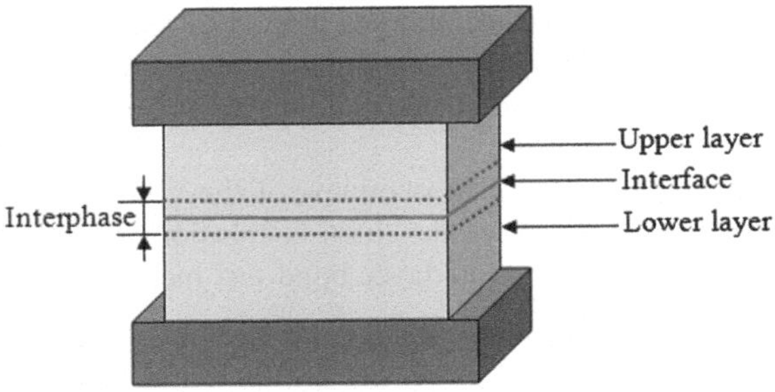


**Fig. 3.9** Tensile Bond Test for a cubical specimen [62] as used at BRRC [70] and the Université de Limoges [63]

A test device comparable to the previous equipment has been designed and constructed at the Université de Limoges [62, 63]. This tensile adhesion test can only be carried out in laboratory. Metal plates are glued to both sides of the specimens (90 by 92 mm). A tensile load is applied in controlled displacement mode (0.5 mm/min) until the specimen fails. During feasibility tests, mechanical data and images speed acquisition are approximately one measurement per second. The experimental temperature is set at room temperature (22 ± 2 °C). A correlation study is realized with a growing up = 0.202 mm/pixel, and the correlation windows size is about 8 × 8 pixels (the shift is equal to 2 pixels) (Fig. 3.8a). Global fields for each method are used and associated to local field studies. Local analysis can be done close to the interface area in order to access mechanical behavior and parameters.

The following paragraph gives an overview of the BRRC working method conditions for the tensile bond test and describes the main conclusions of a study in the laboratory and on site in which this interlayer bond test method was used.

*BRRC working method* The specimens are conditioned for a minimum of 4 h and tested in a climate chamber at 10 °C. From the tensile adhesion test, the tensile force F (in N) as a function of the tensile extension (in mm) is obtained. From both laboratory and field studies described in Destrée et al. [67] and Bergiers et al. [71], the following conclusions can be drawn:

- The direct tensile test developed at BRRC has a discriminating character to study the bonding of both thin (thickness ≥ 30 mm) and ultra-thin wearing courses (thickness ≤ 20 mm) to both asphalt and cement concrete base courses.
- The tensile strength can be obtained in general with the same precision for thin and ultra-thin wearing courses.
- The adhesive strength is much lower in case of a cement concrete base layer. Only for this type of base layer, a significant improvement is seen when using a tack coat. Some latex emulsions and especially the alkaline emulsion have an important effect.

- For tests on specimens prepared in the laboratory, the cubical geometry turns out to be more advantageous than the cylindrical geometry. However, for tests on samples taken on site this cubical geometry will cause more damage and necessitate more extensive subsequent repairs.

#### 3.2.1.3 In Situ Tensile Bond Test

In situ testing allows the determination of bond strength directly on site (see Fig. 3.10). Before applying the gripper system on the pavement surface and applying a tensile force, a ring-shaped-groove is made beneath the surface layer.

*The Layer adhesion measuring instrument—LAMI* The tensile adhesion test described in the Quebec method [72] can be used either in situ or in laboratory to evaluate the tensile bond strength between road construction layers. Two gripper systems can be used according to the thickness of the tested surface course: a plunger or an adhesive-free gripper system (thickness surface course > 35 mm). The test is stress-controlled by applying a tensile force (240 ± 40) N/s to a 100 mm diameter specimen until failure (Principle in Fig. 3.10). In the laboratory, three specimens are conditioned and tested at (20.0 ± 1.0) °C. On site, the applicability domain is between 10 and 25 °C and six tests are performed. For site testing, a correction factor is then necessary to calculate the tensile bond strength at 20 °C. For a segment to be considered as adequately bonded, the average of the three results in a wheel path and that of the three results outside the wheel path must be equal to, or greater than, 0.20 MPa at 20 °C if the tested interface is situated 40 mm or deeper below the surface, and equal to, or greater than, 0.30 MPa at 20 °C if the tested interface is

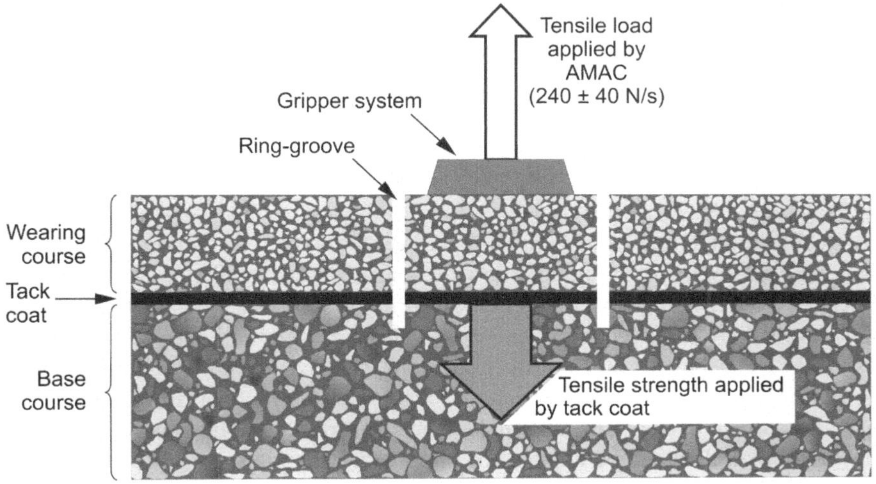


**Fig. 3.10** Principle of the in situ tensile bond test

situated less than 40 mm below the surface. In addition, only one result in three may be lower than the former or the latter of these values. The fate of a segment can, therefore, be determined after only three tests, if the first two results in a wheel path or outside the wheel path are individually lower than the former or the latter of the values mentioned above [73]. For site testing, unlike the torque bond test (see Mode III section) that always requires an adhesive to glue the metal plate to the surface of the core (a both manual and time-consuming action and complicated of a technician), the LAMI with an adhesive-free gripper system is relatively simple, fast and well-defined automatic test method [74]. This test can also be done with limited damage to the pavement layers if the bond is sufficient but from a practical point of view, its main disadvantage is the time consuming coring process and the subsequent cleaning of the cut groove.

*EMPA working method* The so-called pull off test is carried out according to the German testing specification ZTV-SIB 90 [75]. Divergent to this specification, the test disc diameter is (100.0 ± 2.0) mm. The test is conducted stress-controlled with a testing speed of 300 N/s, before starting a pre-force of 100 N is applied. The test can be conducted in the field and in the laboratory. For laboratory testing in a first step, cores are glued onto a concrete plate. After that the steel test disc with a diameter of 100 mm and a thickness of 60 mm is fixed on top of the core. For laboratory use the test device was modified for testing cores by using an aluminium plate with a hole in the middle on top of a steel ring. The pull-off device is then placed on top of the steel ring and connected to the test disc. Testing in the laboratory allows conducting all tests at a constant temperature of (20.0 ± 1.0) °C. In a research project on laboratory prepared specimens, it was found that pull-off testing and LPSD testing (see Sect. 3.3) led to qualitatively similar results and ranking. Another study on the adhesion of rehabilitated concrete pavements [8] revealed a linear correlation between the maximum LPDS shear force and the pull-off strength.

### 3.2.2 Tensile Notch Bond Test (TNBT)

To localize the fracture energy on the interface, the mode I tests proposed here are made on specimen with an initial interface notch. They are called in this document Tensile Notch Bond Tests (TNBT). Regarding static tests, the main parameter indicator is the interface fracture energy, $G_f$ coming from the Load—Crack Mouth Opening Displacement curves of the test results. $G_f$ is calculated by determining the area under the curve and by normalizing it by the fracture area. Regarding cyclic tests, fatigue parameters of the Paris law are fitted on the fatigue curves assuming that this simple law can be applied for bituminous materials.

In the pavement community, this is still an open question.

#### 3.2.2.1 Interface Bond Test (IBT)

Like the disk-shaped compact tension geometry—DC(T) [76], the IBT utilizes the standard fracture mechanics approach of using a notched specimen and crack mouth opening displacement—CMOD-based test control to create a controlled, propagating crack (Fig. 3.11). In this test, the plane of the fabricated notch is made to coincide with the plane of the interface to be tested. IBT testing involves the application of loads through pins inserted into the loading holes located inside the aluminum platens and the measurement of crack mouth opening displacement (CMOD) with a clip-type extensometer, as shown in Fig. 3.11b. In order to provide stable post-peak fracture, the test is controlled through a constant CMOD rate.

The University of Illinois had developed a fracture energy based approach with the interface debonding test (IBT). Three sets of specimens were used in this study. After preparing the specimens in the laboratory, the required edges of the specimens were cut, and notches and grooves were fabricated on each specimen. After gluing the platens using an epoxy with 17.2 MPa (2500 psi) tensile strength, gage points were mounted onto the specimens. The specimens then were placed in the cooling chamber for 2 h at the testing temperature of −12 °C, and the IBT test was performed with a CMOD rate of 0.5 mm/min. With increased wet tack coat application rate up to an optimum application rate, the interface fracture energy increases. The coefficients of variation of the results range between 4 and 19% showing that the test provides acceptable repeatability. It should be noted that the obtained results are based on one tack coat type and one HMA mixture. Further investigation on different tack coat types, wider range of application rates, and different HMA mixtures is recommended.

#### 3.2.2.2 Wedge Splitting Test (WST)

Initially developed for fracture tests [78, 79], the wedge splitting method has been adapted for pavement materials by Tschegg [80–82]. The principle of this method for bi-material interface-samples is shown in Fig. 3.12. Besides cubes, other specimen shapes such as cylinders and beam scan also be used [83, 84]. For the

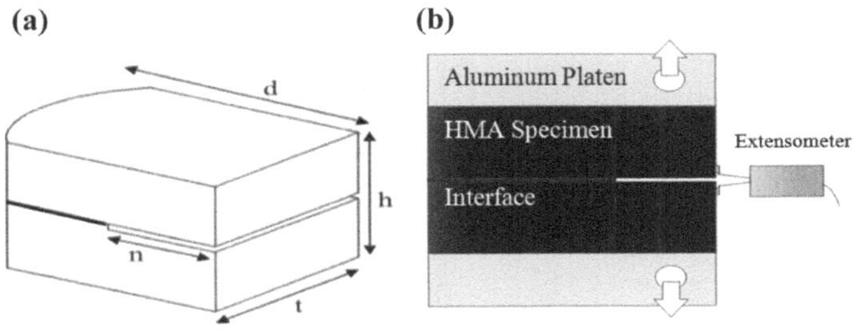


**Fig. 3.11** IBT test **a** Final specimen dimension **b** IBT test setup [77]

interface specimens, the transition from base-to top-layers has to be in the middle of the sample. A symmetrical groove on top of the sample can either be prepared by cutting or, which is more common, by gluing stone cubes on top of the specimen in order to form the groove. To assure crack initiation at the interface or the middle of the specimen, a so called "starter notch" has to be cut at the bottom of the groove. From here, the crack starts to grow stably through the sample. Crack growth is driven by a load; a vertical compressive load FM is generated by a pull–push-testing machine. It is transformed into a horizontal force FH by a wedge and transmitted to the groove of the specimen by load transmission pieces (Fig. 3.12).

To minimize friction, roll bodies are used for the load transmission between the wedge and the sample. Splitting is then caused by the horizontal force component FH (Fig. 3.13a) that can be calculated easily from the equation FH = FM/(2 tg(α/2)), where α is the angle of the wedge. This wedge angle shall be 15 ± 5° for a specimen

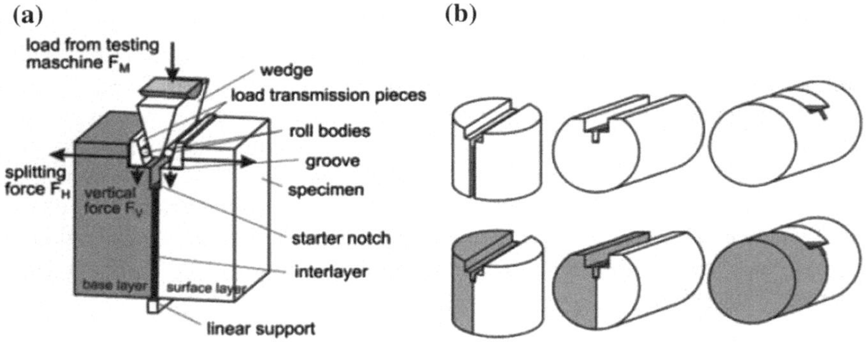


**Fig. 3.12** Wedge-Splitting Test (WST) **a** Wedge-splitting method according to Tschegg [85, 86] **b** Possible specimen geometries [82]

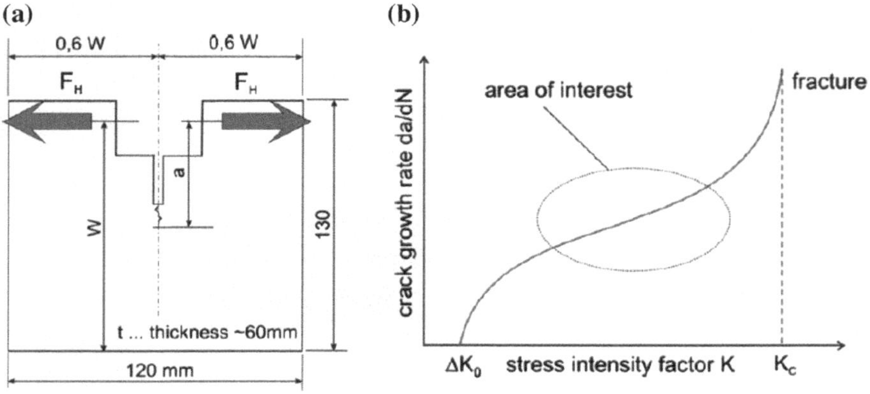


**Fig. 3.13** **a** Profile dimension **b** Schematic Paris law [88]

which is stiff on both sides of the ligament (concrete) and 25 ± 5° for a specimen which is ductile at least on one side of the ligament (asphalt concrete) [87].

In static conditions [80], the specific fracture energy $G_f$ is obtained from the area of the load-displacement curve G divided by the nominal fracture area A. $G_f$ is assumed to be independent of the specimen shape and size if the specimen diameter is 150 mm, the total length 160 mm and the notch 55 mm (Fig. 3.13a). For these dimensions and a small angle value of the slim wedges, this test allows quasi "pure" mode I fracture opening. Specimens coming from pavements may also be tested with a geometrical adaptation of the WST groove [89].

According to [83], this test can also be used for fatigue crack growth testing of asphalt and asphalt and is an Austrian standard test [87]. For fatigue crack growth testing of asphalt and geosynthetic interfaces, Tschegg published that if slim wedges ($\alpha = 5 - 12°$) are used, the vertical force component FV is small in comparison with the horizontal force FH and helps to constrain the crack to propagate within the area between starter notch and linear support [82]. The Stress-Intensity-Factor-concept (K-concept) is used in fracture mechanics to predict the stress state near the tip of a crack caused by a load. To describe the Paris-Law at the area of interest (Fig. 3.13b), crack growth rate and stress intensity factor are calculated discontinuously every 5 mm crack length increment. Results are given at a speed rate of 2 mm/s for asphalt and asphalt interface type and 0.5 mm/mm for cement and asphalt and for temperature of −10, 0, 10, 22 °C. Below 0 °C, asphalt exhibits linear elastic behavior. In order to make experimental results comparable within a temperature range of −10 °C to +10 °C, it is favorable to use the K-concept for +10 °C as well, even if the process zone is larger compared to the crack length and it would thus probably be more suitable to use the J-Integral-concept instead. The magnitude of K depends on the sample geometry, the size and the location of the crack, the magnitude, and the modal distribution of loads on the material. As a conclusion from a study on fatigue crack growth in asphalt and asphalt-interfaces [82, 83], the following was stated: beyond all restrictions of evaluating the crack progress, the wedge splitting method managed to achieve good results for fatigue crack growth analysis for sample temperatures of −10, 0 and +10 °C and for different asphalt-interface samples. Overall, the wedge splitting method has proved to be a practical method in fatigue crack growth testing due to its sensitivity when measuring differences in interface treatment and temperatures.

## 3.3 Mode II—In Plane Shear Mode

Interface Mode II in plane shear mode is a debonding or an interface damage phenomenon created by means of a shear loading (see Fig. 3.4a).

In this type of loading, it is assumed that roughness properties play a large role in the quality and durability of interlayer bonding.

The construction of shear testing devices for asphalt pavements was originally derived from shear testing in soil mechanics and already in the late 1970s different equipment such as the Leutner test in Germany [90] or similar tests in the USA were developed for road pavements [91]. There are two fundamentally different systems: the direct and the simple shear test. In order to introduce the effect of traffic loading some shear testing devices allow for the application of a normal load.

These two types of Mode II—in plane shear mode tests (SBT) are presented below. The first paragraph gives an overview of SBT without additional application of normal stress while the second paragraph presents test methods using an additional normal stress loading condition.

### 3.3.1 Shear Bond Test (SBT) Without Normal Stress

The direct shear test (called Shear Bond Test (SBT) in the European standard [60] is shown in Fig. 3.14.

In general, the SBT is a guillotine-type test where the shear force is induced directly at one side and not at the front surface of the specimen [6] (Fig. 3.15). For the direct shear test, a stress concentration in the point of load introduction can be found. However, the load distribution in the simple shear test is more parabolic but strongly depends on specimen diameter and shear device construction. In the simple shear tests, the upper part of the test specimen is sheared against the bottom part of the test specimen and the shear force is induced at the specimen front surface of the specimen (Fig. 3.15b).

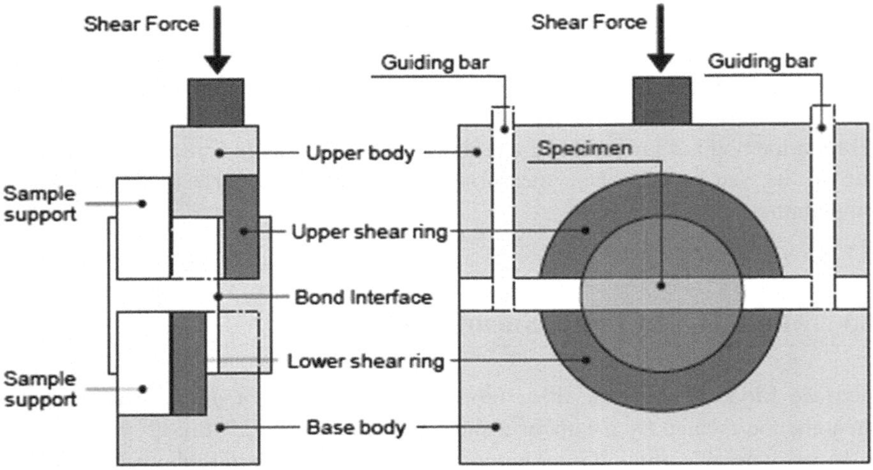


**Fig. 3.14** Pure direct shear device (Leutner) [30]

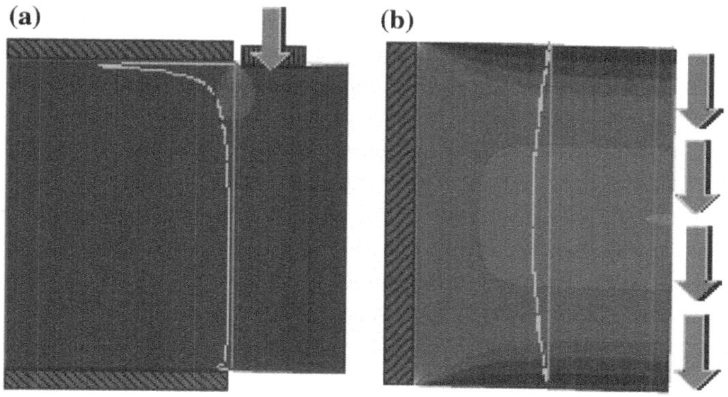


**Fig. 3.15** Shear stress distribution **a** in the direct shear test **b** in the simple shear test [6]

The direct shear testing devices can be divided into devices which use a clamping or fitting system to hold the test specimen [6, 24, 92, 93] and devices which utilize a bending mechanism (three or four point shear tests) to apply the shearing [94, 95]. This test is called Compressed Shear Bond Tests (CSBT) in the European standard (only 2 methods are mentioned in the standard yet). In case of three-layered specimens, as in the double shear test [96], the middle part is sheared against both outer parts. For the simple shear test the mechanism of the different devices is similar, differing mainly in the way the shear forces are applied and how both parts of the test specimen are moved against each other [97, 98].

As opposed to the direct shear tests, where the test specimens can either be clamped or fitted into steel molds, the test specimens in the simple shear test are always fitted into the shear mold by glue or tight fixtures. Therefore, the application of a normal force vertical to the shear plane is always an option, whereas in shear test devices with clamping mechanisms, normal forces are often not taken into consideration. While some shear tests, mainly those used for quality assessment, only allow for static testing, others can be used either in a static or a dynamic mode [98–100].

The shear bond test described in the European pre-standard [60] can only be carried out in the laboratory. The specimens are cores with a diameter of (150 ± 2) mm or (100 ± 2) mm. Although both core diameters are allowed it is worth noting, that in an inter-laboratory study the shear stress of 100 mm cores was found to be 14% higher than the shear stress of 150 mm diameter cores [101]. The minimum thickness of the layers above the interface to be tested shall be 15 mm and below the interface 70 mm, respectively. In the case of ultra-thin surface layers (thickness < 20 mm), a metallic extension plate is glued with a suitable adhesive to the surface layer in order to avoid deformation of this layer and the correct distribution of the shear load at the interface. The specimens are conditioned at (20 ± 1) °C and tested at room temperature using a shear test device (Fig. 3.14) with a recommended gap width between the shearing rings $\leq 5$ mm. This ratio (0–5 mm) is a European compromise allowing for testing with the original Leutner device with a gap with of 0 mm and with the English modification [12, 102, 103] with a gap width of 5 mm which tries to overcome the difficulty with the specimen alignment but which also may lead to an increased eccentricity resulting in a combined bending-shear stress situation and a greater scattering of shear test results [104]. The test is conducted with displacement controlled at a rate of (50 ± 2) mm/min. The obtained result is a shear force versus shear displacement graph (see Fig. 3.16) that allows the calculation of the maximum shear stress (2) and the shear stiffness modulus (3) and the determination of the shear energy determined from the area below the shear force-deflection graph as measured between the displacement at maximum shear stress and the intersection of the linear shear force slope $F_{SBT}$' with the displacement-axis in (Nm) to the nearest 1 Nm.

$$\tau_{SBT,\max} = \frac{F_{\max}}{\pi \times \left(\frac{D}{2}\right)^2} \tag{3.2}$$

where:

$\tau_{SBT,max}$ is the maximum shear stress of the interface in (MPa), expressed to the nearest 0,01 MPa;
F is maximum vertical shear force (in N);
D is the initial diameter of specimen in the interlayer (in mm).

$$k_{SBT,\max} = \frac{F'_{SBT}}{\pi \times \left(\frac{D}{2}\right)^2} \tag{3.3}$$

where:

$k_{SBT,max}$ is the shear stiffness modulus of the interface in (MPa/mm), expressed to the nearest 0.1 MPa/mm;
FSBT′ is the slope of the linear part of the force-deflection graph (in kN/mm);
D is the initial diameter of specimen in the interlayer (in mm).

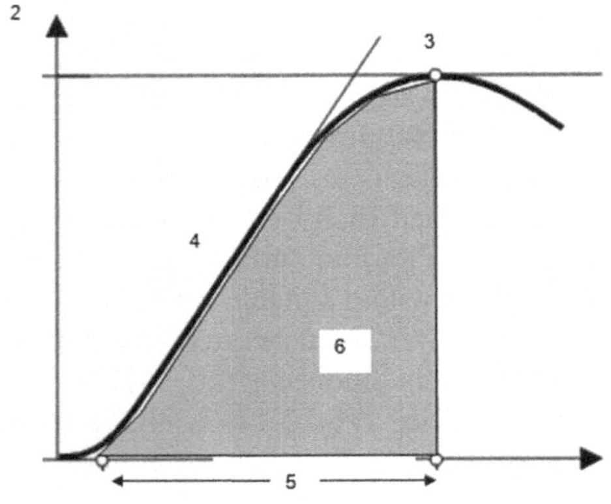


Key
1 displacement δ, in mm
2 shear force F in kN
3 maximum shear force $F_{SBT,\ max}$
4 slope of shear force $F_{SBT}$ (shear stiffness)'
5 displacement at peak shear stress $\delta_{SBT,max}$
6 shear energy $E_{SBT}$

**Fig. 3.16** Example for a shear stress versus shear displacement graph according to [60]

According to the pre-standard, at least two specimens shall be tested.

Here, tests conducted at different laboratories have already shown that testing two specimens is definitely not enough and that at least three specimens should be taken for the determination of the average value. For special constructions such as ultra-thin top layer, even a number of 6 specimens should be taken [6, 67, 101].

Few countries specify shear bond strength limits. Switzerland (CH) and Germany (D) are two countries that do and use a shear bond test as described in the European pre-standard on 150 mm diameter cores and a temperature of 20 °C (Table 3.3).

These requirements were set already during the 1980s in Germany and stated by findings from Codjia [105] (see also Table 3.4).

Some Swiss cantons and administrations carried them over for their applications before they were officially taken over into Swiss standard requirements at the end of the 1990s. Soon after the implementation of requirements in the Swiss and German standards, these values were criticized, and some authors such as Raab and Partl [6] and Stöckert [106] made suggestions for higher limits and recommendations based on their experience and findings opposed to the ones from Codjia [105]. Also, development of tack coats and construction methods has shown [107] that especially for the bond between surface and binder course, higher bond values can be reached and are desirable even though they have not yet been accepted by standardization. For Austria (A), the requirements are dependent on the type of tack coat used (non-polymer modified versus polymer modified). The minimum limits as a function of the tested interface are given in Table 3.5. Here, testing is done according to the European pre-standard on 100 mm diameter cores and a temperature of 20 °C and the requirements are given as shear stress (N/mm$^2$).

**Table 3.3** Shear bond test (SBT) force specifications for D and CH (20 °C, v = 50 mm/min)

| Source | Shear force in kN for Ø 150 mm (shear stress in N/mm$^2$) | |
|---|---|---|
| Standard requirements | Surface/binder | Binder/base or base/base |
| Germany, FGSV | 15 (0.85) | 12 (0.68) |
| Switzerland, SN 640430 | 15 (0.85) | 12 (0.68) |

**Table 3.4** Recommendations for shear bond test requirements based on experience and research

| Source | Shear force in kN for Ø150 mm (shear stress in N/mm$^2$) | |
|---|---|---|
| | Surface/binder | Binder/base or base/base |
| [6] for AC and SMA on binder | 23 (1.3) | – |
| [106] | 25 (1.41) | 20 (1.13) |
| [105] | 15 (0.85) | 10 (0.57) |

Table 3.5 Shear bond test (SBT) stress specifications for Austria ÖNORM B3639-1 [108] (Ø 100 mm cores)

| Source | Shear stress in N/mm$^2$ for Ø100 mm (shear force in kN) | |
|---|---|---|
| Standard requirements | Surface/binder | Binder/base or base/base |
| Non modified tack coat | 0.8 (6.3) | 0.5 (3.9) |
| Polymer modified tack coat | 1.2 (9.4) | 1.0 (7.9) |

#### 3.3.1.1 Layer-Parallel Direct Shear (LPDS)

The Layer-Parallel Direct Shear (LPDS) test device (Fig. 3.17) is an EMPA modified version of equipment by Leutner, being more versatile in geometry and more defined in the clamping mechanism. One part of the pavement core (up to the shear plane to be tested) is laid on a circular u-bearing and held with a well-defined pressure of 0.5 MPa by a semi-circular pneumatic clamp. The other part, the core head, remains unsuspended. Shear load is induced to the core head by a semi-circular shear yoke with a deformation rate of 50 mm/min, thus producing fracture within the pre-defined shear plane. The LPDS has a long clamping and supporting length such that the clamping of the specimen is simple, fast and well-defined. This is particularly beneficial when testing different shear layers in a core or when testing pre-heated test specimens as it minimizes temperature loss [109].

In order to achieve a homogeneous load distribution, the yoke and clamps are adjustable to accommodate core diameters from 148 to 155 mm (normally 150 mm) as well as rectangular specimens. The LPDS test device fits into an ordinary servo-hydraulic testing machine [6]. The gap width between the shearing rings is 2 mm and therefore different from the original Leutner device with a gap

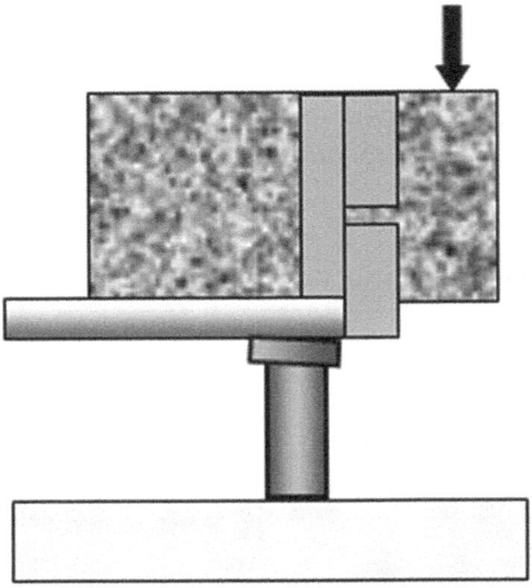

Fig. 3.17 Layer-Parallel direct shear device LPDS [110]

width of 0 mm. The modified gap width was selected to provide for a better alignment of the specimen. For quality assurance and testing in Switzerland, 4 cores are taken and tested at a temperature of 20 °C. The mean value of these 4 tests has then to fulfill the requirement [111]. In addition to the values determined according to the European Pre-standard, the following characteristics as shown in Fig. 3.18 and Eqs. (5.4)–(5.7) are determined and used mainly for research purposes.

The so-called secant stiffness M, which is calculated as the ratio between the maximum shear force and the deformation $w^\circ_{max}$ at $F_{max}$ as given in Eq. (3.4) where $F_{max}$ is the maximal force and $w^\circ_{max}$, the deformation at $F_{max}$

$$M = \frac{F_{\max}}{w^\circ_{\max}} \tag{3.4}$$

The change of the secant stiffness during the test, compared to the original one, can be taken as damage measure (Eqs. 5.5 and 5.6) [110].

$$\frac{M(w^\circ)}{S_{\max}} = 1 \quad \text{no damage} \tag{3.5}$$

$$0 < \frac{M(w^\circ)}{S_{\max}} < 1 \quad \text{damage} \tag{3.6}$$

where $M(w^\circ)$ is the secant stiffness at shear deformation $w^\circ$ and $S_{max}$, maximum stiffness (kmax according to EU standard).

The deformation at maximum shear force $w^\circ_{max}$ is calculated as follows (Eq. 5.7):

$$w^\circ_{\max} = w_{\max} - w_t \tag{3.7}$$

where $w_t$ is the intersection point of $S_{max}$ and x-axis.

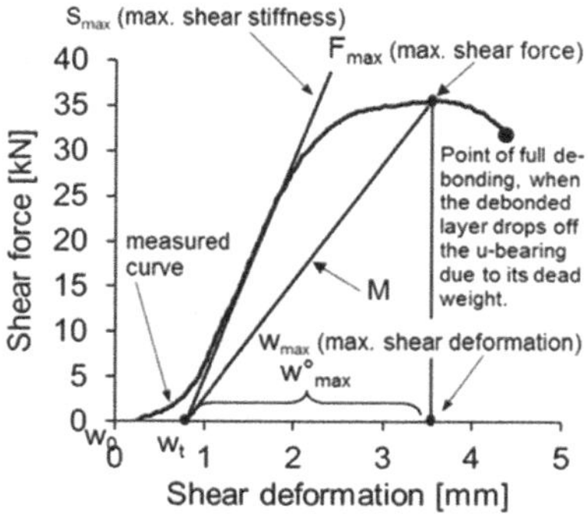


**Fig. 3.18** Schematic shear force-deformation curve with parameters from LPDS test [110]

#### 3.3.1.2 Laboratorio de Caminos de Barcelona Shear Test (LCB)

The Laboratorio de Caminos de Barcelona shear test device known as the LCB shear test (Fig. 3.19) was developed at the Technical University of Catalonia, Spain [94]. Here, cylindrical test specimens are tested under static loading conditions at a deformation rate of only 1.27 mm/min. In the LCB, shear stress is generated in the bond of the asphalt layers to be tested, until it separates them. The specimen is then used as a beam located over two supports, with the test section very close to one of the supports so that the bending moment is almost zero and the specimen is effectively only subjected to shear stress (Fig. 3.19).

#### 3.3.1.3 Double Shear Test (DST)

The double shear test (DST) developed at the Université de Limoges involves a specimen consisting of three layers bonded two-by-two with the same tack coat or interlayer. The test device (Fig. 3.20) allows the investigation of interface behavior in shear monotonic, cyclic and fatigue loading, under controlled temperature and frequency conditions. The shear loading is applied through a servo-hydraulic machine, allowing controlled displacement or controlled force modalities.

The specimen ($70 \times 120 \times 50$ mm$^3$) is composed of three parts (layers), which can be extracted from a two-layer, in situ core (Ø 150 mm) or laboratory produced slabs, including thin or ultra-thin asphalt overlays. As shown in Fig. 3.21, the two upper side layers (AC or CC #1 and AC or CC #3) are fixed during the test, and the central layer (AC or CC #2) is subjected to the monotonic or cyclic load [101].

From dynamic cyclic tests, complex shear stiffness is obtained and can be introduced for conventional pavement modeling. In addition to fatigue tests, fatigue law parameters can be identified [20, 101, 112, 113] (Fig. 3.22).

With the use of finite element modeling, these laws could be implemented in pavement design methods. It would help to open the discussion about the durability of the interface under traffic conditions [20].

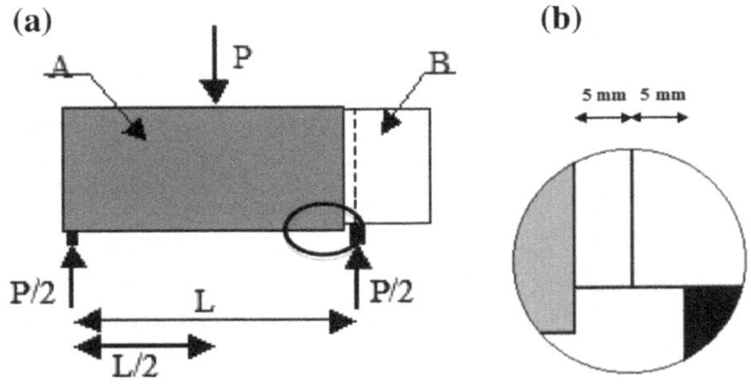


**Fig. 3.19** Laboratorio de Caminos de Barcelona shear device **a** Diagram of forces and stresses acting on the specimen in the LCB shear test **b** Position of the specimen in the press during the LCB shear test [94]

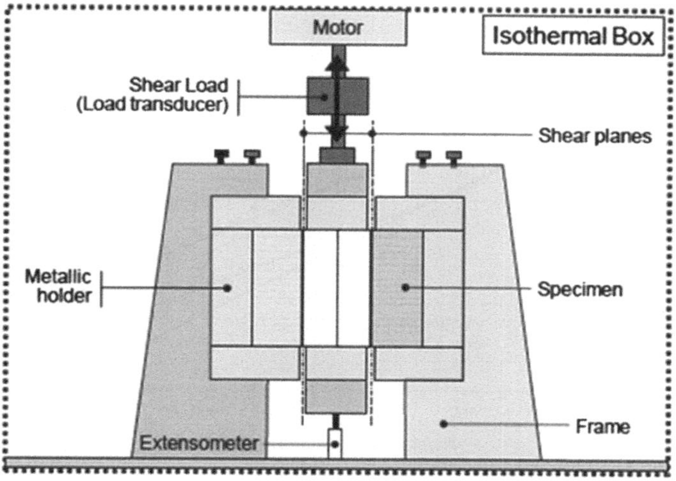


**Fig. 3.20** Double Shear Test (DST) [30]

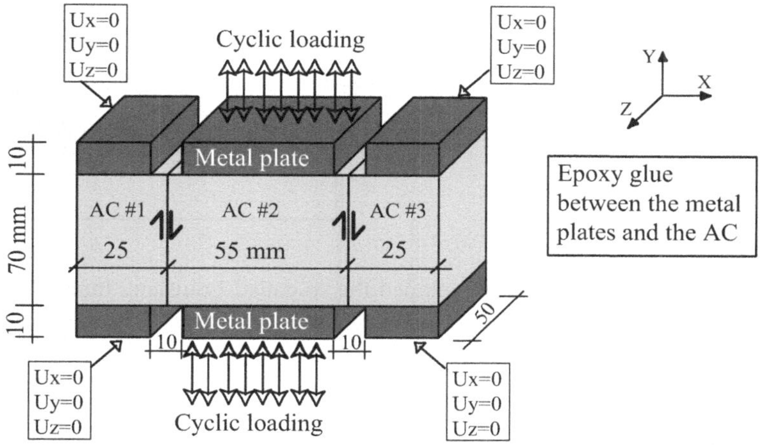


**Fig. 3.21** Schematic diagram of the double shear test [101]

## 3.3.2 *Shear Bond Test (SBT) with Normal Stress*

Researchers in the United States have taken an interest in shear properties since the early 1990s with the development of Superpave Shear Tester (SST) as a shear bond test (Fig. 3.23). Many of the current designs for shear strength testing equipment produce highly variable results or require complex bi-axial load frames.

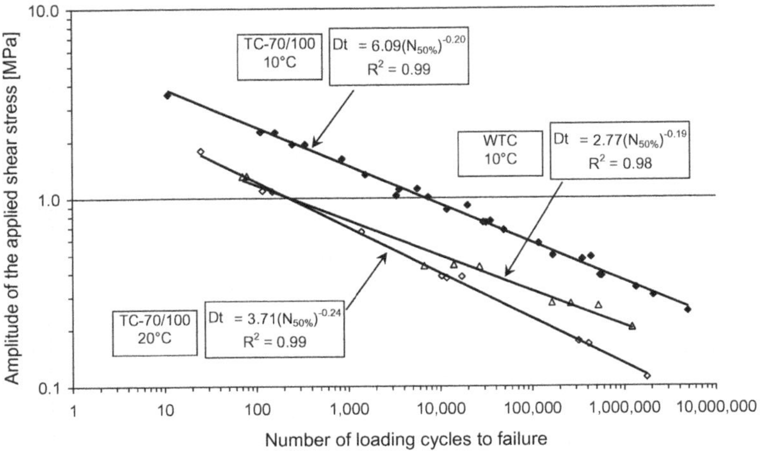


**Fig. 3.22** Conventional fatigue laws

**Fig. 3.23** Shear Bond Test with normal stress [110]

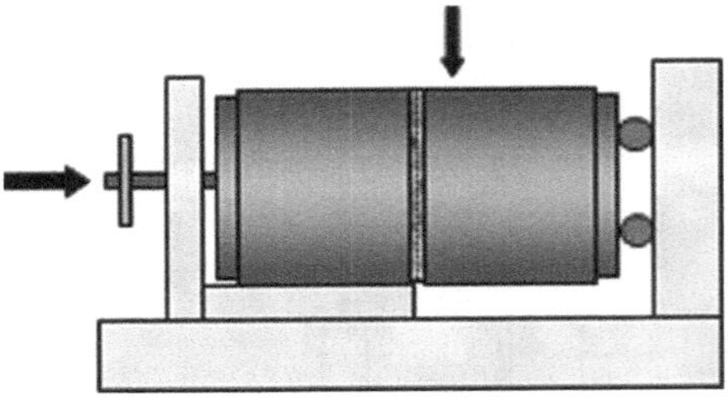

Mohammad et al. [26, 114] developed the so-called Louisiana Interlayer Shear Strength Tester (LISST) consisting of two main parts, a shearing and a reaction frame (Fig. 3.24). While the reaction frame remains stationary, the shearing frame applies the loading. Cylindrical specimens of either 100 or 150 mm diameter can be placed in the frames and locked with collars. The shearing load is applied with a constant rate of 2.54 mm/min. Application of a normal load is also possible. The gap between the shearing and the reaction frame measures 12.7 mm. The LISST device was developed for the characterization of interface shear strength of cylindrical specimens in the laboratory.

The advanced shear tester (AST) was designed to investigate shear properties of the uniform asphalt mix samples as well as interlayer interfaces [116]. It is capable of testing 150 mm specimens in the direct shear. The AST fixture was designed to fit in the computer controllable environmental chamber (40 by 40 mm) and to be attached to a closed-loop servo-hydraulic load frame. The AST consists primarily of two parts, stationary holder and moving collar, with a variable gap size between them. In the current AST version, the gap size can vary incrementally between

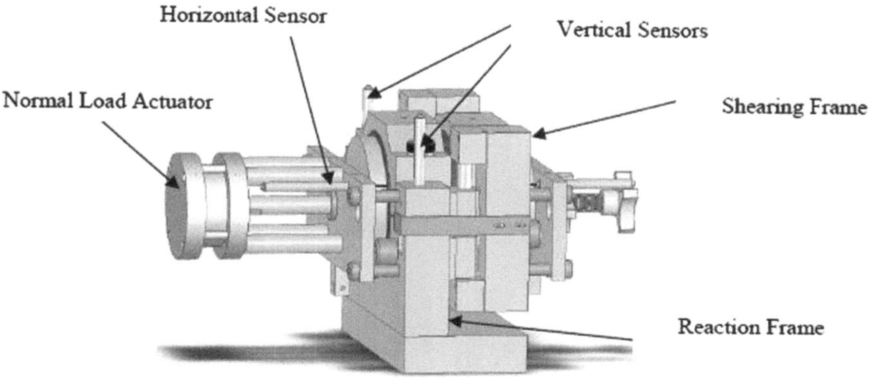


**Fig. 3.24** The Louisiana Interlayer Shear Strength Tester (LISST) [115]

5 and 12 mm, which may influence the results [29]. The AST design allows for up to 15 mm sliding in either direction (up or down) during the testing [117]. This allows for various combinations of the static and cyclic loading conditions in the load- or displacement-control mode to be applied to specimens across various temperature conditions (Fig. 3.25). The moving part of the fixture holds the sheared part of the sample in a collar-like mount and the part slides up and down on the vertical rods via slide bearings. With the collar design of the AST fixture, fatigue tests can be performed in both directions allowing for cyclic loading.

The dilation/contraction of the sample is also measured by strain gauge deflectometer mounted independently from the fixture. The arm of the deflectometer rests on the front specimen stopper that travels when the sample dilates/contracts during the test. The main results may be found in Zofka et al. [116, 117].

From all the devices presented in this section, there is a need to compare existing shear bond experimental setups not only in terms of the shear bond strength but also in terms of the fatigue resistance to the cyclic/dynamic shear conditions.

## 3.4 Mode III—Out of Plane Shear Mode

Torque testing is another way to determine the bond between pavement layers [93, 102, 103, 118]. This method has not been as extensively used as direct shear testing. This might be due to the fact that torsion testing is in most cases more cumbersome compared to shear testing since the specimens often must be glued before testing. Furthermore, torque testing provides a non-uniform shear stress distribution, varying between zero in the center of the interface cross section to a maximum value at its outside. Torque test devices are quite similar compared to one another. In principle, they consist of a plate that is glued or fixed to the test specimen and a torque cell applying the torque moment. The devices mainly differ

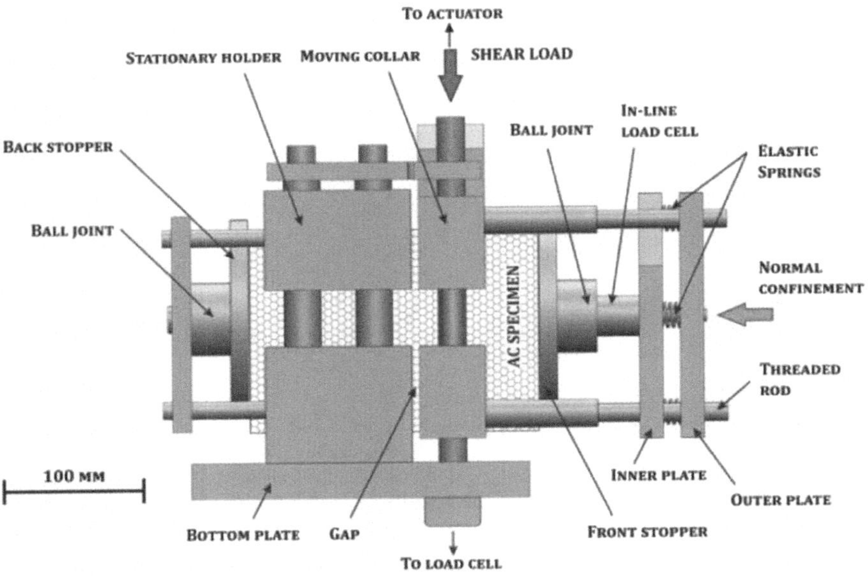


**Fig. 3.25** The AST device [117]

in the way the test specimens are hold (gluing, gripping or clamping) and in the construction of the test frame.

The test specimens for laboratory testing are cylinders with diameters between 100 and 200 mm. Torque testing offers a good possibility for in situ testing. Here, various devices and test set-up have been developed by different institutes and universities in Europe and North America [25, 119]. For in situ testing, the upper layer of the pavement has to be cored below the upper layer to ensure that the specimens separate at the interface. According to the pre-standard prEN 12697-48, the torque bond test is carried out either in situ or in the laboratory using cores. The test procedure consists of applying a torque with a dynamometric key at a steady rate so that the torque wrench sweeps an angle of 90° within (30 ± 15) s. The torque is applied until failure occurs or a torque of 400 Nm is exceeded. For both site and laboratory test methods six torque bonding tests are conducted to evaluate on result. This latter is expressed as the arithmetic mean of the torque bond strength measured on the six specimens (laboratory)/locations (on site), [120].

For laboratory testing (Fig. 3.26), according to prEN 12697, 2015 [60] the core has a diameter of (100 ± 2) mm for top layer thickness < 15 mm and (200 ± 2) mm or top layer thickness ≥ 15 mm. The core is placed and fixed in a mold, so that the top layer and the interface to be tested is (20 ± 10) mm above the rim of the mold. The core is conditioned in a climate chamber at a temperature of (20 ± 2) °C for a minimum of 4 h, and all tests are normally conducted at room temperature according to the test procedure explained above.

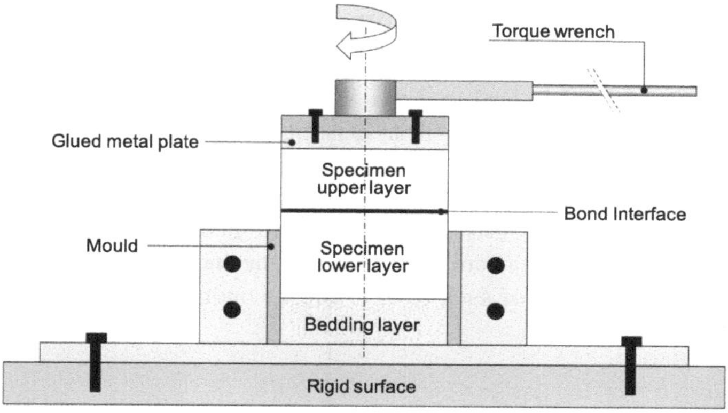


**Fig. 3.26** Torque bond tester [30]

The torque bond strength $\tau_{TBT}$ is calculated for each specimen using the following formula (Eq. 3.8):

$$\tau_{TBT} = \frac{12 \times M \times 10^{-6}}{\pi \times D^3} \quad (3.8)$$

where:

$\tau_{TBT}$ is the inter-layer torque shear bond strength, in kilopascal (kPa).
M is the peak value of applied shearing torque, in Newton meter (Nm),
D is the diameter of core in millimeter (mm).

For site testing [121], for a surface layer thickness $\geq$ 15 mm, a 100 mm diameter groove shall be cored to a depth of (20 $\pm$ 5) mm below the interface to be tested, but the core shall not be removed from the pavement. When the layer thickness of the surface layer is less than 15 mm, no coring is needed. A technician who applies the torque and reads the value at failure performs the torsion test manually. This requires a certain degree of experience and strength; it is physically difficult to exceed a torque of 400 Nm [66, 67]. Unlike the shear bond test (SBT), the torsion test induces shear stresses at the interface as well as in the surface course of the specimen. The torsion test is therefore less precise than the shear test and also more difficult to interpret in terms of type of failure, since it induces both a torque at the interface and in the upper layer itself, unlike the shear test which concentrates the shear loading exclusively on the interface. The idea had been that this test would be applicable directly on site, but in view of the precision, combined with variable environmental conditions and the practical difficulty to glue the metallic plates to the surface, we doubt that this will be a reliable and practical method.

## 3.5 Mixed Mode—Combination of Modes I and II

Some results from the field lead to the conclusion that the interface fracture mode type may not be pure especially near existing vertical cracks through a layer thickness (see Fig. 3.27). In this section, in addition to most of the previously presented test methods and with the aim of deepening the knowledge of complex conditions, some research tests are proposed to investigate interface bonding (behavior) in mixed mode fracture conditions. At the end, the objective should lead to a comparison of results of such tests to those conducted in "pure" opening or shear mode ones.

### 3.5.1 *Three Point Bending Test*

The three point bending test on concrete-concrete composite specimens supplied experimental data to validate a discrete crack model [123, 124]. Then, this numerical tool was exploited to evaluate factors expected to govern the durability of the overlay-substrate interface [125, 126]. Concerning the overlay material, incident parameters analyzed are autogenous shrinkage, Young's modulus, tensile strength, and the strain softening that can be significantly modified notably by a fibre-reinforcement. For the curing conditions, the relative humidity of surrounding ambience was taken into account. Related to preparation of the substrate surface, the tensile strength perpendicular to the interface was considered [127, 128].

Based on the cohesive crack concept, a model has been built and validated by the comparison with experimental data in three-point static bending tests on multilayer specimens. The numerical predictions show that the proposed model is

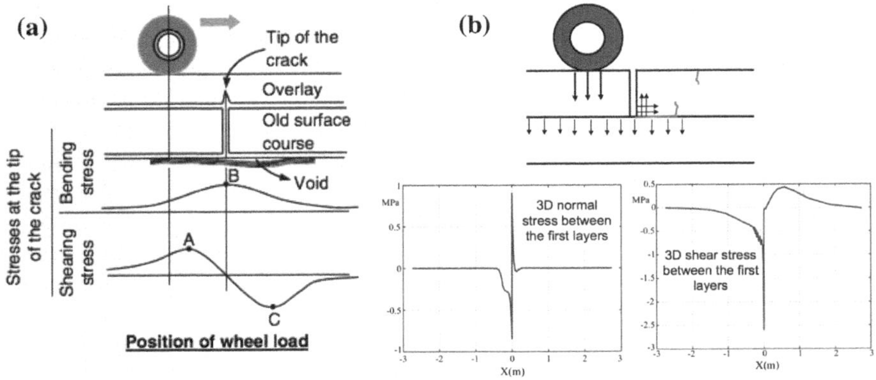


**Fig. 3.27** Interface stress state in pavements near a vertical crack **a** Bending and shear stresses in layers of classical composite or old asphalt pavements [122] **b** Interface M4-5nB stress distribution in UTW pavements [36, 37]

efficient in predicting cracking and interfacial delamination behavior. It also allows the influence of major parameters on the durability of such composites to be studied. From the results obtained, the following findings can be drawn.

- The cracking of the overlay and debonding propagation along an interface are governed by the interlocking phenomenon.
- Shrinkage significantly influences the durability of thin bonded cement based repairs. Overlay materials with low amplitude of autogenous shrinkage are expected to be more durable. The interface debonding propagation can be limited by controlling moisture diffusion from the overlay to the surroundings.
- Fiber reinforcement appears as a solution to improve the durability of thin bonded cement-based overlay. It delays crack opening growth and debonding propagation. Among the commonly used fibers, the high-bond ones are the most efficient for the purpose.
- Overlay materials with low rigidity and high tensile strength contribute to the durability of thin bonded repair systems.

### 3.5.2 Four Point Bending Test

In order to investigate in mixed mode fracture condition the characterization of the crack initiation and propagation at the interface between layers of composite pavements, a four-point bending test on bi-layer structures is adapted to pavement materials [129, 130]. The test is controlled in displacement by a LVDT (Linear Variable Differential Transformer) sensor located on the top middle of the specimen (Fig. 3.28a). To favor and analyze the fracture mechanisms and one edge only, the geometry of the bi-material specimen is chosen non-symmetrical (Fig. 3.28b).

Both experimental and analytical results have been compared on bi-layered Aluminum/PVC and pavement specimens [129, 131]. For Ultra-Thin White-topping (UTW) specimens [132], the cement concrete layer is cast directly onto the prefabricated bituminous slab.

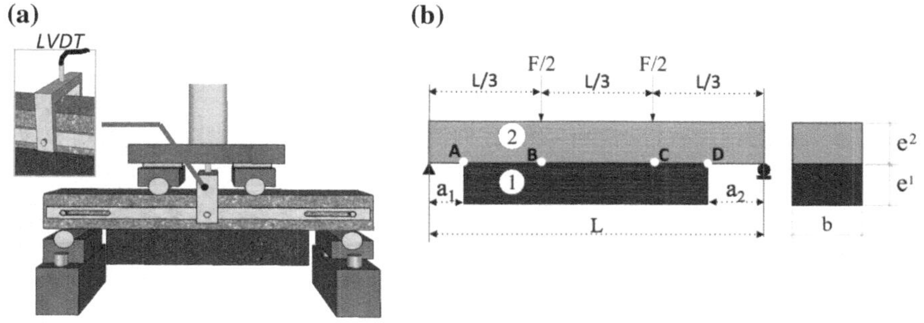


**Fig. 3.28** A 4-point bending test to analyze the bond of UTW interfaces **a** Location of the controlled displacement sensor **b** Specimen geometry [129]

The first tests (L = 420 mm, e1 = e2 = 60 mm, a1 = 40 mm and a2 = 70 mm) have been conducted in static displacement condition fixed to 0.7 mm/min according to other static fracture tests that are done in other Rilem teams. Most of the specimens have been tested at ambient temperature (around 20 °C) and few of them at a cold temperature around 5 °C.

Applied to the 4-point bending test of Fig. 3.28b under plane strain assumptions, the specific M4-5n semi-analytical solutions computed into the Scilab software are obtained very quickly. This modelling allows making several parametric studies for the design of the specimen [130] as illustrated on Fig. 3.29 for the stress field (beam length: L = 420 mm; width: b = 120 mm, Young modulus ratio: E2/E1 ≈ 17, 4, a1 = 40 mm) without any problem of interface stress singularities. Before failure, these quasi-analytical calculations obtained by a specific elastic modelling match the experimental results [129, 133, 134]. Depending on the ratio of modulus of the materials and the geometry of the beam, debonding phenomenon may happen due to a high concentration of interface stresses near the edges of the bi-material beam at points A and D (Fig. 3.29). Due to the bending effects and a high level of tension stress, cracking may also happen at the bottom of the cement layer under point B and C of the beam (Fig. 3.29).

To study water effects as they has been observed in real roads [135] previously saturated by water, specimens have been also submerged in a specific aquarium built for the test. The first results show that water favors debonding phenomenon for UTW interface [129, 131, 136] (Fig. 3.30).

The strain energy release rate is provided by the modelling [137, 138] and compared with some results from tests and from the literature [129]. For the bending test conditions presented in the Fig. 3.31a, the analytical expressions of strain energy releases rates are given in Eq. (3.9) [129, 133].

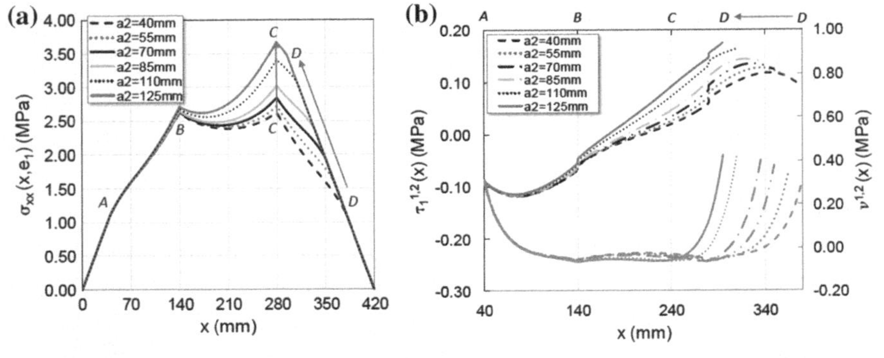


**Fig. 3.29** M4-5n results (F = 5kN) **a** tension stress at the bottom of the cement concrete layer2 **b** interface stress fields [129]

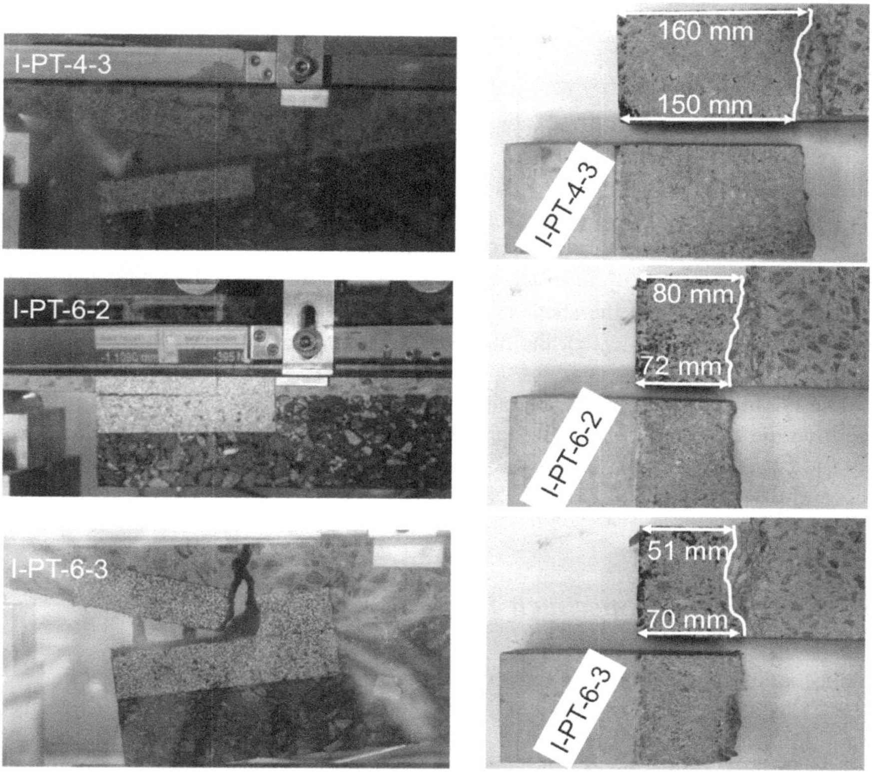


**Fig. 3.30** Failure observations of UTW specimens tested into water (0.7 mm/s, around 20T °C) [129]

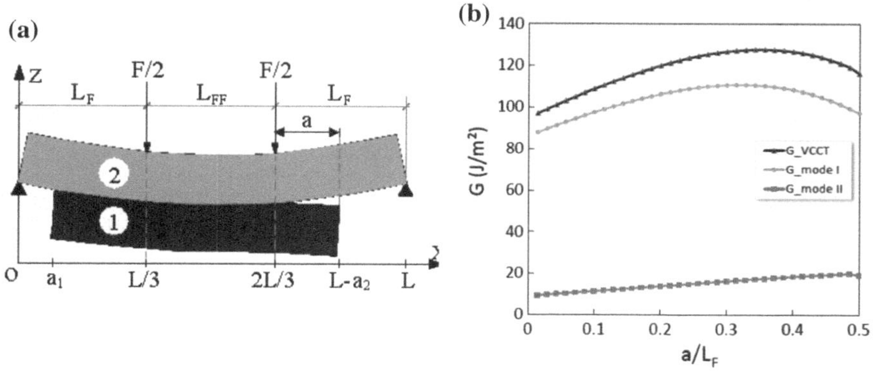


**Fig. 3.31** **a** Debonding configuration **b** Energy release rate simulations (a1 = a2 = 70 mm, F = 12 kN) [129]

$$G_{VVCT}(a) = G_I(a) + G_{II}(a) \text{ with} \begin{cases} G_I(a) = \frac{1}{2b} \cdot \frac{13\left(e^1E^2 + e^2E^1\right)}{35E^1E^2}\left(\nu^{1,2}(a)\right)^2 \\ G_{II}(a) = \frac{1}{2b} \cdot \frac{4e^1\left(1+\upsilon^1\right)E^2 + e^2\left(1+\upsilon^2\right)E^1}{15E^1E^2}\left(\tau^{1,2}(a)\right)^2 \\ -\frac{1}{2b} \cdot \left(\frac{\left(1+\upsilon^1\right)}{5E^1}Q_1^1(a) + \frac{\left(1+\upsilon^2\right)}{5E^2}Q_1^2(a)\right) \cdot \tau^{1,2}(a) \end{cases} \tag{3.9}$$

Only GI is a pure quadratic function of the normal interface stress $\nu^{1,2}$ between layer 1 and 2. GII takes into account the combined terms containing the shear interface stress $\tau^{1,2}$ and the shear layer forces $Q^1$ and $Q^2$ of layer 1 and 2 respectively. For the material tested, the mode I should be recognized as the main failure mode (Fig. 3.31).

Digital Image Correlation techniques linked to TG3 have been performed to improve the determination of the length of the delamination and the displacement field [129] (see Chap. 4). Energy release rates of the M4-5n have been validated successfully compared to Dundurs's approach [129, 131, 134, 139].

### 3.5.3 Composite Specimen Interface Cracking Test (CSIC)

A mechanistic cracking test, termed composite specimen interface cracking test (CSIC), is developed at the University of Florida to evaluate the effect of interface conditions on cracking performance [140]. The test specimen consists of two separately prepared asymmetric specimens, each of them made up of two layers bonded by a tack coat (composite specimens). The testing specimen geometry and loading configuration are shown in Fig. 3.32.

This mechanistic test is designed to initiate and propagate a crack from one material layer through the interface into other material layer(s) using specially prepared test specimens created from either Superpave gyratory-compacted specimens or field-cored specimens. The developed system involves repeated tensile loading and monitoring of number of cycles to failure and rate of damage (reduction in stiffness) to optimize bonding agents and application rates for enhanced cracking performance [141]. The principal features of the test are as follows:

- Asymmetric specimens are glued (with epoxy) together at the surface layer to form a symmetric composite specimen so that bending and/or twisting is prevented during pulling [142].
- A 9.5-mm hole in the center of the specimen serves both as a stress concentrator to initiate cracking (top-down or reflective) and as a platform for load application.
- Repeated haversine load for 0.1 s followed by a 0.9 s rest period is applied to simulate traffic conditions. The rest period allows dissipation of stresses accumulated near the interface during the loading period.

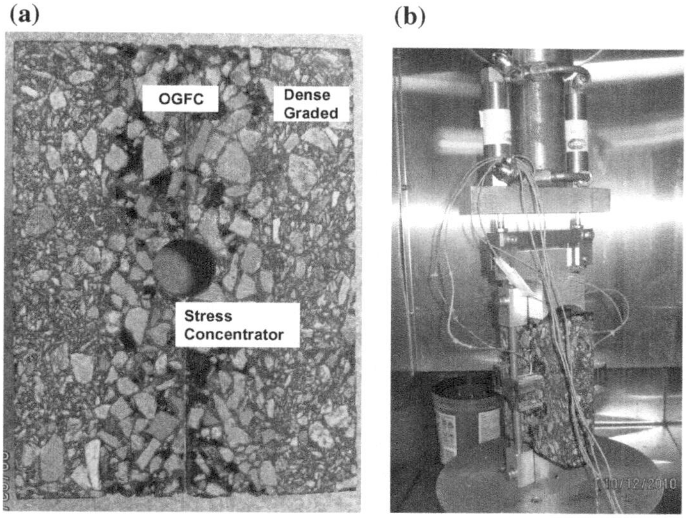


**Fig. 3.32** **a** Composite specimen **b** Test setup [140]

- Load level is selected such that damage and fracture develop gradually (i.e. not catastrophically) to optimize the ability to distinguish the potential effects of the different components on damage and fracture.
- Test temperature is selected such that cracking performance correlates well with that observed in the field and the hypothesis of the HMA-Fracture Mechanics model developed by Roque are satisfied.

The two parameters used to evaluate cracking performance in the CSIC test are: (1) number of load cycles to fail the composite specimen, and (2) rate of damage (Fig. 3.33). The number of load cycles to failure is a straightforward cracking resistance comparison parameter for specimens with different interface conditions subjected to the same loading conditions. However, this parameter provides only the fracture resistance of the whole specimen without any information regarding the damage evolution in the specimen. It has been well recognized that damage induced in the specimen can be estimated by the reduction in stiffness. The specimen stiffness is inversely proportional to the total recoverable deformation, in which results are more convenient for monitoring of damage. The total recoverable deformation is calculated as the average reading of the four extensometers.

A typical total recoverable deformation versus time plot is shown in Fig. 3.33. As can be seen, the total recoverable deformation versus time curve can be divided into three stages: the initial stage, which is known to involve changes in temperature and local damage adjacent to the loading yokes; the second stage, which involves steady-state damage along and through the interface; and the final stage, when the crack propagates rapidly and the specimen breaks. The damage rate is defined as the slope of the steady-state response portion of total recoverable deformation curve, as represented by the straight line in Fig. 3.33.

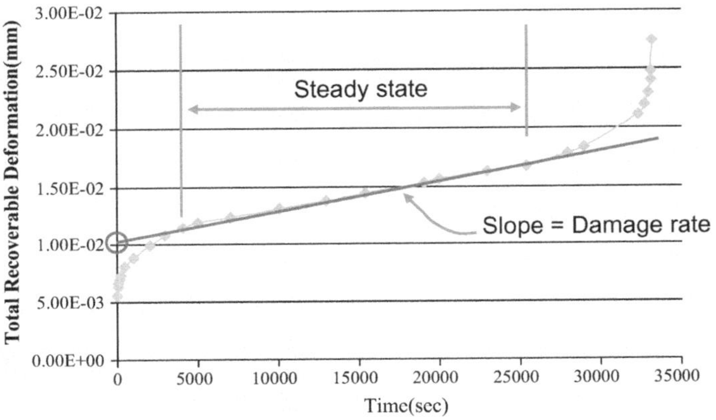


**Fig. 3.33** Typical total recoverable deformation and damage rate [143]

## 3.6 Summary

In this chapter, a synthesis of more than ten contributors working in the field of pavement bonding has been presented. A number of studies, particularly laboratory tests in various loading conditions, exist in the literature and attempt to bring new insight towards bonding performance [30, 144, 145]. However, interface behavior between layers is still not yet thoroughly understood. In this last section, an overview of the existing bonding tests are listed, followed by a summary of key scientific knowledge gaps and future research suggested by contributors of RILEM TC 241-MCD, TG-02.

### *3.6.1 Summary Tables*

The following tables (Tables 3.6, 3.7, 3.8 and 3.9) give a synoptic overview of proposed interface bond tests and characterization methods. Different laboratories have investigated common failure parameters (tensile and shear strength and/or stiffness, fracture or debonding energy, and fatigue parameters) in order to identify the key environmental or design parameters. For example, tack coat content, moisture and dust effect, macrotexture effect, can be identified as material design parameters or environmental conditions affecting interface behavior. Most of the tests can be performed on cylindrical cores with a diameter of 150 mm, which is more convenient for in situ tests, while other tests require slab or rectangular specimens (3PBT, 4PBT).

#### 3.6.1.1 Summary of Mode I Configuration

See (Table 3.6).

**Table 3.6** Different configurations of the Mode I bond tests

| Acronym and loading mode | Reference | Test characteristics (and modifications) | Obtainable results |
|---|---|---|---|
| TBT 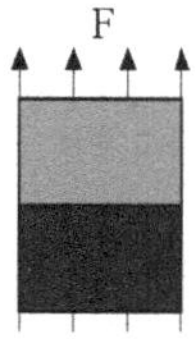  Monotonic loading | TAT: prEN 12697-48 [60] BRRC (Belgium) U Limoges (France) LAMI | – Specimen size: ϕ150 mm for a core or between 80 × 80 mm and 100 × 100 mm for a cube | Tensile strength, normal stiffness, interphase thickness |
| TNBT 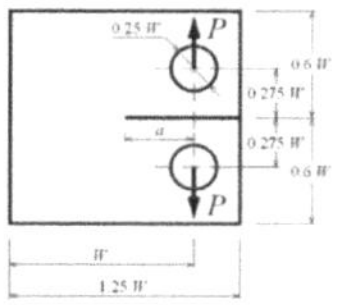 [146] | IBT U Illinois (USA) | – Specimen size ϕ150 mm or holes in glue plates notch Monotonic loading | Debonding fracture energy |
| 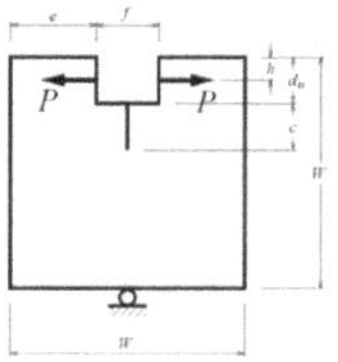 Monotonic and cyclic loading | WST (Austria) [83, 85] Ifsttar (France) [84] | Monotonic and cyclic loading | |

#### 3.6.1.2 Summary of Mode II Configuration

See (Table 3.7).

**Table 3.7** Different configurations of the Mode II bond tests

| Short name and loading mode | Reference | Test characteristics (and modifications) | Obtainable results |
|---|---|---|---|
| SBT 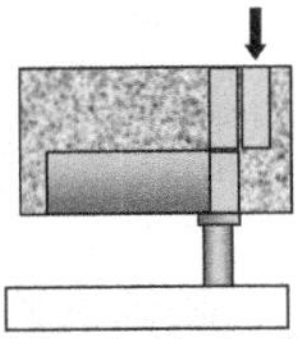 Monotonic loading (without normal force) | Leutner [90] | – Specimen size ϕ150 mm<br>– Gap width: 0 mm<br>– Deformation rate: 50 mm/min<br>– Temperature: 20 °C | Shear strength, Force/deformation diagram, max. force (stress) |
| | European standard EN 12697-48 [60] | – Specimen size: ϕ 100–150 mm<br>– Gap width 0–5 mm | $k_{SBT,\max} = \frac{F'_{SBT}}{\pi \times (\frac{D}{2})^2}$<br>Stiffness modulus |
| | Modified Leutner Nottingham University, UK [12, 103] | – Gap width: 5 mm | |
| | LPDS EMPA (Switzerland), [6, 29] | – Specimen size:<br>– ϕ148 mm ϕ155 mm<br>– 150 mm × 130 mm<br>– Gap width: 2 mm<br>– Temperature for research: 40, −10 °C | Shear Stiffness S (max. force/max slope of the force/deformation curve) |
| LCB 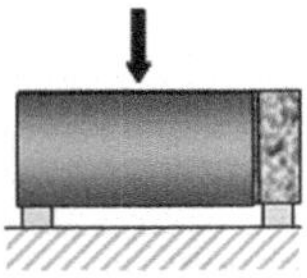 Monotonic loading | Barcelona University, Spain [94] | – Specimen size: ϕ100 mm<br>– Deformation rate: 1.27 mm/min<br>– Temperature: 5 −45 °C | |
| DST 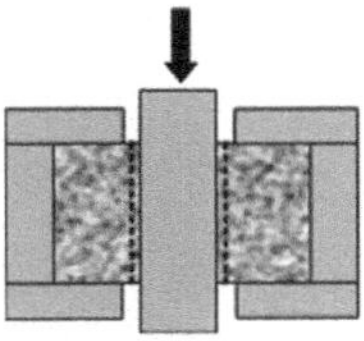 Monotonic and cyclic loading | U Limoges (France) [96] [19] | – Specimen size: 3 layered specimens 70 × 150 × 50 mm<br>– Specimens are placed in a metal frame where the side parts of the sample are fixed while the central part is subjected to a sinusoidal displacement<br>– Loading function: Sinusoidal displacement, 1–10 Hz<br>– Temperature: 5, 10 °C | Shear strength, shear stiffness Ks, fatigue parameter (A, m) |

(continued)

**Table 3.7** (continued)

| Short name and loading mode | Reference | Test characteristics (and modifications) | Obtainable results |
|---|---|---|---|
| SBT with normal stress 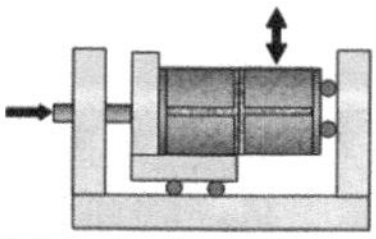 Monotonic and/or cyclic loading | Ascher device University of Dresden, Germany [100] | | |
| | AST Warsaw University, Poland, [117] | – Specimen size: ϕ150 mm<br>– Gap width: 5–12 mm<br>– Specimens are mounted<br>– Loading function: Sinusoidal with amplitudes on three levels of 1, 2 and 4 kN with frequency of 1 Hz, rest periods between amplitude levels: 30 s | Shear failure with coulomb criteria |
| | European standard prEN 12697-48 [60] | – Specimen size: ϕ 100 mm,<br>– Specimens are glued into two steel semicircles<br>– Gap width: 1 mm<br>– Normal force: 0–1.11 N/mm$^2$<br>– Loading function: Sinusoidal with amplitudes of 0.005 to 0.1 mm and frequency of 1–15 Hz<br>– Temperature: −10–30 °C | Force time diagram and deformation time diagram, AK = relative deformation between layers/Shear stress between layers [m3/N] |
| | LISST Louisiana TRC, USA, [115] | – Specimen size: ϕ 100–150 mm<br>– Gap width: 12.7 mm<br>– Deformation rate: 2.54 mm/min | |

#### 3.6.1.3 Summary of Mode III Configuration

See (Table 3.8).

**Table 3.8** Different configurations of the Mode III bond tests

| Short name and loading mode | Reference | Characteristics (and modifications) | Obtainable results |
|---|---|---|---|
| TBT 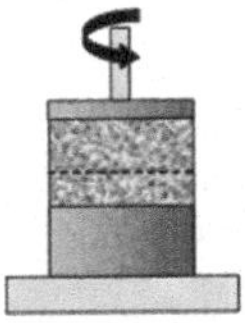 Monotonic loading | European standard prEN 12697-48 [60]<br>BRRC, Belgium Institute of Asphalt Technology, UK [120] | – Laboratory and in situ test<br>– Specimen size: ϕ 100, 150 or 200 mm<br>– Laboratory temperature: 20 °C | – Torque moment at failure<br>– Angular displacement |

#### 3.6.1.4 Summary of Mixed Mode Configuration

**Table 3.9** Different configurations of Mixed-mode bond tests

| Short name and loading mode | Reference | Characteristics (and modifications) | Obtainable results |
|---|---|---|---|
| 3PBT | Univ Toulouse I (France) [123, 128] | – Specimen size: $400 \times 100 \times 100$ mm$^3$ Monotonic and cyclic bending load | Location of internal damages, Fracture mechanics for cracking analysis |
| 4PBT | IFSTTAR (France) [129, 131, 133] | – Specimen size: $420 \times 100 \times 100$ mm$^3$ Monotonic bending load<br>Tested in and out of water | Decoupled debonding<br>Fracture energies GI, GII |
| CSIC test | Univ Florida (USA) and Parma (Italy) [140, 141] | Monotonic and cyclic loading. | Total Fracture energy, Nf number of cycles at failure |

### *3.6.2 Scientific Knowledge Gaps & Future Research Suggested*

In all of these studies, tack coat type and content are the main parameters studied by researchers and engineers. Although roughness surface, water, moisture, freezing, dirt concerning tack coat interface are additional parameters that significantly affect bonding performance [147], these parameters are not yet introduced in the standard methods and pavement design tools. Mainly due to the size effect of specimens made with pavement heterogeneous materials and the loading conditions, significant differences between laboratory results and field performance have also been identified. In addition, the assumptions of Linear Elastic Fracture Mechanics (LEFM), applicable to homogeneous elastic materials, only apply to asphalt if the process zone is small compared to geometric dimensions and crack length. The Tensile Notch Bond Text (TNBT) methods may be a step forward in providing important data for lifetime prediction of roads with a compact testing method. But an open research question surrounding these tests still is: are the dimension of the specimen and the fracture length size enough long to reliably produce one or several material parameter(s)? Concerning the adhesion tests and the use of emulsions as tack coats, there are still knowledge gaps and remaining research questions. A better physical understanding is needed, which should involve:

- Measuring the roughness of interfaces;
- Evaluating the adhesion of interfaces;
- Determining mechanical parameters of interfaces;
- Evaluating the temperature influence on interface behaviour;
- Determining and quantifying the energy evolution during interface damage, and;
- Identifying the mechanical behavior of interfaces under cyclic, mixed-mode conditions.

#### 3.6.2.1 Summary of Scientific Knowledge Gaps Given Specifically in the Ten Contributions

- In the prEN 12697-48 [60], there are three adhesion tests proposed (tensile, shear and torque bond tests) but there is no information concerning the utility (or advantage) of using one or the other test. The precision of the adhesion tests methods is not given either.
- There is not a common amount of tack coat that can be used to optimize the long-term bonding performance of pavements.
- The influence of some parameters on the interlayer bonding is most of the time neglected: milling and cleaning of the interface between layers, age and type of the underlayer.
- For the existing interlayer bonding specifications, there are in general no requirements for the traffic category.
- Compatibility of different interlayer bond devices: do they measure the same intrinsic properties of the interface? Do they at least rank interfaces in the same order? Mixed mode lab tests for mechanical characterization of initiation and propagation of debonding are suggested to improve these comparisons with "pure" debonding tests and the identification of delamination criteria (need also to study size effects in these laboratory tests).
- There exists a need to improve links between static and cyclic test results with the aim of understanding the correlation between laboratory results and field performance. What testing mode (simple shear, cyclic) should be used, under what conditions testing should be performed (temperature, rate, number of replicates) and what parameter should be used?

- There is a need for a clear state of the art from each part of the world on pavement deterioration linked to debonding phenomena (type of pavements—materials and structures, type of climate, type of loading) to better understand the initiation and evolution of delamination: crack geometry, crack propagation speed under traffic and climate conditions.
- Tools for crack and debonding detection (Non Destructive Techniques see Chap. 4) and measurement of their geometry, in the lab and in the field are expected.
- Needs simple modelling tools for prediction of interface debonding initiation and propagation in pavement structures.

#### 3.6.2.2 Future Research Needs

These experimental studies aim to highlight some field of research needed for answering the previous general remarks.

1. Advanced measurements can contribute to better understanding and further the design of interfaces with good bonding properties (see Chap. 4):
   - Measurements of roughness with image analysis can lead to the introduction of this main parameter in pavement bond design and then in HMA design.
   - DIC strain analysis close to the interface can lead to improved pavement modelling and design.
   - Based on the results of NCHRP 9-40 [115], a need exists to evaluate failure mechanisms in tack coat field applications and the influence of shear and normal stresses originating from vehicular and thermal loadings on the performance at the interface. A link between field measurements at the interface using instrumentation and latest sensor technology with advanced three-dimensional FE methods is suggested.
2. Advanced loading testing can enable the study of interface behavior under controlled loading conditions. Using mixed mode tests, actual mixed mode loading calculated in pavement may be simulated to evaluate bond performances in the lab. Shear and normal stresses originating from vehicular and thermal loadings are clearly influences on the performance at the interface.
3. The influence of surface thermal loading on multilayered pavement clearly induces interface loading. Thermal cracking between two-layered systems is assumed as thermal debonding. The fundamental material properties involved in the development of interfacial thermal stresses can be measured such as viscoelastic material parameters E(t,T), ν(t,T) and thermal expansion coefficient (TEC). An experimental program using two-layered systems (AC-AC, AC-PCC, AC-Steel) could be prepared in the laboratory and their thermo-mechanical behavior will be used to calibrate the models. In this phase Digital Image Correlation (DIC) measurements will be extremely useful to detect the deformation field in the interface zone.

From the previously identified scientific knowledge gaps, other future research can be suggested:

In terms of advances in mechanical modelling, debonding energy and surface temperature gradient could be pertinent approaches.

Advanced modelling is possible with information due to advanced measurements. The responses to such critical research problems can be achieved by linking field measurements at the interface using instrumentation and sensoring technology with advanced three-dimensional FE methods. In the near future, models will also have also to include new parameters measured by DIC as primarily parameters [148] (see the following Chap. 4). On the other hand, construction process, water and dirty conditions before spreading the mix could be included as parameters in models.

Further investigation on different tack coat types, a wider range of application rates, and different HMA mixtures under different temperature and moisture conditions and different modes of failure including mode I, mode II and mixed mode is recommended. Data coming from interface bonding tests should also be utilized in numerical pavement models, in order to gain additional insight towards the mechanisms of interface debonding and its prevention. We are also in the process of testing field cores such as done for instance for the WST by Gharbi et al. [89], and coupling the experiments and model with field performance data. The validation in the field of these tests always needs accelerated loading facilities. Both types of studies can be performed:

(1) Comparison of existing interface bonding devices on different materials (reference materials such as polyethylene, and lab- and field-prepared asphalt samples) and under different conditions (loading rate, temperature, conditioning).
(2) Comparison of accelerated pavement test results on different interfaces with laboratory standard test results on cores taken from these pavement sections.

Finally, it should be mentioned that besides fatigue crack data, certain other phenomena such as healing of cracks should be also taken into account in the test methods and calculation models in order to improve the prediction of the pavement lifetime.

Further research should aim not only to predict results for interlayer bond problems using Artificial Neural Network (ANN) modelling [149–151], but should also emphasize the importance of modelling and data collection for the field of pavement engineering, especially interlayer bonding. In this context, more research and effort is clearly needed, so that openly accessible databases will be available providing a basis for future artificial intelligence modelling and machine learning. In this context, it is imperative to continue to bring data and information from different countries and laboratories together to advance the complex field of pavement interface characterization and design.

## References


1. Vanelstraete A, Francken L (1997) State-of-the-Art Report of RILEM technical committee 157 PRC, systems to prevent reflective cracking in pavements. In: Vanelstraete A, Francken L (eds) RILEM Report 18
2. Burmister DM (1943) The theory of stresses and displacements in layered systems and applications of the design of airport run ways. Proc Highw Res Board 23:126–148
3. Collop AC, Thom NH (2002) The importance of bond between pavement layers, Final summary report. School of Engineering, University of Nottingham, UK
4. Vandenbossche JM, Fagerness AJ (2002) Performance, analysis, and repair of ultrathin and thin whitetopping at Minnesota road research facility. Transp Res Rec J Transp Res Board Natl Res Counc Washington, DC 1809:191–198
5. Brown SF, Brunton JM (1984) The influence of bonding between bituminous layers. Highw Transp 31(5):16–17
6. Raab C, Partl MN (1999) Methoden zur Beurteilung des Schichtenverbunds von Asphaltbelägen,Eidgenössisches Departement für Umwelt, Verkehr, Energie und Kommunikation, Bundesamt für Strassen. Report No 442
7. Raab C, Partl MN (2004) Interlayer shear performance: experience with different pavement structures. In: The proceedings of the 3rd Eurasphalt & Eurobitume Congress, Vienna, Austria, 12–14 May, 1:535–545. ISBN 90-802884-4-6
8. Raab C, Partl MN (2006) Rehabilitation of Concrete pavements with Asphalt and Intermediate Layers. In: Third Gulf conference on roads (TGCR06), in Muscat, Sultanate of Oman, March 6–8, pp 48–52. ISBN 1817-4310
9. Khweir K, Fordyce D (2003) The influence of layer bonding on the prediction of pavement life. In: Proceedings of the institute of civil engineers transport, UK, vol 156
10. Pouteau B, Balay J-M, Chabot A, De Larrard F (2004) Fatigue test and mechanical study of adhesion between concrete and asphalt. In: 9th international symposium on concrete roads, 3–6 April, Istanbul, Turkey
11. Chabot A, Pouteau B, Balay J-M, De Larrard F (2008) FABAC accelerated loading test of bond between cement overlay and Asphalt layers. In Taylor & Francis group proceedings (ISBN 13: 978-0-415-47575-4), Sixth international RILEM conference on cracking in pavements, June 16–18, Chicago, US, pp 13–23. https://doi.org/10.1201/9780203882191.ch65
12. Sutanto M (2009) Assessment of Bond between Asphalt Layers. Ph.D. Thesis, Nottingham University, Great Britain
13. Autret P, Baucheron de Boissoudy A, Marchand, JP (1982) ALIZE III Practice. In Proceedings of the 5th international conference on structural design of Asphalt pavements, Delft, Netherlands, pp 174–191. [Alize-LCPC (www.alize-LCPC.com)]
14. Corte JF, Goux MT (1996) Design of pavement structures: the French technical guide. Transp Res Rep 1539:116–124
15. Highways Agency (2008) Specifications for highway works, Manual of contract documents for highway works, vol 1 (MCHW_1), London, UK
16. NCHRP (2001) Guide for mechanistic-empirical design of new and rehabilitated pavement structures. [http://onlinepubs.trb.org/onlinepubs/archive/mepdg/home.htm]
17. Hammoum F, Chabot A, St. Laurent D, Chollet H, Vulturescu B (2010) Accelerating and decelerating effects of tramway loads moving on Bituminous pavement. Mater Struct 43:1257–1269
18. Ambassa ZD, Allou F, Petit C, Medjo Eko R (2013) Fatigue life prediction of asphalt pavement subjected to multiple axle loadings with viscoelastic FEM. Constr Build Mater 43:443–52. http://dx.doi.org/10.1016/j.conbuildmat.2013.02.017
19. Diakhaté M, Phelipot A, Millien A, Petit C (2006) Shear fatigue behavior of tack coats in pavements. Road Mater Pavement Design 7(2):201–222

20. Petit C, Diakhaté M, Millien A, Pouteau B (2009) Pavement design for curved road sections. Road Mater Pavement Design 10(3):609–624. https://doi.org/10.3166/rmpd.10.609-624
21. Vulcano-Greullet, N, Kerzreho JP, Mauduit V, Chabot A (2010) Stripping phenomenon of top layers of thick pavements. In: Proceedings of the 11th international conference on Asphalt Pavements, 1–6 August 2010, Nagoya Aichi, Japan. Curran Associates 1:552–561. Print ISBN: 978-1-61839-073-8
22. D'Andrea A, Tozzo C (2016) Dynamic tests on Bituminous layers interface. Mater Struct 49(3):917–928
23. Mauduit V, Mauduit C, Vulcano-Greullet N, Coulon N, Hammoum F, Hamon D, Kerzreho JP, Piau JM, Chabot A (2013) Dégradation subite des enrobés bitumineux par période de gel/dégel: analyse de cas de terrain et recherche exploratoire en laboratoire. Bulletin des Laboratoires des Ponts et chaussées (279):47–63. http://hal.archives-ouvertes.fr/docs/00/85/09/49/PDF/doc00014196.pdf. (in French)
24. Romanoschi SA, Metcalf JB (2002) The characterization of pavement layer interfaces. In: Proceedings of the 9th international conference on asphalt pavements, Copenhagen
25. Kruntcheva MR, Collop AC, Thom NH (2005) Effect of bond condition on flexible pavement performance. J Transp Eng, ASCE Publ 131(11):880–888
26. Mohammad LN, Bae Elseifi A, Mostafa A (2009) Effect of tack coat materials and application rate on the interface shear strength. In: International conference on maintenance and rehabilitation of pavements and technological control (MAIREPAV6), Torino, sixth proceedings, vol II, pp 636–645
27. Roffé JC, Chaignon F (2002) Characterisation tests on bond coat: worldwide study, impact, tests, recommendations. In: Proceedings of the 3rd international conference bituminous mixtures and pavements, Thessaloniki
28. Pouteau B (2004) Durabilité mécanique du collage blanc sur noir, Ph.D. thesis from ECN Nantes (in French). http://media.lcpc.fr/ext/pdf/theses/rou/throu_pouteau.pdf
29. Raab C, Partl MN Abd, El Halim AO (2009) Evaluation of interlayer shear bond devices for asphalt pavements. Baltic J Road Bridge Eng 4(4):176–195
30. Partl MN, Bahia HU, Canestrari F, de la Roche C, Di Benedetto H, Piber H, Sybilski D (eds) (2013) Advances in interlaboratory testing and evaluation of bituminous materials, RILEM State-of-the-Art Reports of the RILEM technical committee 206-ATB, vol 9, XII, 453 p
31. Bissonnette B, Courard L, Fowler DW, Granju J-L (eds) (2011) Bonded cement-based material overlays for the repair, the lining or the strengthening of slabs or pavements. State-of-the-Art Report of the RILEM Technical Committee 193-RLS Series, vol 3, 1st edn, XIII, 177 p 76 illus edited by Publisher Springer, Netherlands. ISBN: 978-94-007-1239-3
32. Kaufmann N (1971) Das Sandflächenverfahren (The Sand-area method). Strassenbau Technik 24(3):31–50. (Germany)
33. BLPC (1979) Concrete pavements: problems raised by the presence of water in their structure (Chaussées en béton: problèmes posés par la présence d'eau dans leur structure). Bulletin de liaison des laboratoires des ponts et chaussées, special issue, 8. (In French)
34. Fuchs F, Jasienski A (1997) Le phénomène du Punch-Out sur les autoroutes belges en béton armé continu – causes, effets et remèdes. Bulletin from the Federation de l'industrie cimentière Belge and the Centre de Recherches routières Belge. (in French). www.brrc.be/pdf/bulletin/bul30t.pdf
35. Bergeron G, Paradis M, Tourangeau G (2014) Réparation des nids-de-poule. Info DLC, Bulletin d'information technique 19(1). (in French)
36. Chabot A, Tran QD, Ehrlacher AA (2005) Simplified modeling for cracked pavements—Modèle simplifié pour le calcul des chaussées. Bulletin des Laboratoires des Ponts et chaussées, ISSN 1269-1496, (258–259):105–120
37. Chabot A, Tran QD, Ehrlacher A (2007) A modeling to understand where a vertical crack can propagate in pavements. In: Taylor & Francis group proceedings, international conference on advanced characterization of pavement and soil engineering Materials, Athens June 20–22 2007, 1:431–440

38. Chabot A, Chupin O, Deloffre L, Duhamel D (2010) Viscoroute 2.0: a tool for the simulation of moving load effects on asphalt pavement. RMPD Spec Issue Recent Adv Numerical Simul Pavements 11(2):227–250
39. Chupin O, Chabot A, Piau J-M, Duhamel D (2010) Influence of sliding interfaces on the response of a visco-elastic multilayered medium under a moving load. Int J Solids Struct 47 (25/26):3435–3446
40. Grellet D, Doré G, Kerzrého J-P, Piau JM, Chabot A, Hornych P (2012) Experimental and theoretical investigation of three dimensional strain occurring near the surface in asphalt concrete layers. In: Proceedings of the 7th Rilem international conference on cracking in pavements, June 20–22, Delft, The Netherlands. RILEM Bookseries, 4:1017–1027. https://doi.org/10.1007/978-94-007-4566-7_97. Print ISBN: 978-94-007-4565-0
41. Tran QD, Chabot A, Ehrlacher A, Tamagny P (2004) A simplified modelling for cracking in pavements. In: Fifth international RILEM conference on cracking in pavements, May 5–8, Limoges. In Rilem Proceedings. ISBN 2-912143-47-0, 299–306
42. Berthemet F, Chabot A (2013) Apports du massif de Winkler dans la construction d'un outil de calcul d'analyse de structure multicouche fissurée reposant sur un sol. 21ème Congrès Français de Mécanique, 26-30Août, Bordeaux, 6 pages (http://hdl.handle.net/2042/52168)
43. Nasser H, Chabot A (2015) 2D fast software analyzing mechanical fields in elastic cracked pavements. In: Proposed to the international conference on soft computing technology in civil, structural and environmental engineering (CSC2015), Soft computing in pavement engineering special session, Prague, Czech Republic, 1–4 September 2015
44. Nasser H, Chabot AA (2017) Half-analytical elastic solution for 2D analysis of cracked pavements. Adv Eng Softw https://doi.org/10.1016/j.advengsoft.2017.06.008
45. Nasser H, Chupin O, Piau JM, Chabot A (2018) Mixed FEM for solving a plate type model intended for analysis of pavements with discontinuities. Road Mater Pavement Design 19(3):496–510. https://doi.org/10.1080/14680629.2018.1418653
46. Fouchal F, Lebon F, Titeux I (2009) Contribution to the modelling of interfaces in masonry construction. Constr Build Mater 23:2428–2441
47. Raous M, Cangémi L, Cocou M (1999) A consistent model coupling adhesion, friction and unilateral contact. Comput Methods Appl Mech Eng 177(3–4):383–399
48. Monerie Y, Raous M (2000) A model coupling adhesion to friction for the interaction between a crack and a fiber/matrix interface', Z.A.M.M. 205–209
49. Lebon F, Raous M (1992) Friction modelling of a bolted junction under internal pressure loading. Comput Struct 43:925–933
50. Kongo Kondé A, Rosu I, Lebon F, Seguin L, Brardo O, Troude F, Devésa B (2011) Thermomechanical couplings in aircraft tire rolling/sliding modeling. Adv Mater Res 274:81–90
51. Barbié L, Ramière I, Lebon F (2014) Strategies around the local defect correction multi-level refinement method for three-dimensional linear elastic problems. Comput Struct 130:73–90
52. Lebon F Ould, Khaoua A, Licht C (1998) Numerical study of soft adhesively bonded joints in finite elasticity'. Comput Mech 21:134–140
53. Lebon F, Rizzoni R, Ronel-Idrissi S (2004) Analysis of non-linear soft thin interfaces'. Comput Struct 82:1929–1938
54. Lebon F, Ronel-Idrissi S (2004) Asymptotic studies of Mohr-Coulomb and Drucker-Prager soft thin layers. Int J Steel Compos Struct 4:133–148
55. Rizzoni R, Lebon F (2012) Asymptotic analysis of an elastic thin interphase with mismatch strain'. European J Mech A Solids 36:1–8
56. Moës N, Dolbow J, Belytschko T (1999) A finite element method for crack growth without remeshing. Int J Numer Methods Eng 46(1):131–150
57. Van der Meer FP, Sluys LJ, Moës N (2012) Toward efficient and robust computation of energy release rate and mode mix for delamination. Compos A Appl Sci Manuf 43(7):1101–1112
58. Duarte CA, Babuška I, Oden JT (2000) Generalized finite element methods for three-dimensional structural mechanics problems. Comput Struct 7(2):215–232

59. Garzon J, Duarte CA, Buttlar W (2010) Analysis of reflective cracks in airfield pavements using a 3-D generalized finite element method. Road Mater Pavement Design 11(2):459–477
60. prEN 12697-48. Bituminous mixtures—Test methods for hot mix asphalt—Part 48: Interlayer Bonding (2015)
61. Sutton M, Wolters W, Peters W, Ranson W (1983) Determination of displacements using an improved digital correlation method. Image Vis Comput 1(3):133–139. https://doi.org/10.1016/0262-8856(83)90064-1
62. Ktari R, Leandry I, Millien A, Fouchal F, Pop O, Phan VTP, Petit C (2013) Interfaces de chaussées: de la caractérisation du comportement mécanique de l'interface à sa modélisation, 21ème Congrès Français de Mécanique, Bordeaux, 26-30 août, (http://hdl.handle.net/2042/52216)
63. Ktari R (2016) Mécanismes et modélisations des dégradations des interfaces entre couches de chausses. Ph.D. thesis from the Université de Limoges. (https://tel.archives-ouvertes.fr/tel-01417551/document). (in French)
64. RVS 08.16.01 (2010) Technical contract conditions-Bituminous base and wearing courses-requirements for bituminous courses, Wien
65. ÖNORM B 3639-2 (1997) Technische Asphalte für den Straßenbau und verwandte Gebiete - Prüfung - Haftverbund von Asphaltschichten. Austrian Standards Institute
66. De Visscher J, Denolf K, Destrée A, Leprince, L, Piérard N, Vanelstraete A, Vansteenkiste S (2014) European test methods for asphalt and bituminous binders: improvements within the framework of sustainable development, Belgian Bureau for Standardisation (NBN)-funded research project, annual research report
67. Destrée A, De Visscher J, Vanelstraete A (2012) Evaluation of tack coat performance for thin and ultra-thin asphalt pavements. In: 5th Eurasphalt & Eurobitume congress, 13–15th June 2012, Istanbul
68. Destrée A, Leprince L, Piérard N, Vanelstraete A, Vansteenkiste S (2014) Belgian Bureau for Standardisation (NBN)-funded research project, annual research report
69. Destrée A, De Visscher J, Piérard N, Vanelstraete A (2015) Field study to investigate the impact of conditions of application of tack coats on the interlayer bond strength. In: 8th international RILEM SIB symposium, October 7–9, 2015, Ancona, Italy
70. BRRC (2012) MM – MPT – 02.02, Tensile adhesion test, Belgian Road research Centre—BRRC working method for the determination of bond strength to underlayers (2012)
71. Bergiers A, de Visscher J, Denolf K, Destrée A, Vanhooreweder B, Vuye C (2014) Test sections to study the acoustical quality and durability of thin noise reducing asphalt layers. In: International conference on noise and vibration engineering ISMA, 15–17th September 2014, Leuven, Belgium, pp 1–10
72. LC 25-010 (2016) Mesure de la force de liaison avec un appareil de mesure d'adhésion des couches. Méthode d'essai LC25-010, Secteur – liants hydrocarbonés, Transports Québec, December 15
73. MTQ Standard Clause (2016) Evaluation of the bond strength between an asphalt layer and its underlying layer. CCDG (Cahier des charges et devis généraux), Ministère des transports du Québec-MTQ
74. Destrée A, De Visscher J, Vanelstraete A (2016) Field study to evaluate different pre-normative interlayer adhesion tests. In: Proceedings of the 6th Eurasphalt & Eurobitume Congress, p 11, 1–3 June 2016—Prague Congress Centre
75. ZTV-SIB (1990) Zusätzliche Technische Vertragsbedingungen und Richtlinien für Schutz und Instandsetzung von Betonbauteilen, Verkehrsblatt-Dokument; B 5230, Verkehrsblatt-Verlag, Abb., Tab., Lit. 59
76. ASTM D7313-07a (2008) Standard test method for determining fracture energy of asphalt-aggregate mixtures using the disk-shaped compact tension geometry. ASTM Volume 04.03 Road and Paving Materials
77. Hakimzadeh S, AbayKebede N, Buttlar WG, Ahmeda S, Exline M (2012) Development of fracture-energy based interface bond test for asphalt concrete. Road Mater Pavement Design 13:76–87 Sup 1

78. Linsbauer HN, Tschegg EK (1986) Fracture energy determination of concrete with cube shaped specimens, Zement und Beton 31:38–40. (in German)
79. Brühwiler E, Wittmann FH (1990) The Wedge splitting test, a new method of performing stable fracture mechanics tests. Eng Fract Mech 35(1–3):117–125
80. Tschegg E.K, Tschegg-Stanzl SE, Litzka J (1996) Fracture behavior and bond strength of bituminous layers. In: Francken L, Beuving E, Molenaar AAAA (eds) Reflective cracking in pavements, RILEM Published by E&FN Spon, 2–6 Boundary Row, London, SE1 8HN. ISBN 0 419 22260 X
81. Tschegg EK, Macht J, Jamek M, Stegenberger J (2007) Mechanical and fracture-mechanical properties of asphalt-concrete interfaces. ACI Mater 104(5):474–480
82. Tschegg EK, Jamek M, Lugmayr R (2012) Crack growth behavior in geosynthetic asphalt interlayer systems. Road Mater Pavement Design 13(1):156–170
83. Tschegg EK, Jamek M, Lugmayr R (2011) Fatigue crack growth in asphalt and asphalt-interfaces. Eng Fract Mech 78:1044–1054
84. Rossi P, Brühwiler E, Chhuy S, Jenq YS, Shah SP (1991) Fracture properties of concrete as determined by means of wedge splitting tests and tapered double cantilever beam tests, Chapter 2. Chapman and Hall, pp 87–128
85. Tschegg EK (1986) Equipment and appropriate specimen shapes for tests to measure fracture values, AT No. 390328, Austrian Patent Office, Vienna, Austria
86. Tschegg EK (1991) New equipments for fracture tests on concrete. Materialprüfung 33(11/12):338–342
87. ÖNORM B 3592 (2011) Determination of notch tensile strength and specific fracture energy of construction materials, their compounds and composites—wedge splitting method, (in German). Bestimmung der Kerbzugfestigkeit und der spezifischen Bruchenergie von Baustoffen, Baustoffverbindungen und Verbundwerkstoffen – Keilspaltmethode. Austrian Standards Institute
88. Lugmayr R, Jamek M, Tschegg EK (2009) Mechanism of fatigue crack propagation and fracture behavior in bituminous roads. In: Loizos A, Partl MN, Scarpas T, Al Qadi IL (eds) Advanced testing and characterization of bituminous materials. Taylor & Francis, London, pp 807–816
89. Gharbi M, Nguyen ML, Trichet S, Chabot A (2017) Characterisation of the bond between asphalt layers and glass fiber grid with help of a Wedge Splitting Test. In: 10th international conference on bearing capacity of roads, railways and airfields (BCRRA 2017), Athens June 28–30. In CRC Press (Verlag) -Taylor & Francis Group proceedings: 1517–1524. ISBN: 978-1-138-29595-7. https://doi.org/10.1201/9781315100333-217
90. Leutner R (1979) Untersuchung des Schichtverbundes beim bituminösen Oberbau. Bitumen 41(3):84–91
91. Uzan J, Livneh M, Eshed Y (1978) Investigation of adhesion properties between asphaltic concrete layers. Asphalt Paving Technolo 47:495–521
92. Sholar G, Page G, Musselman J, Upshaw P, Moseley H (2004) Preliminary investigation of a test method to evaluate bond strength of bituminous tack coats. Assoc Asphalt Paving Technolo 73:771–801
93. West RC, Zhang J, Moore J (2005) Evaluation of bond strength between pavement layers. National Center for Asphalt Technology, NCAT Report 05-08
94. Mirò R, Pérez Jiménez F, Borras Gonzalez J M (2003) Evaluation of the effect of tack coats. LCB shear test. IN: 6th RILEM symposium PTEBM"03, Zurich, Switzerland, pp 550–556
95. De Bondt AH de (1999) Anti-reflective cracking design of (Reinforced) asphaltic overlays. Ph.D. thesis, Delft University of Technology
96. Millien A, Petit C, Rosier J (1996) Comportement au cisaillement des couches d'accrochage dans les chaussées. Rapport interne Laboratoire 3MsGC, Université de Limoges, France
97. Canestrari F, Ferrotti G, Partl MN, Santagata E (2005) Advanced testing and characterization of interlayer shear resistance. Transp Res Record J Transp Res Board 1929(1):69–78

98. Sanders PJ, Brown SF, Thom NH (1999) Reinforced asphalt for crack and rut control. In: 7th conference on asphalt pavements for Southern Africa, CAPSA '99, Victory Falls, Zimbabwe, Document transformation technologies pp 847–855
99. Crispino M, Festa B, Giannattasio P, Nicolosi V (1997) Evaluation of the interaction between the asphalt concrete layers by a new dynamic test. In: 8th international conference on the structural design of asphalt pavements. Washington State University, Seattle, pp 741–754
100. Ascher D, Wellner F (2007) Untersuchungen zur Wirksamkeit des Haftverbundes und dessen Auswirkungen auf die Lebensdauer von Asphaltbefestigungen [Investigation of the effectiveness of bonding and its influence on the service life of asphalt pavements]. Technical University of Dresden, Germany, Report No. 13589 BR/1
101. Piber H, Canestrari F, Ferrotti G, Lu X, Millien A, Partl MN, Petit C, Phelipot-Mardelle A, Raab C (2009) RILEM Interlaboratory test on interlayer bonding of asphalt pavements. In: 7th international RILEM symposium ATCBM09 on advanced testing and characterization of bituminous materials, Rhodes, Greece vol 2, pp 1181–1189
102. Choi Y, Sutanto M, Collop A, Airey G (2005) Bond between asphalt layers. Project Report to the UK Highways Agency, Scott Wilson Pavement Engineering LtD
103. Choi Y, Collop A, Airey G, Elliot RA (2005) comparison between interface properties measured using the Leutner test and the torque test. J Assoc Asphalt Paving Technolo 74B. ISSN 1553-5576
104. Raab C, Partl MN, Abd El Halim AO (2010) Effect of gap width on interlayer shear bond results. Int J Pavement Res Technolo IJPRT Ref No IJPRT-09544(3), 3(2):79–85
105. Codjia H (1994) Erarbeitung eines Bewertungshintergrundes für das Prüfverfahren Schichtenverbund nach Leutner und Bestimmung der Präzision. Dissertation, Institut für Straßen- und Eisenbahnwesen der Universität Karlsruhe, Heft 43
106. Stöckert U (2001) Schichtenverbund – Prüfung und Bewertungshintergrund. Straße + Autobahn, 11:624–631. (in German)
107. Raab C, Partl MN (2015) In situ service capability of tack coats. In: Conference proceedings, published by CRC Press/Balkema, 6th ICONBMP international conference bituminous mixtures and pavements, Thessaloniki, Greece, 10–12 June
108. ÖNORM B 3639-1 (1997) Technische Asphalte für den Straßenbau und verwandte Gebiete - Prüfung - Schubverbund von Asphaltschichten. Austrian Standards Institute
109. Partl MN, Raab C (1999) Shear Adhesion between top layers of fresh asphalt pavements in Switzerland. In: Proceedings, 7th conference on asphalt pavements for Southern Africa, CAPSA '99, Victory Falls, Zimbabwe, pp 5.130–5.137
110. Raab C (2010) Development of a framework for standardisation of interlayer bond of asphalt pavements. Ph.D. thesis, Department of Civil and Environmental Engineering, Carleton University, Ottawa, Canada
111. Swiss Standard (2000) Schweizer Norm SN 671961. Bituminöses Mischgut, Bestimmung des Schichtenverbunds (nach Leutner), Verein Schweizerischer Straßenfachleute VSS
112. Diakhaté M, Petit C, Millien A, Goacolou H (2008) Interface fatigue cracking in multilayered pavements: experimental analysis. In: Taylor & Francis group proceedings (ISBN 13: 978-0-415-47575-4), Sixth international RILEM conference on cracking in pavements, June 16–18, Chicago, US, pp 281–290
113. Diakhaté M, Millien A, Petit C, Pouteau B (2011) Experimental investigation of tack coat fatigue performance: Towards an improved lifetime assessment of pavement structure interfaces. Constr Build Mater 25(2):1123–1133
114. Mohammad LN, Raqib MA, Wu Z, Huang B (2002) Measurement of interlayer bond strength through shear tests. In: 3rd international conference bituminous mixtures and pavements, Thessaloniki, Greece
115. Mohammad L, Elseifi MA, Bae A, Patel N, Button J, Scherocman JA (2012) Optimization of tack coat for HMA placement. NCHRP report 712
116. Zofka A, Bernier A, Josen R, Maliszewski M (2014) Advanced shear tester for solid and layered samples. In: Proceedings of 2014 international society for asphalt pavements (ISAP) conference, Raleigh

117. Zofka A, Maliszewski M, Bernier A, Josen R, Vaitkus A, Kleizienė R (2015) Advanced shear tester for evaluation of asphalt concrete under constant normal stiffness conditions. EATA conference, June 2015, Stockolm, Sweden
118. Zahw MA (1995) Development of testing framework for evaluation of rutting resistance of asphalt mixes. Ph.D. Thesis, Carleton University, Ottawa, Canada
119. Abd El Halim A, Rickards IR, Haas R, Nabi R (1997) Evaluation of design and construction effects on asphalt pavement performance through a portable in-situ shear test device IN:. Eighth international conference on asphalt pavements, proceedings vol 2
120. British Standard (2004) Guidelines document for the assessment and certification of thin surfacing systems for highways. SG3/05/234, British Board of Agreement, Watford, UK
121. Sonier A, Ramberg J, Leppard B (2007) Mill 'n' Fill Operations, CEE 599–Pavement Construction, May 16
122. Lytton RL, Tsai FL, Lee SI, Luo R, Hu S, Zhou F (2010) Model for predicting reflection cracking of hot-mix asphalt overlays. NCHRP report 669. ISBN 978-0-309-15505-2
123. Tran Q-T, Toumi A, Granju J-L (2006) Experimental and numerical investigation of the debonding interface between an old concrete and an overlay. Mater Struct 39(3):379–389
124. Tran Q-T, Toumi A, Turatsinze A (2007) Modelling of debonding between old concrete and overlay: fatigue loading and delayed effects. Mater Struct 40(10):1045–1059
125. Tran Q-T, Toumi A, Turatsinze A (2008) Thin bonded cement-based overlays: numerical analysis of factors influencing their debonding under monotonic loading. Mater Struct 41(5):863–877
126. Tran Q-T, Toumi A, Turatsinze A (2008) Thin bonded cement-based overlays: numerical analysis of factors influencing their debonding under fatigue loading. Mater Struct 41 (5):951–967
127. Tran Q-T, Toumi A, Turatsinze A (2011) Delamination of thin bonded cement-based overlays: analytical analysis. Mater Struct 44(1):43–51
128. Turatsinze A, Beushausen H, Gagné R, Granju J-L, Silfwerbrand J, Walter R (2011) 'Chapter Debonding'. State-of-the-Art Report of the RILEM Technical Committee 193-RLS Series, Bissonnette B, Courard L, Fowler DW, Granju J-L (Eds), 3:107–139
129. Hun M (2012) Water effect on interface debonding of a bilayer urban road structure subjected to bending/Influence de l'eau dans le décollement par flexion de bicouches de chaussée urbaine. Ph.D. thesis of ECN, http://hal.archives-ouvertes.fr/tel-00777011/. (in French) (2012)
130. Hun M, Chabot A, Hammoum F (2012) A four point bending test for the bonding evaluation of composite pavement. In: Proceedings of the 7th Rilem international conference on cracking in pavements, June 20–22 2012, Delft, The Netherlands. RILEM Bookseries, 4:51–60. https://doi.org/10.1007/978-94-007-4566-7_6
131. Chabot A, Hammoum F, Hun M (2017) A 4pt bending bond test approach to evaluate water effect in a composite beam. Eur J Environ Civil Eng 11:54–69. https://doi.org/10.1080/19648189.2017.1320237 sup1
132. Cole LW, Mack JW, Packard RG (1998) Whitetopping and ultra-thin whitetopping the US experience. In: 8th international symposium on concrete roads
133. Chabot A, Hun M, Hammoum F (2013) Mechanical analysis of a mixed mode debonding test for composite pavements. Const Build Mater 40:1076–1087. https://doi.org/10.1016/j.conbuildmat.2012.11.027
134. Chabot A, Hun M, Hammoum F (2013) Determination of energy release rate for a mixed-mode debonding test for «composite» pavements. In: 6th international symposium on defect and material mechanics, 31–32, July 1–5, Centrale Nantes, France
135. Vandenbossche J, Barman M, Mu F, Gatti K (2011) Development of design guide for thin and ultra-thin concrete overlays of existing asphalt pavements. Task 1 report: compilation and review of existing performance data and information. Technical report, University of Pittsburgh, Department of Civil and Environmental Engineering, Swanson School of Engineering

136. Chabot A, Hammoum F, Hun M (2016) Mixed—mixed-mode debonding approach to evaluate water sensibility in bi-layer composite pavements. In: 8th international conference on mechanisms of cracking and debonding in pavements (MCD2016), Nantes, France, June 7–9. Chabot A, et al. (eds) Springer RILEM Bookseries, 13:613–618, ISBN: 978-94-024-0867-6. https://doi.org/10.1007/978-94-024-0867-6_86
137. Chabot A, Cantournet S, Ehrlacher A (2000) Analyse de taux de restitution d'énergie par un modèle simplifié pour un quadricouche en traction fissuré à l'interface entre 2 couches. Comptes-rendus aux 12ème Journées Nationales sur les Composites (JNC12), ENS de Cachan, 2:775–784. (ISBN 2-9515965-0-2)
138. Caron JF, Diaz A, Carreira RP, Ehrlacher A (2006) Multi-particle modelling for the prediction of delamination in multi-layered materials. Compos Sci Technol 66(6):755–765. https://doi.org/10.1016/j.compscitech.2004.12.022
139. Dundurs J (1969) Edge-bonded dissimilar orthogonal elastic wedges under normal and shear loading. Trans ASME J Appl Mech 650–652. https://doi.org/10.1115/1.3564739
140. Chen Y (2011) Composite specimen testing to evaluate the effects of pavement layer interface characteristics on cracking performance. Ph.D. Dissertation from the University of Florida
141. Chen Y, Tebaldi G, Roque R, Lopp G (2013) Development of a composite specimen interface cracking (CSIC) test for top-down cracking. J Test Eval 41(4):625–634. https://doi.org/10.1520/jte20120002
142. Jones RM (1980) Mechanics of composite materials. McGraw-Hill, New York
143. Chen Y, Tebaldi G, Roque R, Lopp G, Su Y (2012) Effects of interface condition characteristics on open-graded friction course top-down cracking. Road Mater Pavement Design 13(1):56–75
144. Partl MN (2016) Characterization and detection of debonding phenomena in asphalt pavements and on concrete bridge decks. In: Keynote lecture at the 8th international conference on mechanisms of cracking and debonding in pavements (MCD2016), Nantes, France, June 7-9, 2016. (https://mcd2016.sciencesconf.org/conference/mcd2016/Partl_Keynote_MCD2016.pdf)
145. Chabot A, Buttlar B, Dave E, Petit, C, Tebaldi G (eds) (2016) 8th RILEM international conference on mechanisms of cracking and debonding in pavements vol 13, 1st ed. Springer Series, RILEM Bookseries. ISBN 978-94-024-0866-9. https://doi.org/10.1007/978-94-024-0867-6
146. Seitl S, Knésl Z, Veselý V, Routil L (2009) A refined description of the crack tip stress field in wedge-splitting specimens—a two-parameter fracture mechanics approach. Appl Comput Mech 3:375–390
147. Chabot A, Petit C (2017) Mechanisms of cracking and debonding in pavements: debonding mechanisms in various interfaces between layers. Eur J Environ Civil Eng 11:1–2. https://doi.org/10.1080/19648189.2017.1361649 sup1
148. Buttlar WG, Hill BC, Kim YR, Kutay ME, Millien A, Montepara A, Paulino GH, Petit C, Pop IO, Romeo E, Roncella R, Safavizadeh SA, Tebaldi G, Wargo A (2014) Digital image correlation techniques to investigate strain fields and cracking phenomena in asphalt materials. Mater Strcutures 47(8):1373–1390
149. Haykin S (1999) Neural networks: a comprehensive foundation. Prentice Hall. ISBN 0-13-273350-1
150. Miradi M (2009) Knowledge discovery and pavement performance, Ph.D. Dissertation, Deft, University of Technology, Delft
151. Raab C, Abd El Halim AO, Partl MN (2013) Utilisation of artificial neural network for the analysis of interlayer shear properties. Baltic J Road Bridge Eng 8(2):107–116

# Chapter 4
# Advanced Measurement Systems For Crack Characterization

**Gabriele Tebaldi, Alex Apeagyei, Denis Jelagin and Augusto Cannone Falchetto**

## 4.1 Introduction

Interpretation and prediction of cracking phenomena necessarily pass through precise and sophisticated models. In the previous chapters, it is possible to find a detailed overview of the available models related to different conditions of crack initiation and crack propagations. All of those models have a common point: they need detailed local measurements of damage and cracking phenomena in order to be calibrated and validated. The mentioned measures are mainly displacements and the related applied loads that allow calculation of stresses and strains associated with cracks. The stress and strain relation is not the only required information: some models also need information about the structure of the materials to make a correct mesh for finite element applications while other models need information about the damage level of the material and other characteristics.

The main point is that the most popular measurement systems like strain gages, LVDT, clip gages (the so-called traditional measurement systems) are frequently unable to provide the data required by models. To give an example, the strain gages are able to give very precise measurement but alone are not always able to provide

G. Tebaldi (✉)
University of Parma, Parma, Italy
e-mail: gtebaldi@unipr.it

A. Apeagyei
University of East London, London, UK

D. Jelagin
Royal Institute of Technology, Stockholm, Sweden

A. C. Falchetto
TU Braunschweig, Braunschweig, Germany



W. G. Buttlar et al. (eds.), *Mechanisms of Cracking and Debonding in Asphalt and Composite Pavements*, RILEM State-of-the-Art Reports 28,
https://doi.org/10.1007/978-3-319-76849-6_4

high quality data when used to measure strains related to cracks. They provide only local information and cannot provide accurate data when applied to an area of material undergoing large strain gradients. In addition, to be efficiently used, strain gages require an accurate prediction of the cracking phenomena to be applied in the right place. Even still, strain gauges do not allow the full study of post-peak material separation behaviour, since gauges placed over cracks stop producing useful measurements once the foil in the gages is fractured.

In the last few years, to have information that is more accurate and to have more detailed measurements of material behavior, several researchers developed advanced measurement systems to have a better understanding of cracking and of all the collateral phenomena related or generated by cracking. In some cases, the proposed measurement systems were an adjustment to the tools already used in other disciplines. These measurement systems may be classified in several different ways: based on the function of the method or the physics principle used to obtain the measure, based on the function of mechanical principle on which the system is based, whether optical systems are used, whether the system has back calculation, and other methods. This large number of possible different points of view led MCD TG3 to make a classification based on a "classification grid" to highlight the most important details of each systems (Table 4.1) instead of a classification based on set categories. The basic idea behind the information classification grid is to provide researchers interested in these measurement system with a quick overview of their scope and applications.

(1) Measurement principle could be one of the following categories.

- Strain field characterization technique (1)
- Local displacement measurement technique (2)
- Acoustic emission technique (3)
- Optical technique (4)
- Seismic, radar, wave based, NDT (5)
- X-Ray CT (6)
- System using measured structural response to back-calculate damage (7)

(2) The short description of the system may be based on the following characteristics or basic principles of a measurement system.

- Seismic, electromagnetic, NDT based (A)
- Contact technique (B)
- Non-contact technique (C)
- Mono-dimensional (D)
- Bi-dimensional (E)
- Tri-dimensional (F)
- Measurement technique based on direct measurement (G)
- Measurement technique based on post-processing analysis (H)
- Characteristic of applied solicitation (Static, Monotonic, Cyclic, Dynamic) (I)
- Detection of crack length (J)
- Detection of crack propagation with time (K)

- Detection of crack propagation with load (L)
- Detection of crack propagation with deflection/displacement (M)
- Detection of crack propagation with temperature (N)
- System that provide crack spatial distribution and/or make crack map (O)

The MCD-TG3 made a census of the advanced measurement systems currently used in the analysis of cracking in asphalt materials. This chapter shows the systems collected during the census with descriptions provided by the researchers that developed and/or are using those systems. In this chapter, 10 contributions presenting advanced measurement techniques for crack analysis were collected from research groups that are working in different Universities or Research Centers across Europe and USA. These are briefly presented in the next following sub-sections.

## 4.2 X-Ray Computed Tomography (CT) with Tension/ Compression Testing System with Climate Control

The relationship between microstructure characteristics that are known to influence the behavior of asphalt mixtures and macroscopic properties of interest to engineers and designers needs to be better understood. The digital X-ray computed tomography (CT) imaging system with tension/compression testing capabilities and environmental chamber can be used to characterize the three-dimensional microstructure of materials, along with the internal damage and fracture evolution of asphalt mixture specimens during static and cyclic loading regimes. According to the proposed classification grid (Table 4.1), this section concerns Table 4.2.

**Table 4.1** Classification grid of the measurement systems evaluated by TG-03

| Technique/device's short name | Mechanical fields measurement using optical full-field | |
|---|---|---|
| Measurement principle[(1)] | (1), (2), (4) | |
| Categories | Yes | No |
| Field system | | |
| Laboratory system | | |
| Specific setup dependent measurement system | | |
| Measurement system associable with different test setup | | |
| Crack initiation | | |
| Crack propagation | | |
| Crack distribution | | |
| Damage characterization | | |
| Short description of system[(2)] (C), (K), (L), (M), (N), (P) | | |

**Table 4.2** Classification grid of the X-Ray computed tomography

| Technique/device's short name | X-Ray—CT | |
|---|---|---|
| Measurement principle[(1)] | (6) | |
| Categories | Yes | No |
| Field system | | X |
| Laboratory system | X | |
| Specific setup dependent measurement system | | X |
| Measurement system associable with different test setup | X | |
| Crack initiation | X | |
| Crack propagation | X | |
| Crack distribution | X | |
| Damage characterization | X | |
| Short description of system[(2)] (C), (F), (H), (J), (O), (P) | | |

## *4.2.1 Introduction*

### 4.2.1.1 Phenomenon to Be Analyzed by the Proposed System

Evaluation of load-induced changes in asphalt mixture internal structure; especially crack initiation and propagation.

### 4.2.1.2 How the System Works

The high resolution micro-CT system (X-View X5000CT from North Star Imaging Inc.) can be used to characterize the asphalt mixture internal structure at different length scales and at controlled loading and temperature. The test procedure includes scanning the specimen, system calibration, and CT data reconstruction. An industrial computed tomography software called efX CT is used for visualization, calibration and reconstruction. After reconstruction, image analysis software, e.g. Avizo Fire, is used to identify different material phases in the CT dataset and determine qualitative and quantitative parameters describing materials structure. Once all three phases (stones, binder and air voids) have been segmented and analyzed with Avizo Fire, finite element (FE) meshes can be generated and exported to FEA packages. Performing the finite element analysis of different loading and thermal scenarios allows the gaining of further insight on the impact of the material's internal structure on its performance. In summary, this CT system can be used for asphalt mixtures to characterize their microstructure, determine their evolution of microstructure during loading, and to provide the input for the CT-data based micromechanical modelling of the fracture initiation and propagation.

### 4.2.2 Theoretical Basis

The main components of an X-ray tomography imaging system are the X-ray source, collimator, rotational specimen manipulator, and the X-ray detector (Fig. 4.1). When X-rays penetrate into a material, its intensity becomes attenuated due to the absorption and scattering of X-rays by the atoms of the material [1].

The intensities of the transmitted X-Rays are recorded on the detectors placed at the other side of the specimen. The scanning is completed after collecting the intensity measurements for a full rotation of the specimen.

X-Rays intensities are measured before and after penetrating through the specimen in different directions for full rotation of the specimen. The intensity values are used to calculate the distribution of the linear attenuation coefficient in order to generate a map representing the density at every point in the material [2]. Brighter regions correspond to dense objects such as aggregates, and dark regions correspond to low-density objects such as voids.

The digital X-ray tomographic imaging system with tension/compression testing capabilities and environmental chamber provides a three-dimensional characterization of the microstructure of materials, along with the internal damage and fracture evolution of asphalt mixture specimens during static and cyclic loading regimes.

### 4.2.3 Description and Test Setup

X-Ray CT consists of an X-Ray source, a detector, and a turntable carrying the test specimen in between the source and the detector. This is a Real Time X-ray Inspection CT System designed specifically for industrial X-ray applications (Fig. 4.2).

This system utilizes an NSI radiation shielded enclosure and NSI X5000 six-axis component manipulator. A High resolution Amorphous Silicon digital detector (200 μm pixel pitch) (Fig. 4.3) and NSI X-View efX-DR Workstation is included.

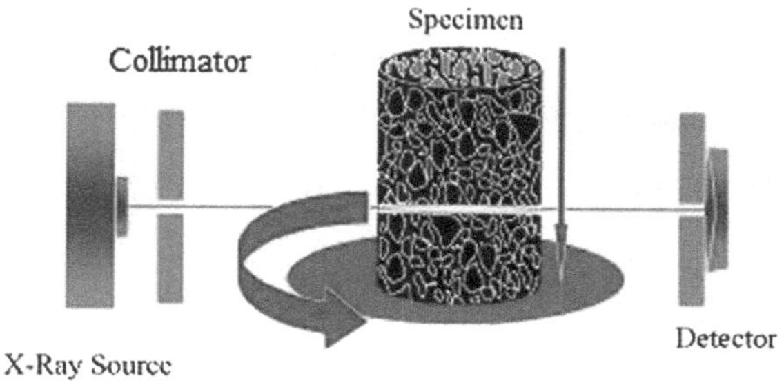


**Fig. 4.1** General mechanism of X-ray tomography scanning

**Fig. 4.2** Computed tomography system

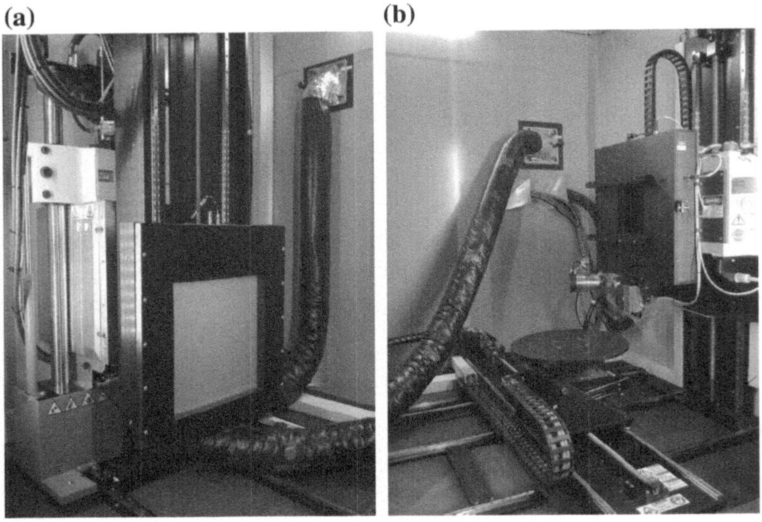


**Fig. 4.3** The X-Ray system **a** Detector; **b** X-Ray source

This imaging system includes an NSI 225 kV Microfocus X-ray tube (5 μm focal spot size) and NSI 450 kV HP X-ray tube with 400 μm focal spot size. This system is state-of-the-art in high-resolution digital X-ray imaging and CT technology. Efx CT software is used for CT calibration, Cone-Beam reconstruction, 3D real time visualization, and analysis tools in the same interface. Depending on the specimen size, degree of the geometrical magnification applied, and the tube used, tomography data with the voxel size down to 5 μm is obtained. The load cell is a MTS Model 370 Uniaxial Test System 100 kN force and 150 mm displacement; the system also includes a EGNU12-6CWL Platinum temperature chamber with temperature range from −20 to 80 °C.

A typical workflow for post-processing with Avizo Fire includes four main steps: image de-noising/correction, materials segmentation, quantitative analysis, and volume meshing. In order to quantitatively characterize the internal structure of asphalt the following parameters can be obtained: stone gradation and orientation, distribution and geometry of the contact zones between stones, variation of the binder, and air void distribution with depth.

### 4.2.4 Examples

Characterization of asphalt's internal structure includes obtaining the following parameters: stone gradation and orientation, distribution and geometry of the contact zones between stones, variation of the binder and air void distribution with depth. As an example, in Fig. 4.4 the reconstructed surface of the effective (interconnected) pore space is shown for asphalt field cores immediately after the core extraction (Fig. 4.4a) and after cleaning the core (Fig. 4.4b). As it may be seen in Fig. 4.4, the cleaning process restores the pore connectivity at the bottom of the field specimen originally reduced due to clogging.

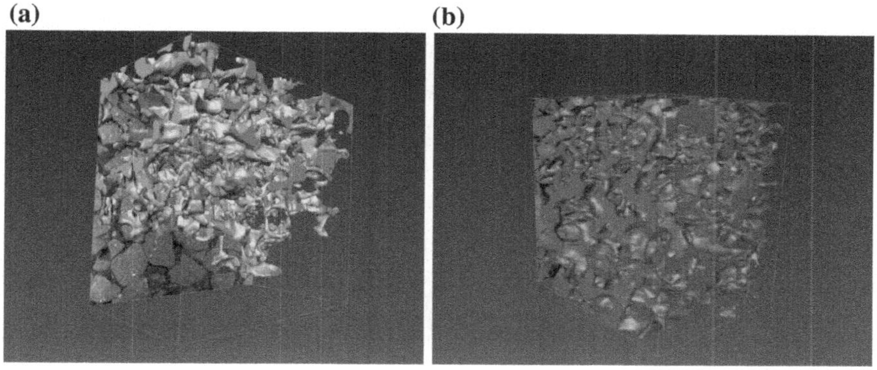


**Fig. 4.4** Air void surface in 'quiet' asphalt **a** Before cleaning; **b** After cleaning

The variation of the air voids ratio with depth is shown in Fig. 4.5 for the cleaned asphalt core.

In Fig. 4.6, the reconstructed CAD surface is shown for the stone phase in asphalt mixture, and the histogram distribution of stone volumes is shown in Fig. 4.7.

The CAD surface representation illustrated in Fig. 4.6 can be used to map the strain field in the specimen provided that the same specimen is scanned before and after loading and correlation analysis is performed.

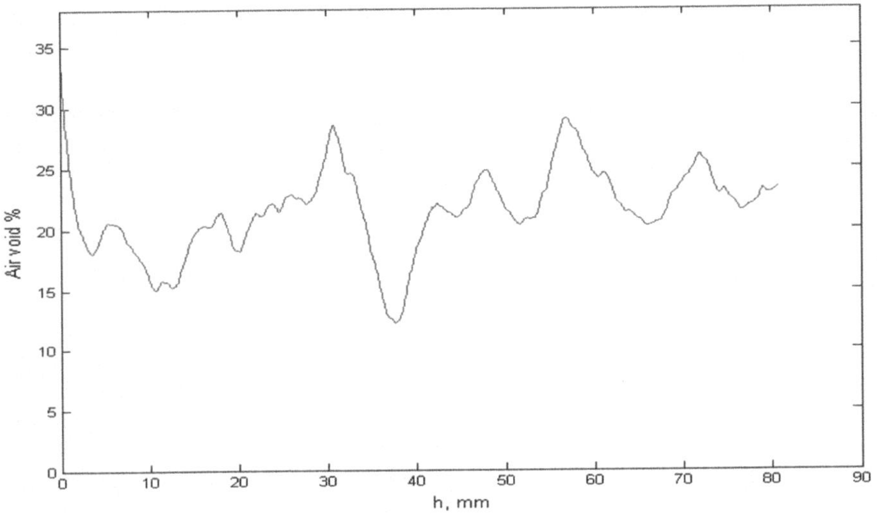


**Fig. 4.5** Variation of air voids ratio with depth

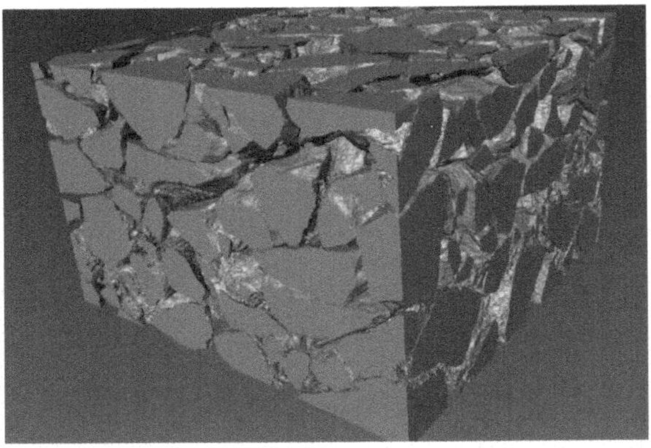

**Fig. 4.6** Reconstructed surface of the stone phase in porous asphalt

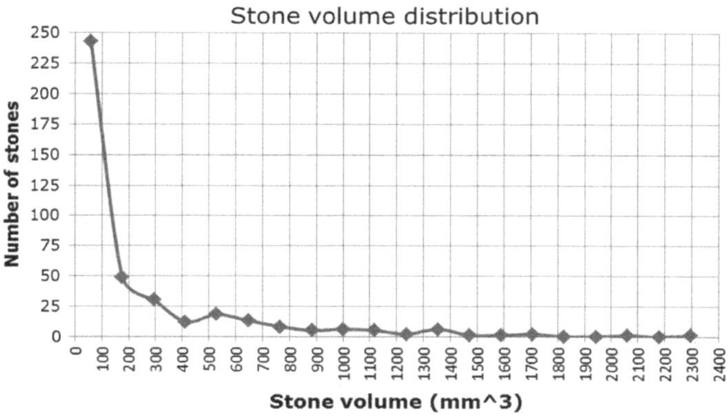


**Fig. 4.7** Volume distribution of stones in the asphalt mixture

### *4.2.5 Benefits and Limits*

#### 4.2.5.1 Benefits

- Static and dynamic load testing capabilities and environmental chamber for X-ray CT
- Tool to develop models for improving the durability of asphalt mixtures
- Entirely non-destructive 3D imaging
- Little or no sample preparation required

#### 4.2.5.2 Limits

- Ability to penetrate objects
- Not mobile
- Large objects
- Image artifacts
- Considerable computer resources for visualization and analysis (Large data files)

## 4.3 High Speed Stereo Vision System (HS-SVS) for Measuring Displacements and Strains During Laboratory Tests

Digital image-based analysis tools are becoming increasingly popular among researchers in the areas of Pavement and Geotechnical Engineering. Optical flow techniques (also called image correlation method) have been used to measure the full-field strain development during testing of materials. In this method, digital images of the specimens are captured during testing, and the analyses of time-lapsed images are performed using optical flow techniques to define strain distributions within specimens. Image-based full-field strain measurement provides much more information than one can obtain from the crosshead extension or an extensometer or an LVDT that may be attached to the sample. This section describes a stereovision system that involves two high-speed (400 frames per second) digital cameras that capture simultaneous images of the samples tested. The system is able to provide three-dimensional (3D) strain distribution (i.e., strains x-, y- and z-direction) as well as the temporal change on the strain (i.e., strain speed or crack speed if applied to a fracture test). It uses the 3D stereo-vision technique (explained in the next section) to calculate the 3D coordinates of points specified on the surface of a specimen being tested. According to the proposed classified grid (Table 4.1), this section concerns Table 4.3.

### 4.3.1 Theoretical Basis

Typical optical flow algorithms search for a constant size of block (called macro block) between successive image frames based on a matching criterion (see Fig. 4.8). Similar patterns are tracked sequentially from one image to the next, and the amount of movement is directly related to the movement of selected patterns. The amount of displacement is calculated for each block, which produces optical flow for the entire image. In the software used here, two matching algorithms are implemented: a commonly used technique called normalized cross-correlation (NCC) [3] and a more accurate a phase-based approach for sub-pixel resolutions

**Table 4.3** Classification grid of the HS-SVS measurement system

| HS-SVS | Mechanical fields measurement using optical full-field | |
|---|---|---|
| Measurement principle(1) | (1), (2), (4) | |
| Categories | Yes | No |
| Field system | | X |
| Laboratory system | X | |
| Specific setup dependent measurement system | X | |
| Measurement system associable with different test setup | X | |
| Crack initiation | X | |
| Crack propagation | X | |
| Crack distribution | | X |
| Damage characterization | | X |
| Short description of system(2) (C), (K), (L), (M), (N), (P) | | |

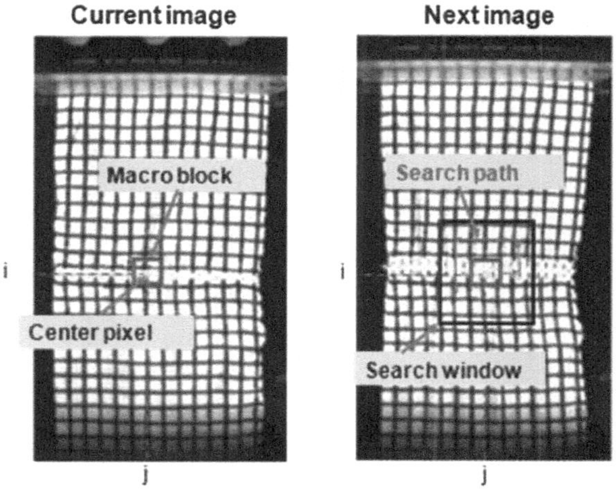


**Fig. 4.8** Illustration of optical flow based on pixel match

where contours of constant phase over time are tracked [4], since these are more robust to variations in lighting conditions and deviations from pure translation than contours of constant amplitude.

Most strain measurement systems that use the method shown in Fig. 4.8 (i.e., NCC) provide only the 2D strains parallel to the camera plane. In order to measure the displacements in 3D, i.e., x-, y- and z-directions, at least two cameras are needed. With the knowledge of relative location of the two cameras as well as the focal length of the cameras, one can theoretically compute the 3D coordinates of an object in space using Eq. (4.1)

$$Z = f\frac{T}{x_l - x_r} \tag{4.1}$$

where Z is the z-coordinate of the point of interest relative to the cameras, T is the horizontal distance between the focal points of two cameras, f is the focal length of the cameras, $x_l$ and $x_r$ are the pixel coordinates of the projection of the point on the left and right cameras, respectively. Figure 4.9 illustrates the parameters in Eq. (4.1).

The most important issue related to the 3D reconstruction is finding the same points in left and right cameras (i.e., $x_l$ and $x_r$). To address this challenge, a robust pixel-matching algorithm is needed, similar to the NCC method described earlier (see Fig. 4.8). Figure 4.10 shows an example reconstruction of 3D surface of a jar using the stereovision system. Another important step is the camera calibration, i.e., finding the focal length and relative distance between the left and right cameras. This is achieved by using a checkerboard as described in http://www.vision.caltech.edu/bouguetj/calib_doc/.

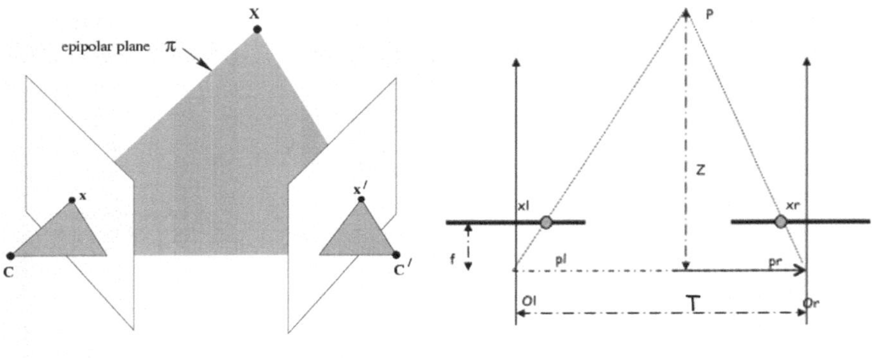


**Fig. 4.9** Illustration of two camera system and parameters used in the calculation of Z

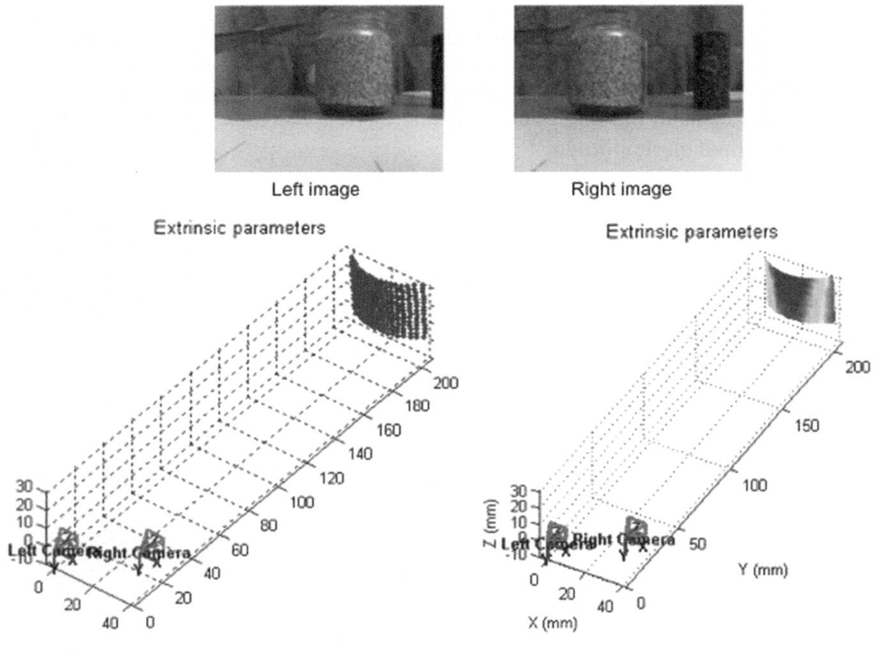


**Fig. 4.10** Example reconstruction of 3D surface of a jar using the stereovision system

## 4.3.2 *Asphalt Mixture Performance Tests (AMPT)*

This system can be typically used to measure the strains during Flow Number (FN) and Dynamic Modulus (|E*|) tests. However, the same system can be used to measure strains during testing of many materials in the laboratory (e.g. concrete, soils, geosynthetics etc.) Axial and radial strains during FN and |E*| tests can be computed. By measuring the displacements on a grid of points allows computation of strain distribution on a specimen (Fig. 4.11).

The system can also be used to compute the strain distribution on the surface of a variety of materials being tested. Figures 4.12a, b show examples of computed strain distribution within a geotextile and asphalt specimen, respectively. In Fig. 4.12b, the location of the crack can easily be seen (colored graph on the right),

**Fig. 4.11** Picture of the MSU's 3D stereovision system used to measure strain during flow number tests

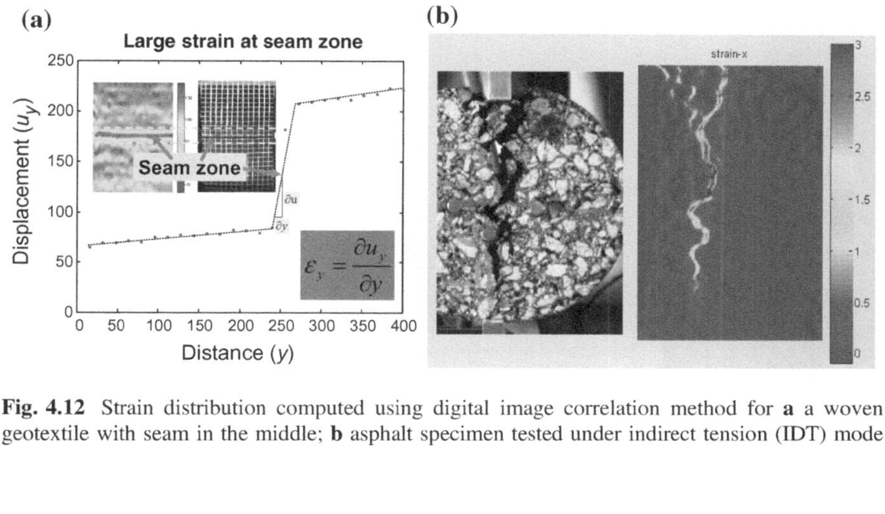


**Fig. 4.12** Strain distribution computed using digital image correlation method for **a** a woven geotextile with seam in the middle; **b** asphalt specimen tested under indirect tension (IDT) mode

and the crack was visible even before it actually appeared on the image. This result was due to the localized large strains that developed around the cracks before the actual crack occurred.

### *4.3.3 Results*

Figure 4.13 shows the validation of the 3D stereo algorithm developed here. During the lab test, the displacement at the bottom of the specimen is measured with the testing machine's (i.e., the AMPT's) actuator LVDT. The top of the specimen is fixed. The displacement values computed by the OF_GUI (the name of the algorithm used) at 4 locations shown in the figure are plotted along with the LVDT measurement, where an excellent match is visible. This concerns a cyclic compressive test run at 1 Hz. Peak strain is about 900 microstrain. Displacements as low as 10 microns have been measured/computed by the system and algorithm.

### *4.3.4 Benefits and Limits of HS-SVS*

Following are the specifications of the camera system used in HS-SVS:

- Frame rate: 400 frames per second (fps)
- Grayscale 1280 by 1024 pixels from each camera
- Regular Lens: Strains as low as 50 microstrains can be measured at 400 fps (Field of view = 150 mm by 150 mm)
- Microscopic Lens: Resolution: 12 micron (10–6 m) at a field of view of 15 mm by 12 mm.

With this system, along with the in-house algorithm, it is possible to capture/compute 40 points on a strain curve caused by a cyclic test with a frequency of 10 Hz.

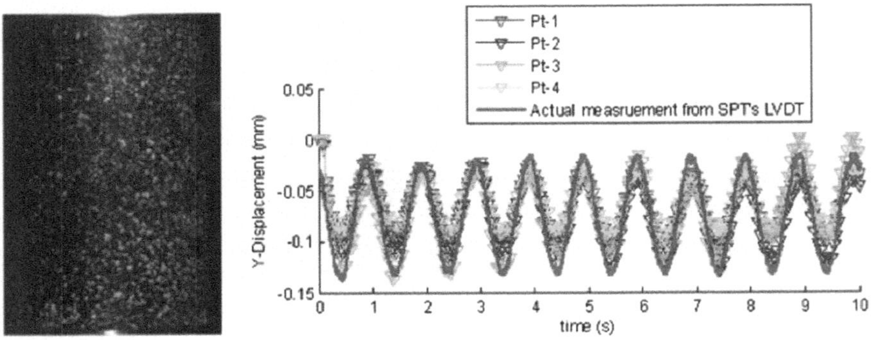


**Fig. 4.13** Validation of the strain measurement during cyclic asphalt mixture testing

Limits of the system:

- A well-distributed random texture needs to be generated on the specimen surface, which requires a special spray painting method. This slows down the sample preparation stage.

### 4.3.5 *Examples*

An example application of the HS-SVS is the measurement of strains and displacement in a Flow Number (FN) Test. Figure 4.14 shows a FN test with 0.1 s load followed by a 0.9 s rest period. The camera-based strains (4 and 6 in the legend) show the entire time history of the displacements. AMPT data is the plastic strain measured by the LVDT of the equipment. The image-based plastic strains (i.e., 4 (red) and 6 (blue) lines shown in the legend) at the end of each cycle match well with the AMPT's LVDT measurements.

## 4.4 Mechanical Field Measurement Using Optical Full-Field Techniques

In this section, an experimental analysis of mechanical behaviour of the bitumen materials from optical full-field measurements is presented. Using the digital image correlation (DIC), two experimental tests are analyzed in terms of displacement and strain fields. This feasibility study shows that by using the optical full field technique, it is possible to identify and characterize the damage process for the complex

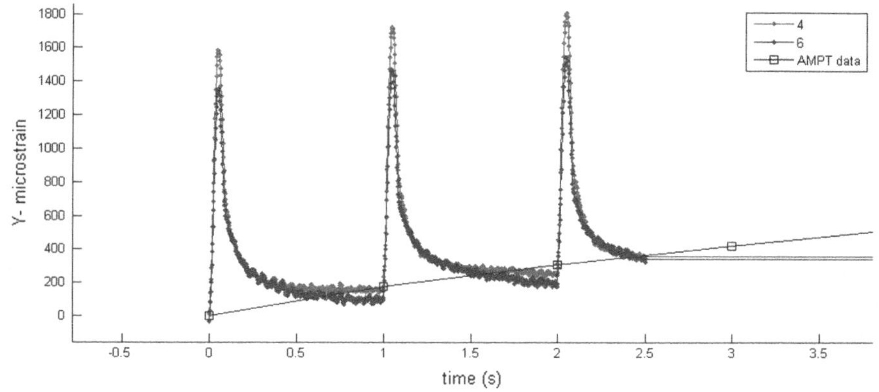


**Fig. 4.14** Measurement of axial displacements in a flow number test

**Table 4.4** Classification grid of the optical full fields techniques

| Optical full fields techniques | Mechanical fields measurement using optical full-field | |
|---|---|---|
| Measurement principle[(1)] | (1), (2), (4) | |
| Categories | Yes | No |
| Field system | | X |
| Laboratory system | X | |
| Specific setup dependent measurement system | X | |
| Measurement system associable with different test setup | X | |
| Crack initiation | X | |
| Crack propagation | X | |
| Crack distribution | | X |
| Damage characterization | | X |
| Short description of system[(2)] (C), (K), (L), (M), (N), (P) | | |

loading. The tests are realized using several materials and loading configurations. According to the proposed classified grid (Table 4.1), this section concerns Table 4.4.

### 4.4.1 Introduction

With the passage of time, optical methods find more applications in the mechanical characterization of materials and structures. Associated with the full fields techniques, the optical methods can be easily correlated with the energetically approaches such as the integral invariants. Today, several methods are developed in order to measure the different mechanical fields. These methods include the interferometry and the moiré to measure principally the out-of plane displacement [5], the photoelasticity to evaluate the principal stress [6], and most recently the DIC [7, 8] and the mark tracking techniques [9] to measure the displacement and strain field in-plane. Using the stereo DIC [10], it is possible to obtain the third component of the displacement fields. Initially developed to characterise the "classical" materials such as the steel, aluminium, polymers or composites, these methods were then adapted for the materials employed in civil engineering (e.g. concrete, bitumen, asphalt …). Among the optical methods summarized above, the DIC and the mark tracking method seem to be the best ways to characterize the mechanical behavior in the case of bitumen materials. Another particularity of the optical full fields' techniques is the possibility of coupling this one with the numerical approach such as the Finite Element Method (FEM) [11–13]. Using the optical methods, the Zone of Interest (ZOI) can be discretized either by the subsets similar to the finite elements of the mesh with the DIC or by the marks similar to the finite element mesh nodes with the mark tracking method. In this section, a feasibility study to characterize the mechanical behavior of bitumen materials by the DIC technique is proposed.

### *4.4.2 Digital Image Correlation*

Related to the optical full field methods, DIC is an optical method that allows measuring in-plane displacement.

Typically for displacement measurement by DIC, the experimental set-up includes the following components (Fig. 4.15): an optical system which groups a CCD camera with lens allowing for the recording of the images of the plane surface of the sample under loading, a light source for illuminating the sample surface, and a computer allowing for the processing and storage of the digitized images. The basic principle of the method is based on the comparison between two images of the sample plane surface acquired at different states: one before deformation and the other one after. Then, the displacements are estimated by comparing the degree of resemblance between these two image subsets. As is shown in Fig. 4.16, in DIC the displacements are calculated into the ZOI discretized by small areas with multiple pixels called subsets (where, m is the subset numbers). Note, that each subset is characterized by a unique light intensity distribution (i.e., gray level). Assuming that during the test the light intensity distribution on the plane surface of sample does not change, the displacements are estimated by searching the subsets' changes (translation + rotation + rigid body motion) between the undeformed and the deformed images. Finally, the displacement field is built from the displacement vectors ($u_1$, $u_2$) corresponding to displacements of all subsets centers.

Let us consider the same rectangular subset, called (m) in undeformed image and ($m^*$) in deformed image. Then, the subset transformation can be written in terms of intensity values at positions (m) and ($m^*$) as in Eq. (4.2):

$$f(m) = f(x_1, x_2) \text{ and } f^*(m^*) = f^*(x_1 + u_1, x_2 + u_2) \tag{4.2}$$

The DIC algorithm, based on iterative method, allowing for estimating the displacements could be described by the cross coefficient in Eq. (4.3):

$$C = 1 - \left[\sum\sum\nolimits_m {}^*f(x_1, x_2) \cdot f^*\left(x_1^*, x_2^*\right) / \sqrt{\sum\sum\nolimits_m [f(x_1, x_2)]^2 \cdot \sum\sum\nolimits_m \left[f^*\left(x_1^*, x_2^*\right)\right]^2}\right] \tag{4.3}$$

With Eq. (4.4):

$$x_1^* = x_1 + u_1 + \frac{\partial u_1}{\partial x_1} \cdot \Delta x_1 + \frac{\partial u_1}{\partial x_2} \cdot \Delta x_2 \text{ and } x_2^* = x_2 + u_2 + \frac{\partial u_2}{\partial x_1} \cdot \Delta x_1 + \frac{\partial u_2}{\partial x_2} \cdot \Delta x_2 \tag{4.4}$$

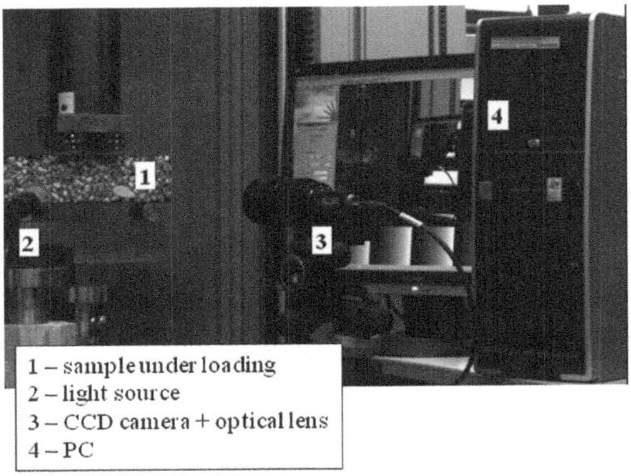


**Fig. 4.15** Typical experimental set-up for DIC measurement

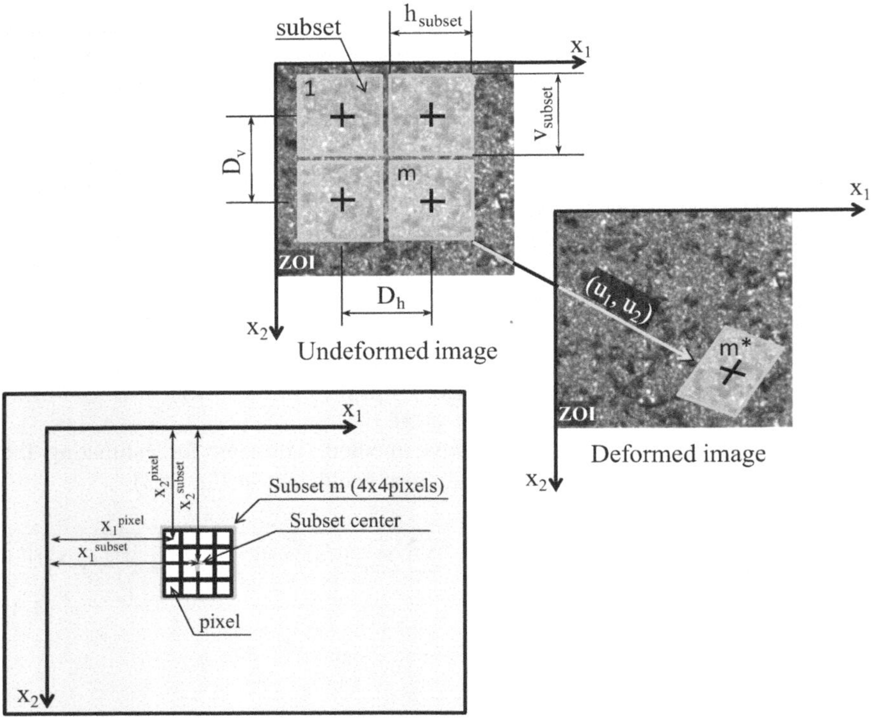


**Fig. 4.16** Principle of DIC

where:

- $f(x_1, x_2)$ is the gray level value at coordinate $(x_1, x_2)$ for undeformed image;
- $f^*(x_1^*, x_2^*)$ is the gray level value at coordinate $(x_1^*, x_2^*)$ for deformed image;
- $u_1$, $u_2$ are the displacements for the subset center in the $x_1$ and $x_2$ directions respectively;
- $\Delta x_1$ and $\Delta x_2$ are the distances from the subset center to points $(x_1, x_2)$.

The correlation is performed by calculating the local deformation values for subset $u_1$, $\partial u_1/\partial x_1$, $\partial u_1/\partial x_2$, $u_2$, $\partial u_2/\partial x_1$, and $\partial u_2/\partial x_2$ which minimize the cross-correlation coefficient.

Another important aspect in DIC is the specimen plane surface preparation (Fig. 4.17). So, in order to obtain a gray level distribution, it is necessary to create a characteristic speckle pattern in which a black speckle is deposited on the white background by spraying the black and white paints.

### *4.4.3 Materials and Methods Used in Two Lab Tests*

In this paragraph, two experimental tests are described (Fig. 4.18).

In the case of the double shear test (Fig. 4.18a) the loading is realized under displacement control and the displacement magnitude is equal to ±30 μm. For this test, several frequencies are tested: 0.01, 0.05 and 0.1 Hz. The tested sample is a 0/

**Fig. 4.17** Sample preparation

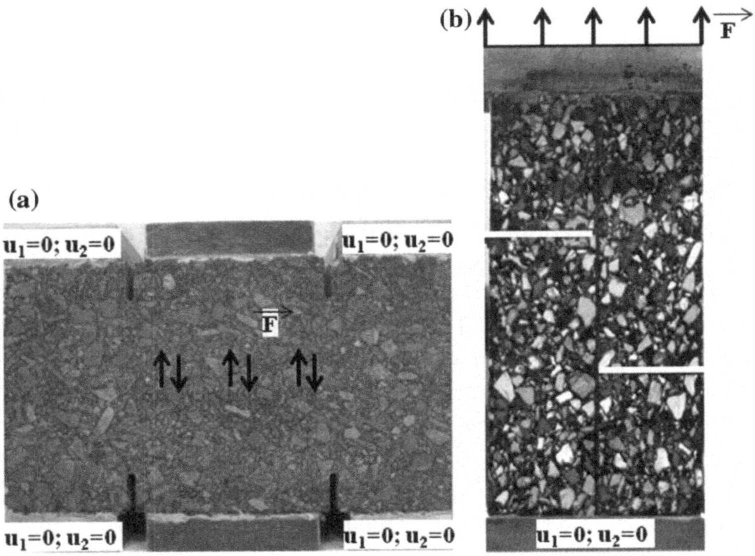


**Fig. 4.18** Geometries and boundary conditions of experimental sample **a** The double shear test; **b** The shear deboning test

6BBC (French Standard). For the shear debonding test, several materials and geometries are tested.

The displacement fields are calculated by the DIC method from the images recorded during the test. The images recorded correspond to different states of deformation of specimen. The evaluation of displacement fields by DIC requires some preparation of the specimen surface. So, in order to obtain the measurements, as explained above, a speckle pattern is applied on the specimen surface. In this case, a black paint is splashed onto a white background applied on the specimen surface. The images of sample during the test are acquired using an 8-bit CCD camera (1392H × 1040V pixels$^2$). The frame rate of camera can vary between 1 and 14 fps.

### *4.4.4 Displacement and Strain Fields Results*

After the test the displacement and strain fields are evaluated in a ZOI discretized by the subsets. For double shear test, using DIC the vertical displacement for three extreme positions (−30; 0 and +30 μm) are represented in Fig. 4.19.

By analyzing displacement fields, we can observe the presence of shear bands at the junction between the central and the ends parts of sample. Note also that for 0.01 Hz the distance between metallic heads glued on the fixed parts is equal to 7.5 mm. For 0.05 and 0.5 Hz the distance is fixed at 4.5 mm. The sample material

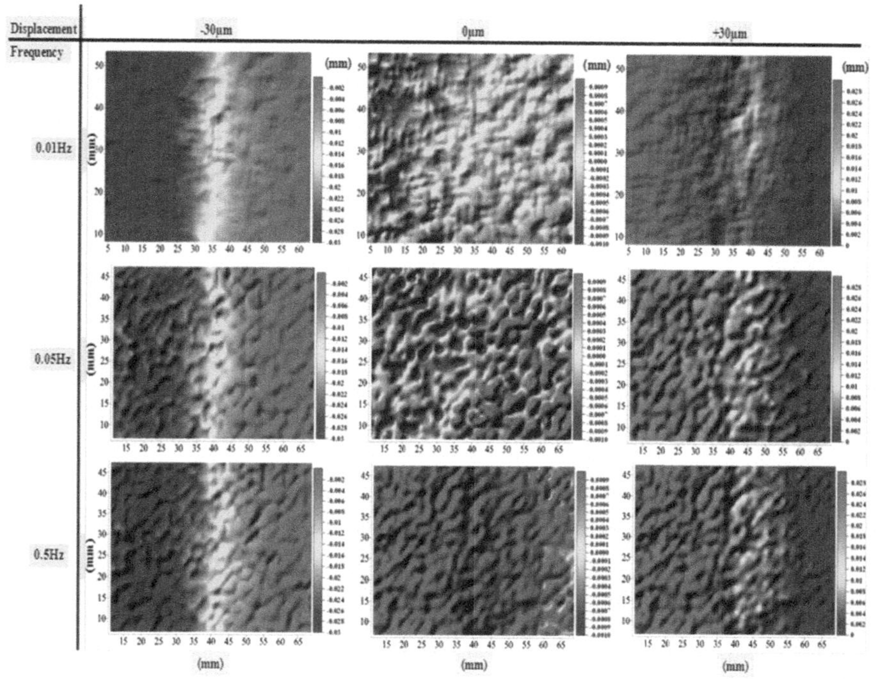


**Fig. 4.19** Vertical displacement field cartographies versus frequency

is the same for all three cases, 0/6BBC. This preliminary analysis allows for estimation of the width of shear band and the study of the influence of boundary conditions and the loading parameters on shear bands characteristics. For 0.01 Hz the ZOI is discretized by subsets of 32 by 32 pixels and horizontal and vertical gap between two subsets is 8, respectively 8 pixels. In the case of 0.05 and 0.5 Hz, the vertical displacements are calculated, this time, in a ZOI discretized by the subsets of $16 \times 16$ pixels$^2$ with a vertical and horizontal gap of 16 pixels. The factor scale is equal to 0.108 mm/pixels.

As shown in Fig. 4.20, the study can be also completed by a finite elements analysis by using the experimental data. According to the geometrical dimensions of ZOI, the mesh is generated using the Cartesian coordinates of subset centers. Taking into account the subsets' disposition, the finite element mesh is generated using four-node quadrilaterals finite elements [14]. Figure 4.20 represents the experimental deformed meshes of ZOI for the test realized at 0.01 Hz. The deformations correspond to two loading levels −30 and +30 μm respectively. The finite element mesh obtains from DIC technique is discretized using 4032 nodes.

The deformed meshes evidence a light noise of the displacement fields, which can be associated with experimental parameters such as light, speckle pattern and vibrations. In this case, using the adjustment procedures the displacement field

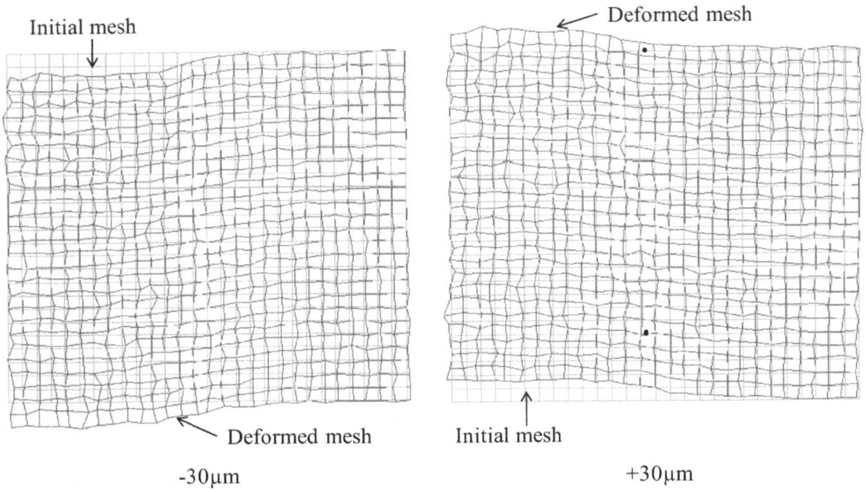


**Fig. 4.20** Experimental deformed mesh (with an amplification factor equal to 100)

optimization may be considered in order to increase the accuracy of the experimental fields [15].

Concerning the shear debonding test, the preliminary results are illustrated in Figs. 4.21 and 4.22. Figure 4.21 represents the characteristic curve load-displacement, and this one allows observing the global mechanical behaviour of the sample during the test.

Using the images acquired during the test, a preliminary study of displacement and strain fields is realized in Fig. 4.22. The cartographies correspond to loading levels represented on the characteristics curve load-displacement. By analyzing the displacement and strain fields, it is possible to observe the damage evolution and to estimate the crack length.

## 4.4.5 Conclusion

In this study, an experimental technique is proposed in order to characterize the kinematic state and implicitly the damage in bituminous mix sample. The DIC is employed in order to measure the displacement and strain fields on the specimen surface. The study of kinematic state allows detecting the presence of the share bands for double shear test. The results allow to estimate the shear band width and to evaluate the influence of material structure and experimental boundary conditions on the shear bands. In the case of shear debonding tests, the results show the possibility to localise the damage and to estimate the crack advance during the test. Moreover using the experimental data the study can be completed by a finite

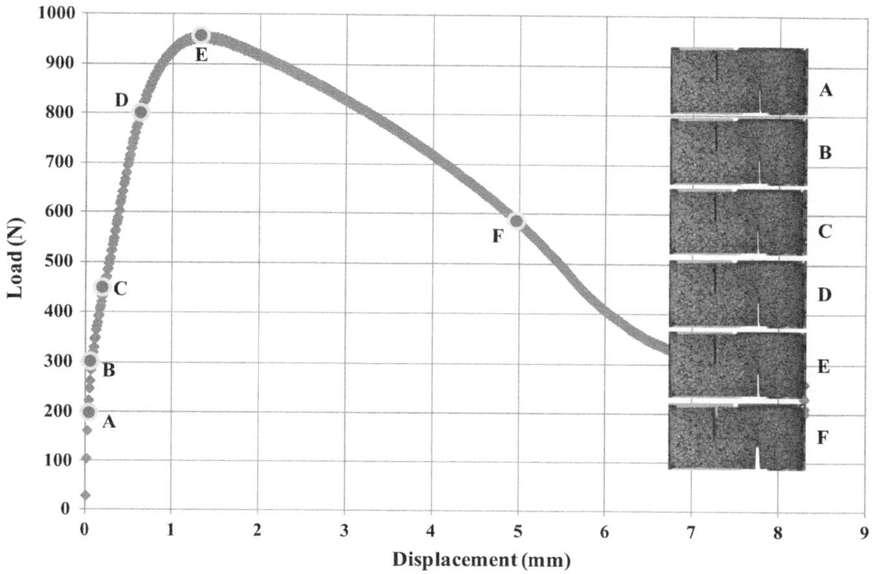


**Fig. 4.21** Load-displacement characteristics curve

elements analysis by allowing visualizing the sample deformations. This possibility can be explored to generate meshes of the complex geometry of specimens.

## 4.5 Characterization of Interface Behaviors Using DIC Technics

Asphalt and cement concrete are heterogeneous materials, wherein, in the case of UTW bi-layer specimen, its interfacial bonding behavior is complicated and the characterization of interfacial bonding parameters becomes difficult. As mentioned in the previous Chap. 3, a four-point bending (4 PB) test is adapted during Hun's Ph.D. research work (2012) in order to investigate the characteristics of the concrete-asphalt-interface in the lab [16]. This test allows generating interface failure in mixed mode (mode I and II) [17]. The DIC technics are used to deepen the knowledge of debonding mechanisms for specimens tested in air and in a water bath. It provides also data for comparing Dunder's values with those of the M4-5n [18]. According to the proposed classified grid (Table 4.1), this section concerns Table 4.5.

In order to favor the delamination at the interface of bilayer materials, the bi-layer specimen geometry is optimized. Using a specific model called Multi-particle elastic Model of Multilayer Materials (M4-5n) a nonsymmetrical geometry is defined (Fig. 4.23) [16].

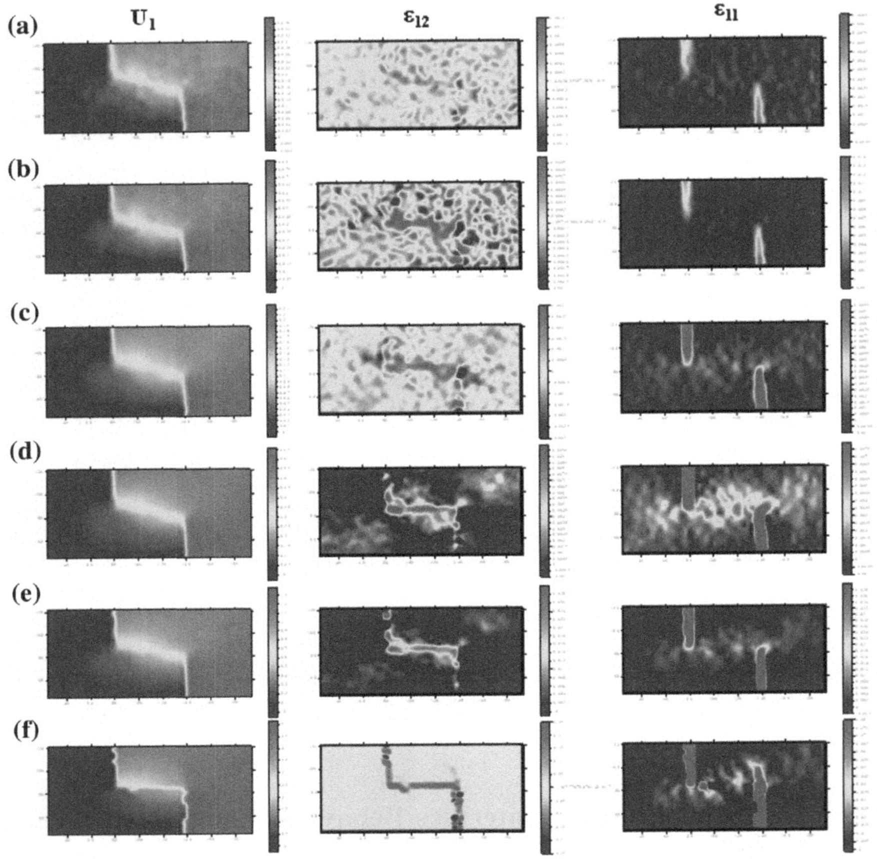


**Fig. 4.22** Displacement and strain fields

## *4.5.1 Use of Digital Image Correlation (DIC) for Characterizing Interface Behaviors*

Numerous papers have been published on the phenomenon of debonding, especially on the possible causes of debonding, methods for predicting the debonding potential of bilayer materials, and the use of DIC technique for characterizing interface behaviors. In order to obtain microscopic information on the failure processes at the interface of bimaterial, a robust full-field measurement method is required [19]. Using DIC, a full-field surface displacement can be measured with high accuracy for specimens with multiple cracks at incremental levels of fracture. This technique can be used to monitor the testing of a wide range of specimen sizes under different loading conditions [20]. The DIC has been used for estimating the stress intensity factors of two and three-dimensional cracks [21]. Hence, in this

**Table 4.5** Classification grid of the optical full fields techniques

| DIC technics | Characterizing interface behavior of bilayers | |
|---|---|---|
| Measurement principle(1) | (1), (2), (4) | |
| Categories | Yes | No |
| Field system | | X |
| Laboratory system | X | |
| Specific setup dependent measurement system | X | |
| Measurement system associable with different test setup | X | |
| Crack initiation | X | |
| Crack propagation | X | |
| Crack distribution | | X |
| Damage characterization | X | |
| Short description of system(2) (C), (K), (L), (M), (N), (P) | | |

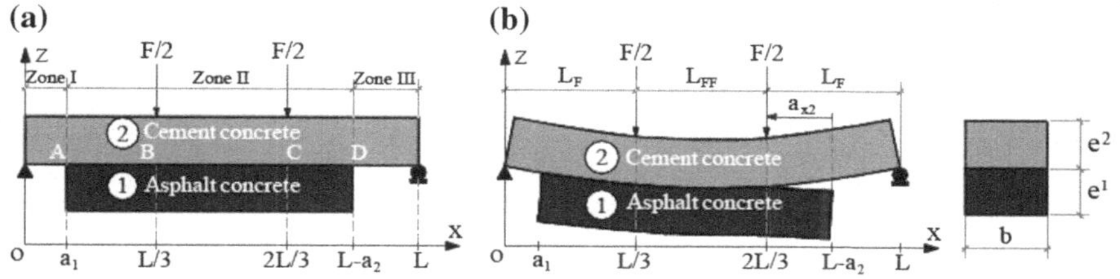


**Fig. 4.23** The 4-point bending test on the bilayer specimen **a** antisymmetric geometry; **b** debonding crack length geometry

work, the DIC is chosen to analyze and determine the displacement, strain components, crack opening and sliding displacements, load-point displacement, crack length and crack tip location. The mode I and mode II stress intensity factors and strain energy release rate for bilayer material interfaces are determined using the crack opening, crack sliding displacements and crack length obtained from DIC [16, 18]. The DIC method was originally developed by Sutton et al. [22, 23]. It is an optical and noncontact measurement technique which is adopted to analyze the displacements on the surface of an object of interest. In DIC, a digital camera from which the displacement at any point of the image is computed takes the surface images before and after the deformation. DIC is based on the following principles: the image of the body is described by a discrete function representing the grey level of each pixel. The grey level is a value between 0 and 255 of its grey levels with the lowest value representing black, highest value white, and values in between representing different shades of gray. The correlation calculations are carried out for a set of pixels, called a pattern. The displacement field is assumed to be homogeneous inside a pattern.

The initial image representing the body before distortion is a discrete function $f(x, z)$ and is transformed into another discrete function $f^*(x^*, z^*)$ after distortion or displacement in Eq. (4.5). The theoretical relation between the two discrete functions can be written as:

$$f^*(x^*, z^*) - f(x + u(x, z), z + v(x, z)) = 0 \tag{4.5}$$

where $u(x, z)$ and $v(x, z)$ represent the displacement field for a pattern (Fig. 4.24a).

In this image correlation, a Normalized Cross-Correlation function that measures how well subsets match is chosen in Eq. (4.6).

$$C_{NCC} = 1 - \frac{\int_{\Delta M} [f(x, z).f^*(x^*, z^*)]^2 \mathbf{dx}\mathrm{d}z}{\left[\int_{\Delta M} f(x, z)^2 \mathbf{dx}\mathrm{d}z. \int_{\Delta M} f^*(x^*, z^*)^2 \mathbf{dx}\mathrm{d}z\right]^{1/2}} \tag{4.6}$$

where $f(x, z)$ is the gray level value at coordinate $(x, z)$ for initial image and $f^*(x^*, z^*)$ is the gray level value at the point $(x^*, z^*)$ of the deformed image. The coordinates $(x, z)$ and $(x^*, z^*)$ are related by the deformation which has occurred between acquisition of the two images. They are given by Eq. (4.7).

$$\begin{aligned} x^* &= x + u + \frac{\partial u}{\partial x}\Delta x + \frac{\partial v}{\partial z}\Delta z \\ z^* &= z + v + \frac{\partial v}{\partial x}\Delta x + \frac{\partial v}{\partial z}\Delta z \end{aligned} \tag{4.7}$$

where $u$ and $v$ are the displacements for the subset centers in the $x$ and $y$ directions, respectively. The terms and are the distances from the subset center to point $(x, z)$. By performing image correlation, the values of coordinates $(x, z)$, displacement $(u, v)$, and the derivatives of the displacements $\frac{\partial u}{\partial x}, \frac{\partial u}{\partial z}, \frac{\partial v}{\partial x}$ and $\frac{\partial v}{\partial z}$ are determined [24, 25].

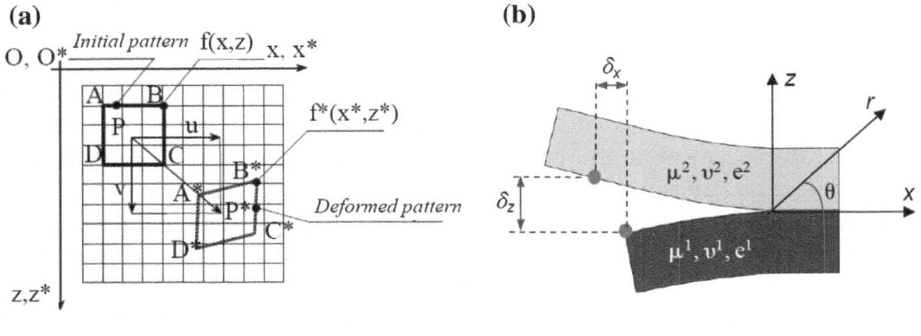


Fig. 4.24 a Initial and deformed pattern [19] b Bimaterial interface crack [16]

### 4.5.2 Experimental Procedure and Obtainable Results

In order to immerse the specimens in water during the tests, a specific aquarium is constructed. r is used to project the load from the hydraulic press MTS on the image (Fig. 4.25).

The double glazing, which allows the control of the temperature in the aquarium and the use of the DIC method, is used for the aquarium wall. The testing temperature is controlled by a climate chamber and two thermal transfer pipes which create a closed circuit. This circuit is assured by means of two fans, one placed in the aquarium and another one placed in the chamber, which pump cold air from the aquarium into the chamber (Fig. 4.25). The warm air from the chamber is thereby moved back to the reservoir. For DIC, the speckle pattern is to be made on the specimen. The specimens are initially black washed, and a speckle pattern is prepared over it using a black spray paint (Fig. 4.26).

The images are captured by a camera CDD with a frame rate of 63 fps, which is installed on a tripod to maintain the distance between the camera lens and the specimen. In order to synchronize the load with the images, a LCD projector is used. The digital images are correlated using an algorithm originally written by

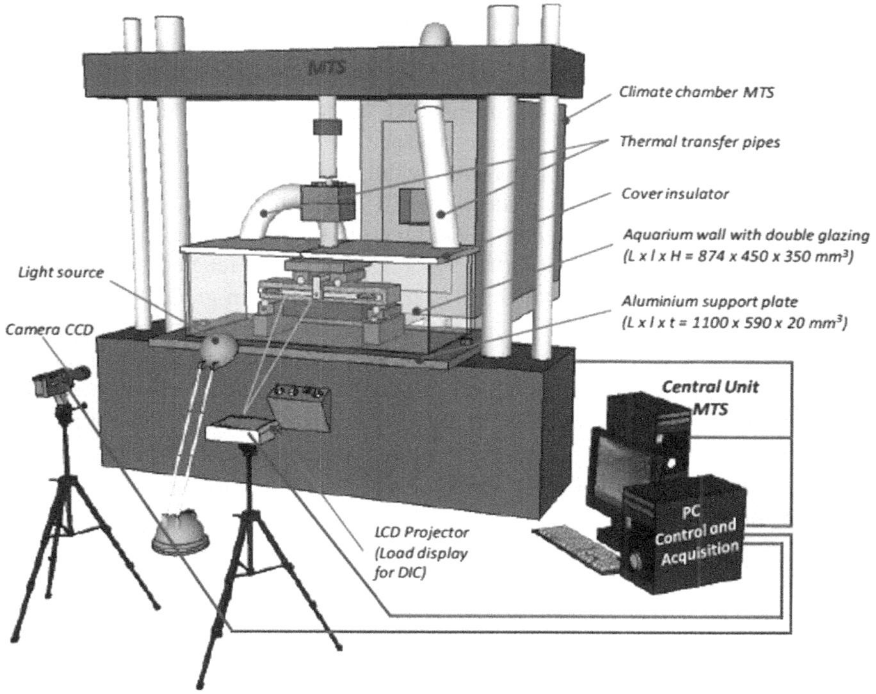


**Fig. 4.25** Four-point bending test set-up and DIC technique [16]

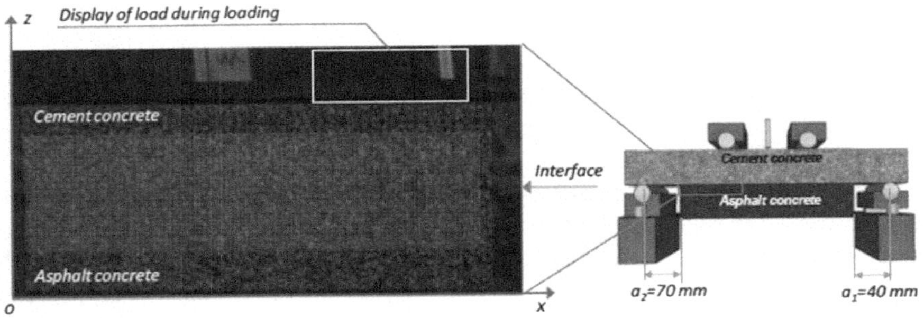


**Fig. 4.26** Zone of Interest (ZOI) on the reference image [16]

Eberl et al. [26] using Matlab. By using bilinear interpolation between pixels, the sub pixel displacement values are obtained. The algorithm applies a normalized cross-correlation function on two images before and after displacement and identifies the peak. The displacement is assumed to be at the location of the peak of the normalized cross-correlation function. Figure 4.26 illustrates the ZOI for the image correlations. The strains and displacements are computed by considering a square grid of pixels in each of $x$ and $z$ directions. A resolution of 16.9 ∓m is obtained for displacement by using 680 pixels on a flat field of 115 mm in length. This resolution is limited in function of the constraints of the ZOI and the capacity of the camera lens. The accuracy of the crack opening displacement (COD) and crack sliding displacement (CSD) obtained from the DIC is about 0.1 pixel.

The dimensions of the specimens are with non-symmetrical geometry (a1 = 40 mm, a2 = 70 mm), 20 mm distance between supports (L), 60 mm thickness of each layer (e1, e2), 100 mm width (b), (Fig. 4.27). The cement concrete material (denoted BC6) is made with a CEM I 52.5R cement (NF EN 206-1) and aggregate size 0/11. Its Young's modulus is 35,000 MPa. The asphalt material (denoted BBSG 0/10; NF EN 13108-1) is a conventional French bituminous mix made using aggregate size 0/10 and 5% of the 35/50 bitumen grade with 9.59% of air void. The equivalent modulus value of the asphalt material equals approximately 2000 MPa at 20 °C [17, 18]. The test is conducted at ambient temperature (around 20 °C). Loading is monotonic with a speed of 0.70 mm/min controlled by a LVDT sensor. By using the image correlation, the displacements in the $x$ and $y$ directions are determined. Figure 4.27 shows the evolution of vertical displacement $uz$ during different stages of loading. This figure illustrates the crack opening at the interface as the test progresses.

Using this profile of displacement, the COD and crack length are determined (Fig. 4.28a). Figure 4.28b shows the pictorial description of strain distribution and vertical displacements obtained from the DIC method and the crack pattern at the interface during the test. It is seen that the vertical displacement profile qualitatively agrees with the crack pattern observed on the specimen.

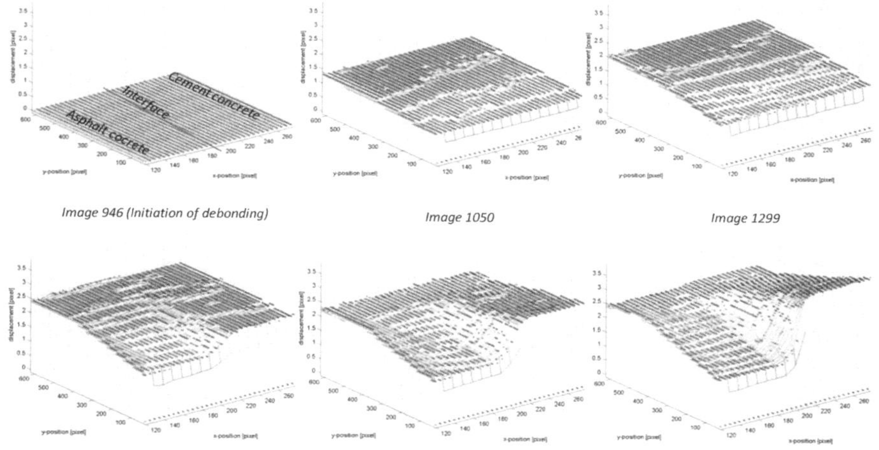


**Fig. 4.27** Evolution of vertical displacement *uz* during different stages of loading [16]

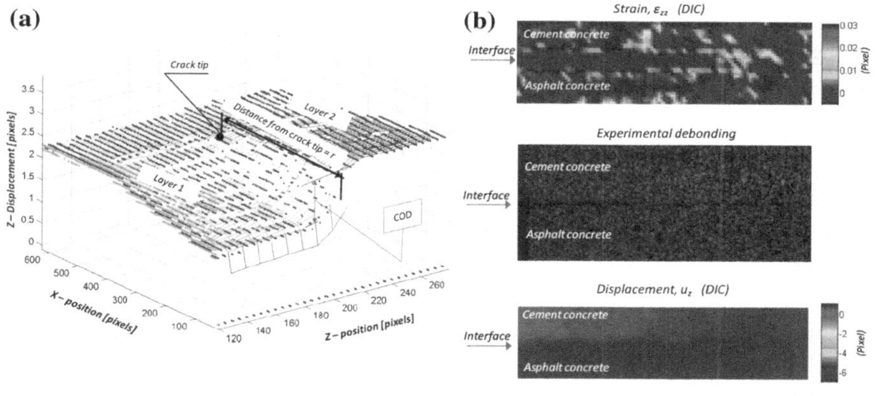


**Fig. 4.28** **a** Determination of crack opening displacement and crack length; **b** Results of strain, crack pattern and displacement [16]

Figure 4.29 illustrates the values of COD and CSD of the point located on the interface during different stages of loading. It is seen that there is a departure of interface crack after the peak load.

### 4.5.3 Benefits and Limitations

The image correlation method allows for the analysis of the surface displacement, surface strains, crack opening displacement, crack sliding displacement, and crack

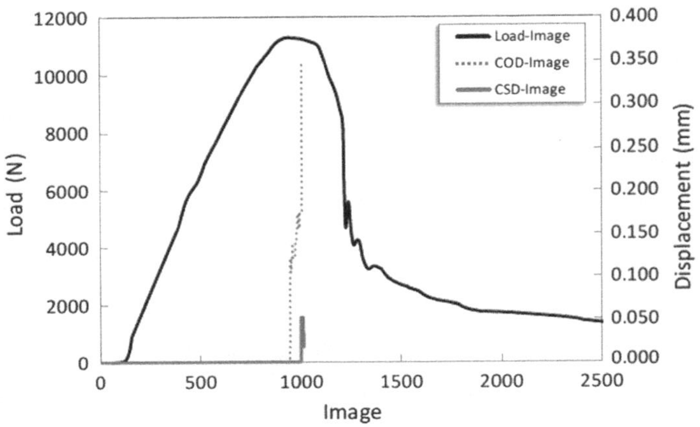


**Fig. 4.29** Load, COD, CSD during different stages of loading [16]

length which are complicated to measure for a mixed mode test. Using DIC technique coupled with Dundurs method, the mode I and mode II stress intensity factors and strain energy release are experimentally determined and compared successfully with M4-5n results. However, for such a mixed mode test that first propagate the crack in an instable way [17] the frame rate of camera used in this study (63 fps) is limited to visualize the exact initiation phase of interface crack [18].

## 4.6 An Optical Strain Measurement System for Asphalt Mixtures

Knowledge of the conditions governing the initiation and propagation of cracks in Hot Mix Asphalt (HMA) mixtures is a prerequisite for a comprehensive understanding of HMA cracking mechanisms. Traditional strain measurement sensors have proved to be not completely adequate since they do not provide point wise measurements, thus not pinpointing the location of crack initiation and not accounting for non-uniform strain distributions. This section presents a DIC system for non-contact and full strain field measurements, conceived for the purpose of investigating the cracking behavior of HMA mixtures. The whole system was developed to account for the special nature of typical HMA testing configurations. An image matching technique (Least Squares Matching) was employed for providing matches with sub-pixel accuracy. The performance of the method was investigated by several tests. The DIC system was shown to overcome the shortcomings of traditional on-specimen strain measurement devices achieving satisfactory accuracy compared to strain gauges.

**Table 4.6** Classification grid of the optical strain measurement system

| Technique/device's short name | Optical Strain Measurement System—OSMS | |
|---|---|---|
| Measurement principle(1) | (4) | |
| Categories | Yes | No |
| Field system | | X |
| Laboratory system | X | |
| Specific setup dependent measurement system | | X |
| Measurement system associable with different test setup | X | |
| Crack initiation | X | |
| Crack propagation | X | |
| Crack distribution | X | |
| Damage characterization | X | |
| Short description of system(2) (C), (E), (H), (I, monotonic), (P) | | |

According to the proposed classified grid (Table 4.1), this section concerns Table 4.6.

## 4.6.1 Experimental Setup

A digital camera Basler AF 101 (resolution 1300 × 1030, focal length 8 mm, pixel size 6.7 μm, 12 fps@max resolution) is currently employed. The optics adopted at maximum magnification allow 30 μm per pixel resolution. The camera, which is directly connected with a personal computer, is located on a support inside a climatic chamber of a material testing system, focusing up to 3.5 cm from the area of most interest of the specimen (i.e., the most stressed area). A lighting system, created for the purpose of providing adequate illumination of the specimen inside the climatic chamber, is composed of 4 white lights, which can be oriented according to the specimen shape and/or dimensions. Two horizontal guide rails allow horizontal movements (Fig. 4.30).

### 4.6.1.1 Specimen Preparation

The specimen requires a preliminary surface treatment to ensure a successful imaging acquisition and the subsequent application of the DIC method. The technique involves measurements of the greyscale level at each pixel location of the image. Thus, very well contrasted images are fundamental for achieving high measurement accuracy. The surface treatment adopted for asphalt mixtures consists of the application of thin white paint overprinted by a speckle pattern of black, resulting in a homogeneous randomly oriented texture. Care must be taken in

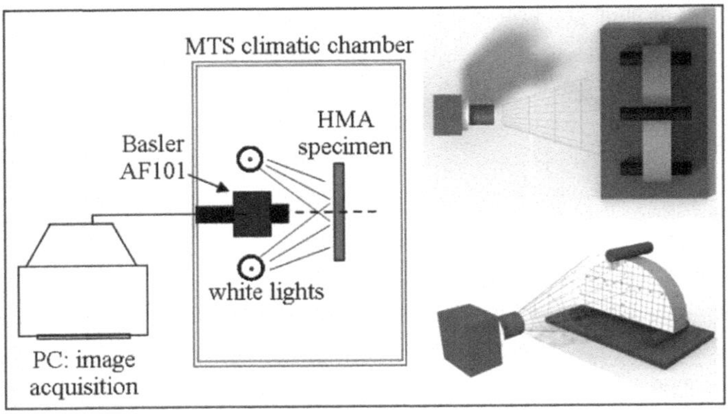


**Fig. 4.30** Experimental setup

ensuring thin enough layer thickness to avoid tracking the deformation of the paint film rather than the specimen deformation. The paint adopted was a water-based paint, which is lightly absorbed by the asphalt mixture so as not to affect the estimation of the real cracking behavior of the specimen.

## *4.6.2 Theoretical Principles*

The feature tracking is achieved using Area Based Matching (ABM), a long established technique for the extraction of image correspondences based on similarities between grey values (g.v.). In ABM, each image point to be matched is the center of a small window of pixels (template) in an undeformed reference image (master image), which usually corresponds to the first image in the sequence of frames. The grey values of the template are statistically compared with those of an equally sized window of pixels (patch) in a deformed search image (slave image), which corresponds to another image in the sequence of frames.

Two different approaches can be adopted for the evaluation of the similarities between the patch and template's grey values: cross-correlation and least squares matching. The former uses a proper correlation function to determine a coefficient, which establishes whether a point in the template corresponds to another in the patch. The latter is based on an iterative least-squares resolution algorithm. The maximum correspondence between the grey values of the two windows is established if the patch and template are exactly the same. However, grey value correspondences always differ since the patch is affected by both radiometric and geometric differences. Radiometric differences are due to sensor response, illumination changes, and object reflective changes. Geometric differences arise from object movements (translations and rotations), object deformation, and perspective effects (camera location and object shape).

Cross-correlation is a method of choice in computer science, since it is faster to implement and results in a more efficient computational performance. However cross-correlation tracks the interested point only by shifting the template pixel by pixel along the horizontal and/or vertical axes, not accounting for rotations. Thus it works well only if geometric and radiometric distortions of the patches are kept at the minimum.

During Hot Mix Asphalt fracture tests, rotations and/or scale changes between two images always occur, thus the cross-correlation approach appears inadequate for this kind of analysis. In contrast, the Least Square Matching technique uses a more complete functional model, providing matches with sub pixel accuracy while accounting for both translation and rotations.

#### 4.6.2.1 Data Extraction

The image correlation technique tracks a dense set of features along the acquired image sequence using an approximate value of the patch position estimated at the previous frame. The displacement is computed as the difference of the feature location between each image frame in sequence and the reference one (which is fixed). Displacement estimation depends mostly on two concurrent factors: the partial or total rigid displacement of the specimen and the displacements occurring in the image system, e.g. camera vibrations. These effects must be minimized to recover only strain field information. Thus, in each sequence, a point is selected as the origin of a reference system attached to the specimen. The origin is selected as a distinct point feature, with high and consistent correlation values between epochs and located on a specimen area not significantly stressed. The region of interest (ROI) is then meshed regularly in both (x, y) directions ($\Delta x$ and $\Delta y$ mesh steps respectively). The displacements are computed at the nodes of the regular grid by linear interpolation of the displacement values estimated by Least Squares Matching over the template bordering the same nodes. Strains are finally estimated by using a finite difference scheme at the double ends of the template window.

The accuracy achievable in the strain tensor components depends on the Least Squares Matching accuracy. Let us assume that displacement measurements are independent. This is true when the mesh step is larger than the template size and the accuracy is uniform ($\sigma_u = \sigma_v$) such as in the case of assumed isotropic specimen texture. Assuming no uncertainty in the determination of the mesh step $\Delta x$, and according an error propagation, the mesh step directly affects the uncertainty in the strain estimation: a small one provides a pinpoint strain description but with less accuracy while a larger one gives an average value of the strain with better accuracy. Thus, considering that both $\sigma_u$ and $\sigma_v$ depend on the accuracy in the image space, the pixel size in the object space directly defines the achievable accuracy.

### 4.6.3 Verification of the Method Accuracy

The performance of the method was investigated by means of several tests in order to assess its accuracy for both displacement and strain measurements. Comparisons between the image method strain measurements and strain gauge ones were also observed.

#### 4.6.3.1 Accuracy in Displacement Measurements

The accuracy in the displacement measurements is affected by both LSM performance and template dimension. Using a small template, the assumption of strain isotropy and affine deformations becomes more realistic, but the measurement redundancy drastically decreases (in terms of number of pixels considered, number of equations involved, and the likelihood of finding well-contrasted pixels). On the other hand, the use of large templates improves the redundancy but adversely affects initial assumptions of affine deformations. Indeed, the use of larger templates makes the approximation of local displacements by affine transformations harder; the implementation of a non-linear shape model for patch transformation is advisable if large deformations are expected [27, 28]. The displacement measurements' accuracy was assessed by performing two different kinds of tests. The first set of tests was performed on synthetic images, re-sampling the original one with different transformations (translation, rotation, non-isotropic scaling and shear): mean accuracies of 0.005–0.007 pixel in patch position were obtained. The second set of tests was performed using a micrometric 10 μm slide. This kind of test was selected to account only for translations, thus neglecting the possible development of local deformations. An enlarged picture of the specimen surface was glued on the micrometer slide and imaged at each 100 μm shift. An appropriate distance between the camera and the target texture was chosen to achieve an object space accuracy of the slide of 1/1000 pixel in image space. The theoretical $\varepsilon_{xx}$ accuracies for locally homogeneous strain fields according to the ROI dimension and the mesh step size are listed in Table 4.7 assuming a sensor with 1300 × 1000 pixels and $\sigma_u$ = 1/100 pixel. For instance, for a 4 cm wide ROI and a 1300 pixel resolution in x direction, the pixel size in object space is 30 μm. Least Square Matching accuracy between 1/100 pixel and 1/150 pixel was achieved corresponding to value between

**Table 4.7** Theoretical image system accuracies

| ROI | 1 cm | 3 cm | 5 cm | 10 cm |
|---|---|---|---|---|
| Displ Accuracy (μm) | 0.08 | 0.23 | 0.38 | 0.77 |
| ε grid 1 mm | 0.011% | 0.033% | 0.054% | 0.109% |
| ε grid 2 mm | 0.005% | 0.016% | 0.027% | 0.054% |
| ε grid 5 mm | – | 0.007% | 0.011% | 0.022% |

0.30 and 0.20 μm in image space. Assuming $\Delta x = \Delta y = 1$ mm (1200 measurement nodes), the σ accuracy of the tensor components is determined as shown in Eq. (4.8):

$$\begin{aligned} \varepsilon_{xx}: \sigma_\varepsilon &= 0.03\text{–}0.04\% \\ \varepsilon_{xy}: \sigma_\varepsilon &= 0.02\text{–}0.03\% \end{aligned} \tag{4.8}$$

#### 4.6.3.2 Accuracy in Strain Measurements

The accuracy achievable in the strain measurements rests on the distribution of the measurement points on the specimen surface. If the nodes are too close to each other, even a small inaccuracy in the displacement measurements may lead to large errors in the strain values as can be observed in Eq. (4.7). On the other hand, if the reciprocal nodes are too distant, the local effects can be lost due to the interpolation process. The accuracy in the strain measurements was assessed performing a test in which results obtained from the image method were compared to results obtained with on-specimen strain gauge measurements. A strain gauge, arranged in a quarter Wheatstone bridge, was mounted on an aluminum bar, machined to a dog-bone shape for an uniaxial tensile test. The strain gauge was painted to obtain a dotted pattern and imaged during the testing. Strain gauge and digital image measurements were recorded. The strain values obtained with the DIC method were averaged over the strain gauge area; their differences were computed during the linear elastic portion of the stress-strain response as well as during the initial yielding stage. The ROI was about 1 cm wide and the displacement measurement grid was 0.5 mm spaced. The root mean square of the differences between the strain gauge measurement and the DIC mean strain value over the strain gauge length was about 0.003%. The result agrees with the variance propagation law shown in Eq. (4.9):

$$\sigma_{mean} = \frac{\sigma_\varepsilon}{\sqrt{n}} \tag{4.9}$$

where $\sigma_{mean}$ is the error in the mean value and n is the number of measurement points. Indeed, 100 measurement points uniformly distributed over the strain gauge area (with expected local accuracy $\sigma_\varepsilon = 0.02 \div 0.03\%$) were used to obtain the mean strain value at each epoch.

### *4.6.4 Tests on Asphalt Mixture Specimens*

The DIC method was employed to investigate the cracking mechanism of HMA mixtures from three of the most common monotonically increasing load strength test configurations: the Superpave Indirect Tensile Test (IDT), the Semi-Circular

Bending Test (SCB) and the Three Point Bending Beam Test (3PB). Each test was performed at 10 °C; the load was applied on a centered loading plate (IDT) or ring (SCB, 3PB) with a displacement-controlled system moving at 0.08 mm/s speed. The measurement grid spacing is set equal to 1 mm; the pattern size is in the range between 30 and 40 pixels, dot size is about 3 pixels. The origin point of the reference system attached to the specimen is selected according to the highest values of correlation between epochs and in an area not significantly stressed during all test long. For each test setup one strain gauge was mounted on one face of the specimen. The same face was then treated to obtain the required texture for digital image analysis. The test was imaged and processed; then the mean strain value of the length covered by the strain gauge was estimated by the digital image system interpolating all the strain values of the grid points located at the strain gauge length. Strains registered by the strain gauge were finally compared with those estimated with the DIC System, as shown in Fig. 4.31 for instance for the Indirect Tensile Test.

Computing the Root Mean Square (RMS) of the differences between measured and estimated strain values, accuracies of 0.015% for the IDT test, 0.034% for the SCB test, and 0.017% for the 3 PB were obtained. These results agree well with the mean error previously estimated ($\sim$0.03%). The DIC System accuracy matches that obtained with strain gauges, while allowing a dense description of the field of interest where the cracking is developing. Results related to first crack detection and to observations of crack propagations for both unmodified and polymer modified asphalt mixtures can be found in [29–31].

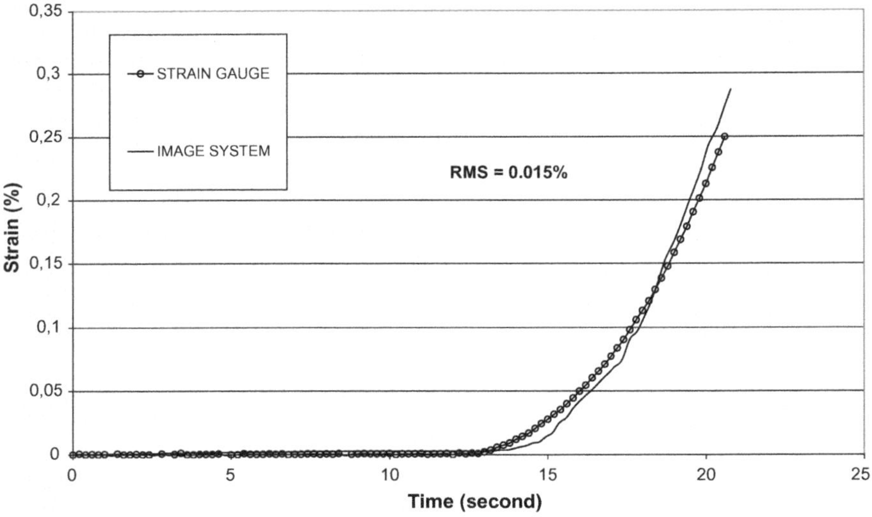


Fig. 4.31 Comparison between image correlation and strain gauge measurements in the HMA indirect tensile testing [29]

### *4.6.5 Description of Software Tools*

The visual graphic interface of the image correlation software was conceived on purpose to pick the point of interest, choosing which information is requested (horizontal/vertical/shear strains, horizontal/vertical displacements). The strain values are then exported to plot stress-strain responses at the specific chosen point. This tool is very convenient for fracture mechanics analysis since it allows for strain response estimation at the accurate point in which a crack initiates. Figure 4.32 shows the measured horizontal strain map for the mixture during the IDT, from micro-crack initiation to major crack opening; high strain values develop within the whole center area of the IDT specimen.

From full field strain maps, it can be observed that the tensile strains are greatly localized in the area in which a crack initiates. The full field strain maps also allow for the observation of tensile strain development around aggregates, while no strains are registered where a coarse aggregate exists. The comparison between horizontal strain map and the correspondent specimen image emphasizes the capability of the DIC method to catch strain values and to visualize crack patterns at crack initiation and propagation. Full field strain estimations allow for a comprehensive analysis of HMA cracking behavior since strains vary within the overall ROI. Traditionally, for wide area analysis, users mount multiple sensors with different lengths and positions, while in the image correlation system a single camera functions as a single sensor from which information from all the grabbed areas can be obtained.

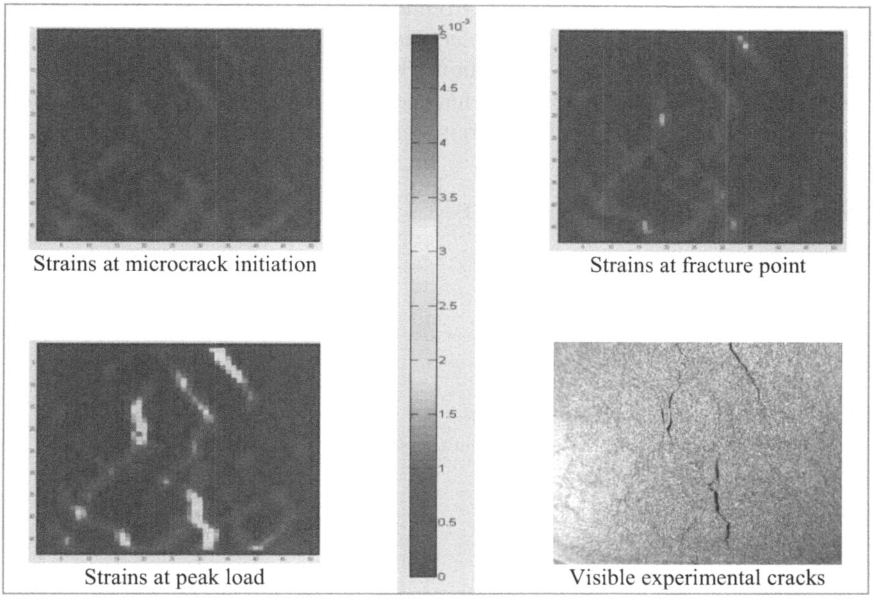


**Fig. 4.32** Full field strain maps during the IDT test

## 4.7 The Colibri Device

A shock applied to a road structure generates different physical phenomena according to the characteristic of the shock (frequency, intensity). When a Falling Weight Deflectometer (FWD) applies the load, static surface deflection of the roadway is measured. When the impact echo method is applied, compression waves propagate through the structure and the surface response is measured at a higher frequency. With an intermediate solicitation, the vibratory response on surface road can be measured and the Frequency Response Function (FRF) can be evaluated. This function is sensitive to the presence of internal damage such as voids or cracks. Sensitive frequency band depends on the investigated structure and also the type and the location of damages. To interpret the results, it is necessary to standardize them using a reference function. A statistical process is usually used to define this reference function, which is representative of the healthy structure. Colibri apparatus applied the method to detect and locate defects on test sites. According to the proposed classified grid (Table 4.1), this section concerns Table 4.8.

### *4.7.1 Colibri Device and Test*

The Colibri device (Fig. 4.33) is a mobile, automated measurement device, which uses a hammer to apply a dynamic solicitation (shock) to the road surface and measures the vertical surface response (vertical acceleration) close to the solicitation with an accelerometer placed at 10 cm from the impact. It analyzes vibration modes that are sensitive to the presence of defect such as delamination or vertical cracks. Colibri is used all along a roadway section. Then, the process aims at comparing the FRF modulus [32] by defining a reference FRF representative of the healthy structure and at identifying FRF, which is significantly different from this reference

**Table 4.8** Classification grid of the Colibri Device

| Technique/device's short name | Colibri device | |
|---|---|---|
| Measurement principle[(1)] | (5) | |
| Categories | Yes | No |
| Field system | x | |
| Laboratory system | | x |
| Specific setup dependent measurement system | x | |
| Measurement system associable with different test setup | | x |
| Crack initiation | | x |
| Crack propagation | | x |
| Crack distribution | x | |
| Damage characterization | x | |
| Short description of system[(2)] (A), (B), (D), (H), (P) | | |

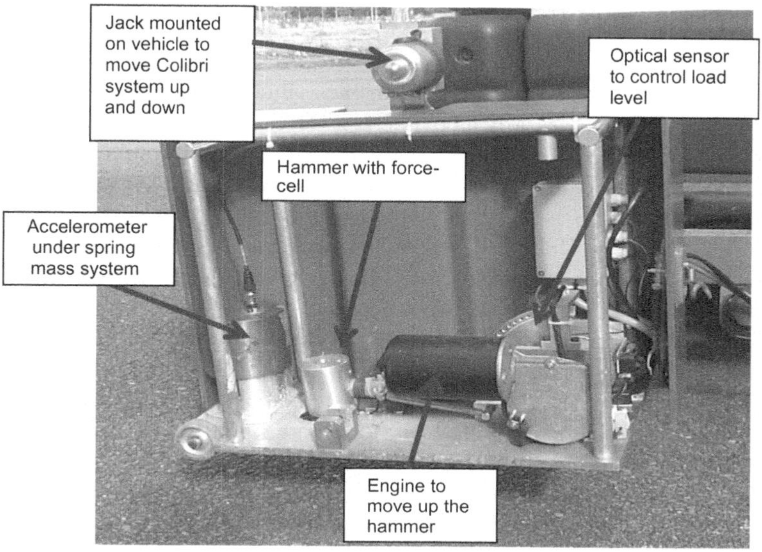


**Fig. 4.33** Colibri apparatus

function. The variations observed are then representative of the presence of damages that lead to a softer structure.

The Colibri apparatus is a complete automated system for dynamic investigation of pavements. Colibri is used to investigate the experimental pavement section. The Colibri device can be mounted on a vehicle or it can be moved manually to go from one measurement point to the next. It includes:

- A hammer with a force-cell to measure the shock application;
- An accelerometer placed at 0.10 m from the impact to measure the surface response and held by a spring mass system;
- An optical sensor to control the level of the shock.
- An engine to move up the hammer and produce the shock
- An electric jack enabling the lowering and raising of the system on the road surface as well as the placement of the sensors when Colibri is mounted on a vehicle;
- Electronics and data acquisition systems and a computer to manage the measurement sequences and to store the data.

Usually, measurements are recorded along longitudinal or transversal profiles. Measurements are made at regularly spaced points. The distance between 2 measurement points is adapted to the length of the profile, and to the size of the defect, which has to be located. Usually, 3 impacts and the acceleration are recorded at each individual measurement point.

### 4.7.2 Theoretical Basis

For a healthy structure, Colibri shock generates vibrations of the whole pavement. When a structure includes a defect (interface or crack), low frequency vibration modes appear which correspond to the vibration of a part of the structure (above the delamination or close to the crack). The inertance modulus estimated for the delaminated structure is higher than the one of the healthy structure. It increases at each eigenfrequency. Thus, a difference of inertance is observed in a sensitive frequency band. This band and particularly the lowest frequency, depends on the characteristics of the defect (extension, depth, nature).

To apply the technic to pavement, the methodology consists of collecting the inertance function all along a roadway section [33]. Then, the process aims at comparing the FRF modulus by defining a reference FRF representative of the healthy structure and at identifying FRF, which are significantly different from this reference function. It has to be noted that the reference function is related to the investigated roadway. So, we suppose that measurements are recorded on a homogeneous structure (materials and layer thicknesses). The variations observed are then representative of the presence of damages, which lead to a softer structure. For each measurement point, $i$, a spectral analysis of a series of tests leads to calculation of the inertance, $A(f,i)$, and the coherence function, $\gamma(f,i)$, between the pavement response and the applied solicitation. These functions depend on the frequency, $f$, and on the measurement point. The coherence function estimates the dependence of the output signal compared to the input signal. It is a real value ranging between 0 (no dependence) and 1 (full dependence). A minimum threshold of coherence (usually 0.8) is chosen to validate or not the calculation of the inertance. This threshold can be adapted according to the studies. For each frequency and each measurement point, FRF is validated if the coherence value is higher than this threshold. Thereafter, the analysis is restricted to the population of validated measurements. On a homogeneous zone, data are then processed in 2 steps:

- Estimation of a reference function representative of the healthy structure;
- Calculation of a normalized damage.

To estimate the reference function modulus, it is assumed that a part of the tests is carried out on a healthy zone. This can be done voluntarily by investigating an un-trafficked zone such as an emergency lane. In practice, we usually consider the set of modulus, $|A(f_k,i)|$, measured at a fixed frequency, $f_k$. The reference value at this frequency, $|A_{réf}(f_k)|$, is defined as a percentile of selected population. We usually adopt the percentile 20, which allows obtaining a low value representative of the healthy structure and eliminating abnormal measurements. The set of reference values are used to build the reference transfer function representative of the healthy structure, $|A_{réf}(f)|$. Inertance modulus increases with frequency. We normalize the FRF modulus $|A(f_k,i)|$ using the modulus of the reference function. For each frequency and each measurement point, the damage, $D(f_k,i)$ is calculated according to Eq. (4.10).

$$D(f_k, i) = \begin{cases} 0 & \text{si} \quad |A(f_k, i)| < |A_{réf}(f_k)| \\ 1 - \frac{|A_{réf}(f_k)|}{|A(f_k, i)|} \end{cases} \tag{4.10}$$

This value varies between 0 and 1. The matrix $D$ represents the damage on the road section for the different frequencies. It can be presented as a "damage mapping" where:

- The X-coordinate is the abscissa along the road section;
- The Y-coordinate is the frequency band;
- Colors represent the level of damage.

### 4.7.3 Obtainable Results

Using the damage matrix $D$ (13), Colibri results are plotted on a map. The map allows one to detect horizontal defects of layer interface (extended zone) or inside vertical cracks (local effect). Damage is then located, and the sensitive frequencies give information about their nature. A damage index can be calculated. The index is useful to survey the evolution of damages [42] or to compare different design of rehabilitation works.

At Ifsttar Nantes, a test section on the carousel location has been investigated using Colibri prototype. Its structure includes 4 layers: a 0.06 m thick AC wearing course, two 0.10 m thick BBGM layers, and a 0.20 m thick layer of unbound granular mixtures (UGM). The structure includes two pieces of Kraft paper (0.5 × 0.5 m, which represent interface defects. One is placed at the AC/BBGM interface; the second is located at the interface between the 2 BBGM layers. Figure 4.34 presents the image applying the dynamic investigation method. The theoretical localization of the defects is indicated on the graph. Frequency band between 3000 and 5500 Hz is sensitive to the interface defects. The first one (nearer to surface), is easier to locate because the stiffness contrast between the healthy structure and the substructure located above the defect is high. Its localization is good. The deeper defect is less readable because the structure under the defect is soft.

Vertical cracks represent variations of limit conditions, which reduce the eigenfrequencies. They also can be detected with dynamic investigation. We used the techniques during an experiment, which tests different technologies to limit reflective cracking. A concrete structure was sawed. A different anti reflective cracking process was applied before covering the structure with a 0.06 m thick AS wearing course. Pavement fatigue machines applied one million loads above each crack. Dynamic investigation was conducted at the beginning and at the end of the experiment. It extended 0.5 m on each side of cracks in 3 longitudinal profiles: in (noted north and south and between (noted center) the wheelpath. Intervals between consecutive measurements were only 0.02 m at short crack distance (<0.20 cm) and 0.05 m at higher distance. Pavement sounding was held over one half-day under

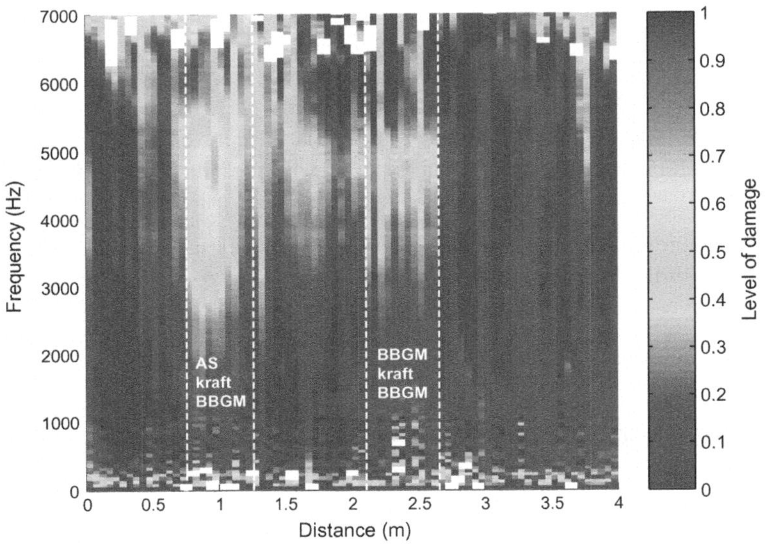


**Fig. 4.34** Mapping from Colibri measurements

homogeneous and constant operating conditions. An unloaded zone has been sounding to evaluate the influence of climatic conditions on measurements. Figure 4.35 compares mapping deduced from Colibri investigation before and after fatigue testing. Crack is located at the null abscissa. Its influence is readable before and after fatigue testing in a wide frequency range. Before fatigue testing, the crack influences results at a distance lower than 0.2 m. During fatigue testing, the crack grows up through the anti-reflective system and part of the wearing course. Thus, the damage image after the fatigue testing is influenced in a longer zone.

We calculated a damage index for each measurement point. It was equal to the damage average in the frequency range 1500–5000 Hz. Comparing the different profiles, is easier using such index. Figure 4.36 presents the index variation along the 6 profiles for the reference construction technique (Fig. 4.36a) and an alternative construction technique (Fig. 4.36b). Before fatigue testing, small differences exist in the longitudinal profiles which correspond to transversal dispersion of the test site. Damage level increases up to 0.5 close to the crack. It influences dynamic results only at a short distance (<0.2 m). After fatigue testing, the damage index increased. It is different in and between the wheel-path. For the reference technique, the index grows up to 1 in the wheel-path. Between the wheel-path, it increases only up to 0.7. The index decrease to a level equivalent to the one observed before fatigue testing at a distance of 0.3–0.4 m. For the alternative technique, the index increased up to 0.9 between the wheel-path near the crack. In the wheel-path, the index is always higher than the initial level.

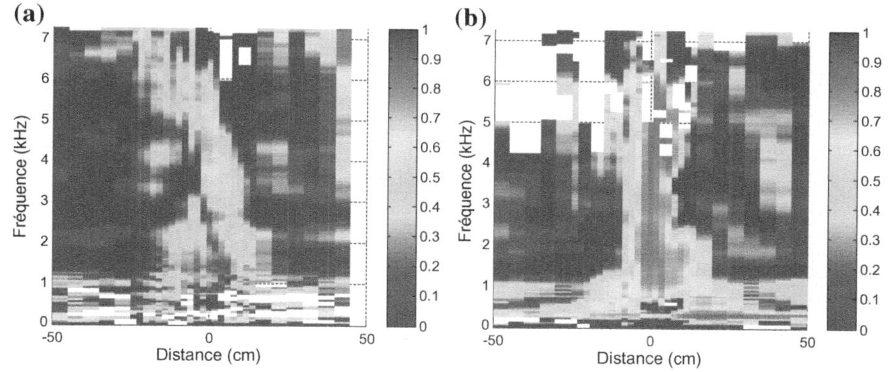


Fig. 4.35 Images of estimated damage over a crack between the wheel-path **a** before fatigue testing; **b** after fatigue testing

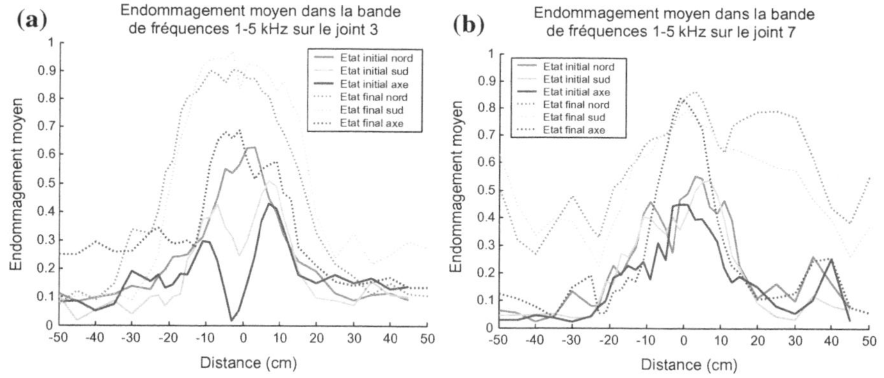


Fig. 4.36 Average damage in the frequency range 1500–5000 Hz estimated by Colibri on the various profiles before (initial) and after (final) fatigue testing; **a** Reference technique; **b** alternative technique

## 4.8 Deflection Devices

This section presents a set of diagnostic methods for flexible pavement based on the surface deflections measurements. These methods (Falling Weight Deflectometer and Rolling apparatus) are able to give an analysis of the bearing capacity of the pavement structure or an estimation of structural defects like cracks or delaminations. According to the proposed classified grid (Table 4.1), this section concerns Table 4.9.

Deflection measurement is a nondestructive method to estimate pavement material properties. It can be used to calculate a structural index of bearing capacity. This index is integrated into the pavement management system. At the project level,

**Table 4.9** Classification grid of the deflection basin measurement devices

| Technique/device's short name | Deflection basin measurement devices | |
|---|---|---|
| Measurement principle[(1)] | (7) | |
| Categories | Yes | No |
| Field system | X | |
| Laboratory system | X (if full scale tests are considered) | |
| Specific setup dependent measurement system | | X |
| Measurement system associable with different test setup | X | |
| Crack initiation | | X |
| Crack propagation | | X |
| Crack distribution | | X |
| Damage characterization | X (requires assumption regarding the existing damage type) | |
| Short description of system[(2)] (B), (E), (H) (I,c), (M), (P) | | |

deflection measurement can also be used to predict the residual pavement service life and to evaluate the need of reinforcement or rehabilitation.

The deflection basin is also sensitive to inside damages such as cracks or delamination. Variation of deflection basin can be helpful to detect and survey the evolution of damage during in situ experiments.

### 4.8.1 Theoretical Basis and Deflection Measurement Apparatus

Deflection is measured on the road surface when a load is applied. This load can be a rolling load (heavy truck) or a falling load (which simulate a heavy truck run). Roadway material behaviour is usually elastic linear or visco-elastic. Static or dynamic models are used to interpret deflection measurement. Multilayered continua model exists in elastic [34] or visco elastic [35] case. In a non-continuous model, finite elements or central finite difference has to be used. In the dynamic case, the motion of the forced regime of a structure submitted to loading is govern by the equation: $M\,\ddot{u}(t) + C\,\dot{u}(t) + Ku(t) = F(t)$ where M, C and K are the $N \times N$ global mass, damping and stiffness matrices respectively, while quantities u(t) and F (t) designate the generalised vector of displacement and the loading respectively.

Several devices are used to measure deflection basins. Punctual sensors can be used such as inclinometer, accelerometer, or geophone put on the road surface, or fit inside road during construction. In this case, a rolling load is usually used to apply the load. Specific apparatus have been developed to measure deflection basins. Falling Weight Deflectometer (FWD), Lacroix Deflectograph, curviamètre, or Benkelman Beam are the most commonly used. The load application is a main difference between these devices. FWD applies an impulse dynamic load and measures deflection at different distances from this punctual load while other systems use a rolling load and measure the deflection bowl in a fixed point.

The load level is usually 50 kN, which corresponds to a heavy vehicle. For a rolling device, the load is applied at the speed of the vehicle. The FWD load reproduces the dynamic impulse load similar in magnitude and duration to a moving truck at 65–85 km/h. The maximum deflection is a common indicator that is measured with the different apparatus. Other deduced parameters cannot be compared. For example, the radius of curvature close to the maximum deflection can be estimated with a Benkelman beam, a Deflectograph, or a curviameter. It cannot be compared with the difference in deflection between the central sensors and those at different distances form the load.

## 4.8.2 Obtainable Results

*FWD Results* [36]

The FWD method is used for a feasibility study for detecting defects in a pavement structure. The numerical simulation is implemented in a finite element program. The model (Fig. 4.37) takes into account the dynamic aspect of the FWD loading and allows a more adapted simulation of the actual solicitations due to the traffic. The multilayered pavement structure modeling was implemented on a 3D finite element model. Absorbing boundaries were used in the model in order to avoid, as best as possible, the waves reflection on the mesh edge. The modeling of the interface flaw can be done considering the interface state as slippery or by creating a thin interlayer with a weak modulus (50 MPa: modulus of the sand). Similarly, crack modeling has been performed by a thin trench.

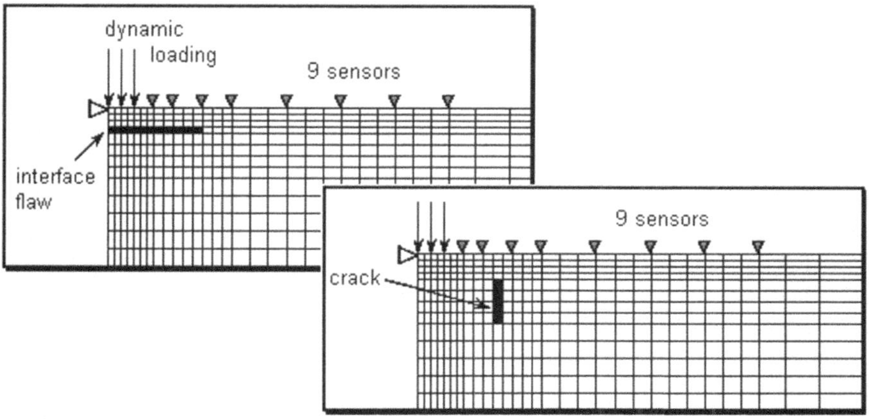


**Fig. 4.37** Modeling of interface flaw and crack

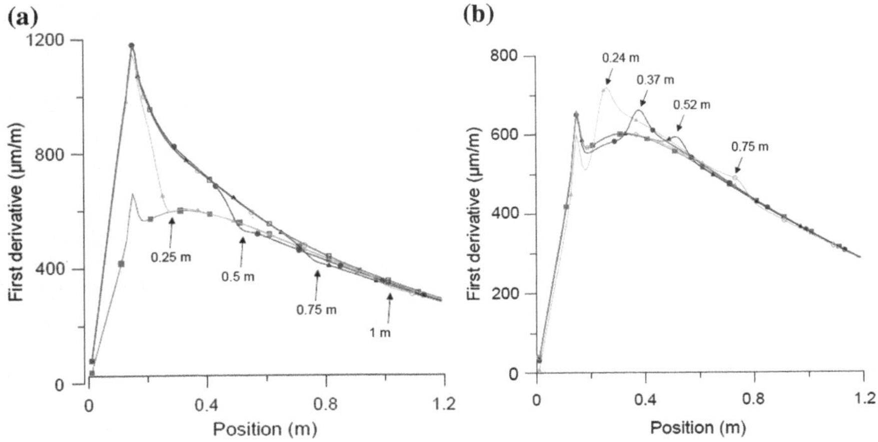


**Fig. 4.38** Numerical results for **a** interface flaws and **b** cracks

Figure 4.38 introduces results for interface flaws and cracks for several positions. The effect of the presence of a crack on the pavement surface detection is minimal even if the crack is located near the loading area. The better exploitation of defection discontinuity on the pavement structure response caused by the presence of the crack is obtained by spatial derivatives. So, the first derivatives of the maximum deflection curve will highlight the crack impact on the pavement response under FWD Loading. It can be observed that the more the crack is close to the load area, the easier it is to locate. With the same method, the presence of delamination is clearly identified and its expense can be "numerically" determined.

#### 4.8.2.1 Rolling Apparatus Results [37]

A structure consists of a 0.08 m wearing course of asphalt concrete (AC), a 0.15 m road base of bituminous bound granular mixtures (BBGM) over a 0.21 m sub-base of cement bound granular mixtures (CBGM) laying on an infinite pavement foundation of unbound mixtures (UM). Numerical models use 3D finite elements to include a 1 mm vertical crack in the sub-base layer and optionally in the base layer. A standard dual-wheel was applied at different distances of this crack. Figure 4.39 shows the deflection basins. On the left, 6 cases are presented: the load center is located at 2, 1 and 0.1 m of the crack. The crack crosses only through the CBGM (solid line) or crosses through CBGM and BBGM (dotted line). On the right, the crack crosses the 2 layers; the graph compares the deflection basin for different distances between load and crack (2, 1 and 0.1 m). For a given load position, deflection basins are similar. On the right, we clearly distinguish variations of the deflection basin. An exhaustive study shows that deflection basin varies only when the distance between load and crack is less than 0.2 m. The maximum deflection

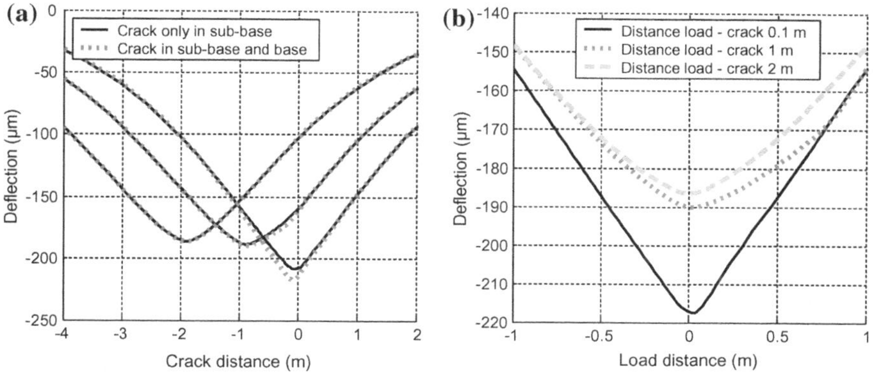


**Fig. 4.39** The deflection basins calculated close to the crack for different position of the load **a** according to the crack position; **b** according to the load position

increases only of about 10% while the radius of curvature increases by more than 50%. A continuous measurement of deflection basin is clearly necessary to detect inside damages such as cracks or delaminations. It is better to use the first or second derivative than the usual maximum deflection.

## *4.8.3 Examples*

### 4.8.3.1 FWD Measurement Over a Delamination

A pavement structure has been built on the pavement fatigue carousel in Nantes. It includes a 0.30 m granular sub-base, a 0.08 m bituminous base layer, and a 0.06 m bituminous wearing course. This test section includes several artificial delaminations, rectangular areas of different size using different techniques (sand, plastic film, absence of tack coat). They are located at the interface between the 2 bituminous layers. Several NDT methods (GPR, Dynamic investigation, Ultrasonic Pulse Echo and deflection) will be used to survey this section.

FWD tests were performed along a longitudinal profile every 0.10 m. The test was repeated 3 times. Figure 4.40 shows maximum deflection measured with FWD over a sand delamination which extends from abscissa 0 to abscissa 1.5 m. The maximum deflection increases from 450 µm before (and after) the delamination zone to about 700 µm over the delamination. Measurement over the zone including geotextile or without tack coat confirms that maximum deflection is sensitive to these artificial damages. However, shorter zones are more difficult to investigate because the sensitivity of maximum deflection is lower, obtaining an accurate position of FWD is always difficult, and some differences exists between design and construction.

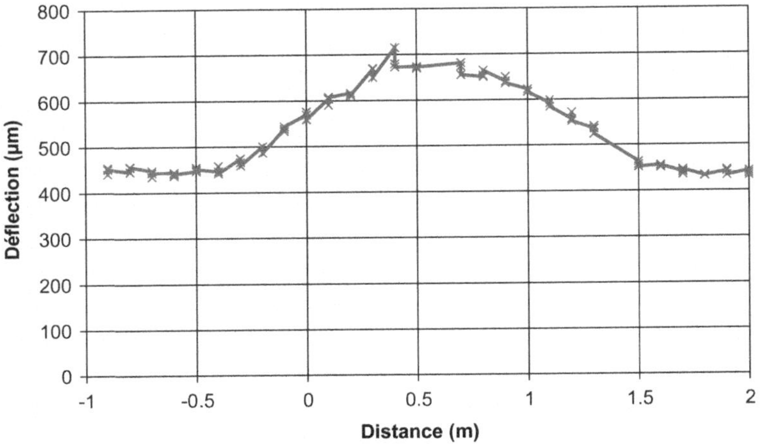


**Fig. 4.40** Maximum deflection measured with a FWD over a sand delamination

#### 4.8.3.2 Evaluation of Radius of Curvature with an Inclinometer Near a Crack

Some tests with an inclinometer (Fig. 4.41) were performed during a full scale experiment and gave interesting results.

In that experiment, a transverse crack was sawn in the base layer (bituminous material) and covered with a bituminous wearing course 5 cm thick. The radius of curvature was estimated with an inclinometer. Tests were performed every 10 cm along a 2 m long longitudinal profile. Figure 4.41 shows the measured values of the radius of curvature. The internal crack is located close to the abscissa 0. The level of measurement clearly decreases at the distance 0 and 0.10 m, which correspond to the influence zone of the internal crack.

### *4.8.4 Benefits and Limits of Deflection Measurement Tools*

Deflection measurements are usually used to estimate bearing capacity of in situ roadway (pavement material properties). They are also sensitive to the presence of cracks or delamination in pavement, therefore they can be used to detect or to survey specific zones including inside damages.

These apparatus are designed to operate on roadways in safe conditions. However, the deflection basin is not continuously measured. An interval of several meters exists between 2 consecutive measurements. Continuous measurement will be time consuming.

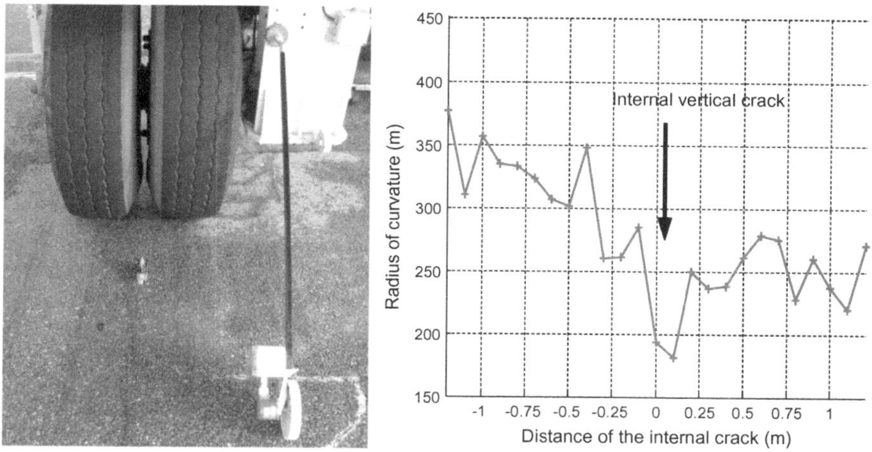


**Fig. 4.41** Maximum deflection measured with a FWD over a sand delamination

The interpretation of results requires the development of a new scheme, which takes into account the in site conditions which are not controlled and have to be measured (specifically material temperature).

## 4.9 Radar Systems

Over the past few years, radar systems have emerged as a powerful nondestructive testing (NDT) technique that features several major advantages, such as a high acquisition rate, global monitoring through quasi-continuous measurements, and production of a mapping of the subsurface. Radar systems take advantage of the penetration capability of electromagnetic (EM) waves to image the dielectric contrasts within the subsurface. According to the proposed classified grid (Table 4.1), this section concerns Table 4.10.

A radar system (Fig. 4.42) consists of a generator that generates EM waves towards the studied road layers and a receiver that receives the reflected part, via one or several antennas, placed above (horn antennas) or just on (dipole antennas) the road surface. The vertical structure of the pavement can then be deduced from radar data by means of echo detection and amplitude estimation. Conventional radar systems (Ground Penetrating Radar GPR) emit EM pulse waves. The actual limit of GPR is the resolution of the thickness measured that depends on the permittivity of the medium, and on the central frequency of the used antenna, that is currently limited to 2.5 GHz. An alternative technology called step-frequency radar (SFR) could emits shorter wavelength and allow thinner thickness determination. This system used a Vector Network Analyzer (VNA) that emits monochromatic waves in a wide band [i.e., 1.4–15 GHz], corresponding to the band pass of specific

**Table 4.10** Classification grid of the radar systems

| Technique/device's short name | Radar systems | |
|---|---|---|
| Measurement principle[(1)] | (5) | |
| Categories | Yes | No |
| Field system | X | |
| Laboratory system | X (if full scale tests are considered) | |
| Specific setup dependent measurement system | | X |
| Measurement system associable with different test setup | X | |
| Crack initiation | | X |
| Crack propagation | | X |
| Crack distribution | X | |
| Damage characterization | X | |
| Short description of system[(2)] (A), (C), (E), (O) | | |

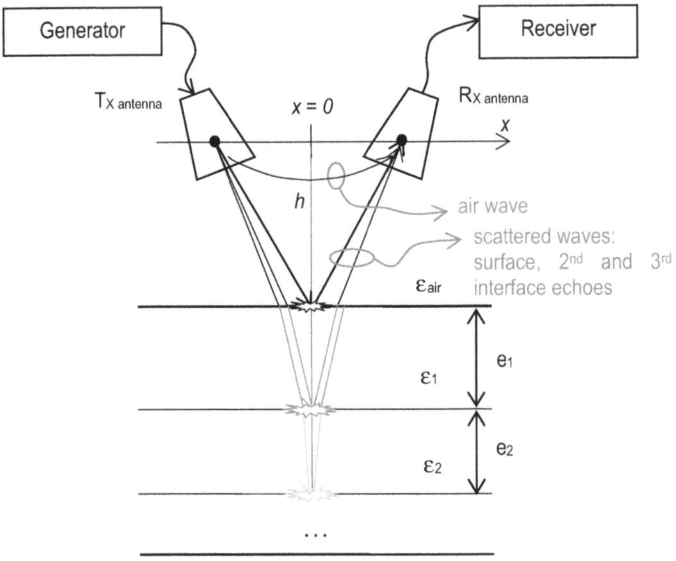


**Fig. 4.42** Radar configuration: monostatic mode (simplest) x = 0; bistatic mode (x > 0)

antennas as Exponential Tapered Slot Antennas (ETSA). Thus, the equivalent central frequency, after IFFT, increase (about 7.5 GHz) and the resolution of the system is improved. The analysis of the amplitude variation allows locating the interface variations that represents the horizontal crack or the debonded area. In theory, signals have to be disturbed by vertical crack.

### 4.9.1 Theoretical Basis

Dielectric contrast within the medium generates EM echoes, which propagate back to the receiver antenna. The travel time along the wave velocity is used to estimate the geometry of the subsurface structure. The layer thickness is given by the following Eq. (4.11):

$$e_i = v_i \Delta t_i / 2 \qquad (4.11)$$

where $\Delta t_i$ and $v_i$ are the travel time and the wave velocity within the interface i, respectively. The wave velocity depends on the relative permittivity $\varepsilon_i$ of the interface as follows in Eq. (4.12):

$$v_i = c_o / \sqrt{\varepsilon_i} \qquad (4.12)$$

The relative permittivity $\varepsilon_i$ can be determined from the echo amplitudes. Permittivity $\varepsilon_1$ can be retrieved from the ratio between the echo amplitude $A_1$ associated with the top surface echo amplitude and the amplitude of the Tx pulse $A_0$ according to Eq. (4.13) [38]:

$$\varepsilon_1 = \left( \frac{1 - A_1/A_0}{1 + A_1/A_0} \right)^2 \qquad (4.13)$$

### 4.9.2 Radar Systems Measurement

Two kinds of systems can be found in the literature, according to the antennas' position with regards to either the ground or each another. The use of air-launched antennas is more common for pavement surveys. Horn antennas or Vivaldi antennas are raised above ground at a certain height to guarantee far field conditions, i.e., plane waves impinging on the medium. Nevertheless, some other commercial devices afford ground-coupled antennas that improve the strength of the scattered echoes. Near-field data may be however more difficult to interpret compared with far field data. Besides, monostatic and bistatic radar configurations are associated with the spatial distance between the Tx and Rx antennas. The bistatic mode implies oblique propagation as the Tx and Rx antennas are moved apart from each other. This configuration may require more physics for the data interpretation especially when the bistatic mode is associated with ground-coupled antennas, i.e., it may include dispersive guided waves. For pavement survey, Tx and Rx antennas are set as close to each as possible, a monostatic mode. This configuration is better matched to the probing of horizontally stratified media and then also for pavement probing.

The radar data consist of three types. A vertical profile (or A-scan) is recorded at each location along the track. The data vectors collected along the track are gathered into a data matrix which is called the B-scan image (Figs. 4.43 and 4.44). C-scan is a 3D data matrix which is obtained from gathering the collection of parallel B-scans over the surface to probe.

For detection purposes, data processing mostly aims at picking the maximum of echoes on the vertical profile (signal amplitude) and at performing accurate time

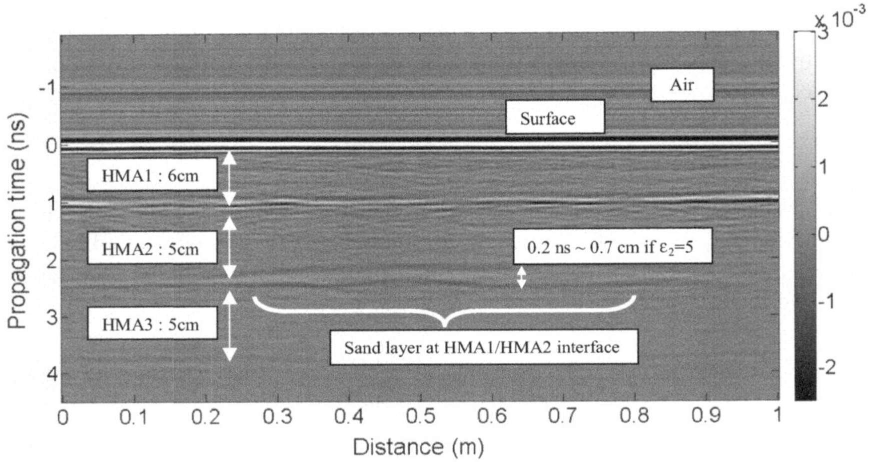


**Fig. 4.43** B-scan obtained with a SFR system from measurement above the sand defect

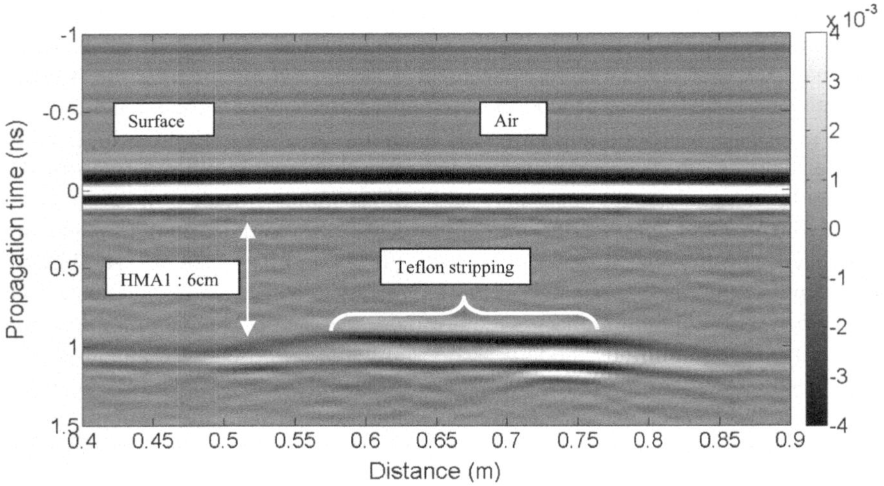


**Fig. 4.44** B-scan obtained with the SFR system from measurements above the teflon interface

delay estimation. Layer estimation is based on the time delay of primary echoes, which have been scattered only once within the medium and compose the first part of the scattered signal. Multiple echoes, if any, cannot disturb this processing but may limit the data detection and analysis of deeper echoes.

## 4.9.3 Obtainable Results

Most actual GPR allows NDT of pavements above 4 cm in thickness. This limitation is due to the frequency bandwidth 2 GHz at the most for conventional GPR. Providing simplified assumptions (the bandwidth equals to twice the central frequency), the vertical resolution is found to be roughly the half of the wavelength. When layers to probe are thinner than this limit, echoes from two successive interfaces are overlapping with each another, merging into an apparent single echo with longer time duration.

The first issue to overcome this limit has been to enlarge the frequency bandwidth as well as the central frequency [39]. This was one of the major motivations behind the use of the step-frequency radar technique.

Another issue has been recently emerged from the signal processing community. As shown in [40], some advanced signal-based signal processing techniques may overcome the bandwidth limitation of conventional radar. As a result, the vertical resolution can be enhanced by a factor 10 at high signal to noise ratio (i.e., above 20 dB).

### 4.9.3.1 Short Background on the Detection of Layer Interfaces

We consider the two HMA layers media with a thin debonded zone between these two layers. This defect could be detected as far as the emitted electromagnetic field from the antenna has a wavelength of the order of half the thickness and the surface area of the defect. The vertical and horizontal resolutions are deduced from Eq. (4.14).

$$r_v = \frac{\lambda}{2} \text{ and } r_h = \sqrt{\frac{\lambda^2}{16} + \frac{\lambda_z}{2}} \tag{4.14}$$

where λ is the wavelength of the emitted electric field that is parallel to the surface and z is the height of the antenna above the surface. For instance, for a HMA layer with a permittivity of 5 and a thickness of 5 cm, and an ETSA radiated in the [1.4–20 GHz] band at a height of 15 cm from the HMA surface, a debonded or damaged interface of 3 mm to 4.5 cm of thickness, and of 2.6–10 cm of horizontal length, is detectable. This is a theoretical approach. Actually, it is better to consider the limit of 7.5 GHz given by the central frequency of the impulsion emitted from the ETSA, that leads to a potential detection of less than 1 cm of thickness, and a horizontal dimension of less than 4 cm.

### 4.9.4 Examples

The measurements were carried out on the pavement fatigue carousel (Ifsttar testing facility). The results presented concern the detection of two defects: a sand interface and a Teflon interface. This one is also characterized in terms of nature, by the measurement of its permittivity. This is possible because there is a significant permittivity difference between the Teflon and bituminous materials. It is less feasible for other natures of interfaces, with sand for instance. Figure 4.43 shows the radargram (or B-scan) of a SFR measurement on the sand interface defect which is clearly detected between the HMA2 & HMA3 layers. Its length is about 0.60 m and its thickness (less than 7 mm) can be estimated by considering the travel times between its top and its bottom (less than 0.2 ns) and an approximated permittivity of 5 in the middle. Nothing allows its characterization in terms of nature because its permittivity is not estimated; the first calculation conducted is currently insufficient and some improvements must be carried out.

Figure 4.44 presents the SFR measurements realized just above the Teflon interface. It is clearly identified by a displacement of the HMA1/HMA2 interface position in time and by a variation of the reflected amplitude.

Permittivities of the HMA layers are estimated according to the classical approach [38, 41, 42]. It consists of measuring the amplitude $A_{HMA}$ in time domain of the reflected pulse and comparing it with the amplitude $A_{fmp}$ of the total reflection on a flat metal plate. The plane wave approximation is assumed and the HMA electrical conductivity is neglected. The reflection coefficient $R_{01}$ (sign positive) between the air and HMA1 and $R_{12}$ (sign dependent on dielectric contrast) between HMA1 and HMA2 are directly proportional to the ratio of these respective measurements as written in Eq. (4.15).

$$R_{01} = \frac{A_{HMA1}}{A_{fmp}} \text{ and } R_{12} = \frac{A_{HMA2}}{A_{fmp}} \tag{4.15}$$

The permittivities of the first and second layer are given by Eq. (4.16).

$$\varepsilon_1 = \left(\frac{1 + R_{01}}{1 - R_{01}}\right)^2 \text{ and } \varepsilon_2 = \varepsilon_1 \left(\frac{1 - R_{01}^2 + R_{12}}{1 - R_{01}^2 - R_{12}}\right)^2 \tag{4.16}$$

The permittivities of the HMA1 layer and the HMA2 layer are presented in Fig. 4.45 for each position of the antenna above the surface, with a step of displacement of 1 cm.

As illustrated in Fig. 4.45, the presence of the Teflon decreases the permittivity value of the second interface. The value for the HMA2 layer shows a strong variation where the Teflon stripping is located. The permittivity calculated in the frequency domain (2.3 ± 0.5) is closed to the theoretical one (2.1).

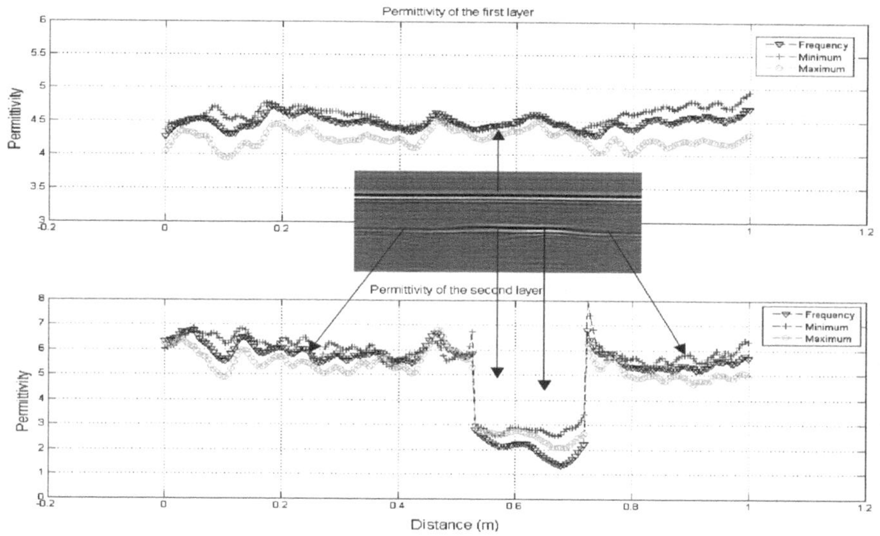


**Fig. 4.45** Results of estimated permittivity for the two HMA layers

### *4.9.5 Benefits and Limits of Radar System*

Radar systems are a NDT technique with a high acquisition rate. So, quasi-continuous measurements (every 1 cm) are done to produce a mapping of the subsurface. Radar systems can be mounted on a vehicle to survey roadways at traffic speed (according to acquisition rate) to obtain global monitoring of the structure. Electromagnetic waves are sensitive to the presence of inside damages such as voids or cracks. Specific signal analysis is performed to locate and characterize such damages. Thus, radar systems are convenient systems that give back thickness measurement of asphalt layers but also detect inside the pavement the damages such as voids, delaminations or cracks.

Electromagnetic waves that are sensitive to the presence of water and may reveal interface damages are also disturbed by metallic elements that are sometimes included into new materials. Radar systems seem to be less sensitive to internal damages than mechanical NDT method. However, the evolution in electronics and in signal processing will allow improving the performance of radar method.

## 4.10 Evaluation of Low Temperature Asphalt Binder and Mixture Behavior Using the Acoustic Emission Technique

In recent years, the acoustic emissions (AE) technique has increased in popularity, particularly for applications such as in situ evaluation of material behavior, structural condition assessment, and damage initiation detection. In the present study, the

AE approach has been successfully used to rapidly and accurately characterize the embrittlement temperature of asphalt materials. The embrittlement temperature is defined herein as the temperature threshold where fracture-related acoustic events are emitted from an asphalt mixture sample subjected to thermal cooling. The embrittlement temperature appears to be strongly related to the temperature at which thermally-induced cracks appear in a pavement with an asphalt concrete surface. This section provides an overview of the AE test procedure for determination of asphalt materials embrittlement temperature. According to the proposed classified grid (Table 4.1), it concerns Table 4.11.

Cracks in asphalt pavements are expensive and difficult to properly treat. In cold climates, thermal cracking is a prevalent cause of asphalt pavement premature deterioration. This type of distress can be developed early in pavement life, thereby contributing to loss of pavement serviceability long before the fatigue-based design life has been reached. A typical thermal cracking pattern occurring in asphalt pavements is shown in Fig. 4.46. This form of distress drastically affects asphalt pavement life span and smoothness, and is associated with billions of dollars of repair and maintenance annually in the United States. If left untreated, thermal cracks will continue to deteriorate and widen over time, allowing moisture to readily infiltrate the pavement system. Detrimental effects of low-temperature cracking have motivated a number of studies in an effort to experimentally design and control asphalt properties related to the low temperature performance of asphalt pavements. However, accurate predictions of thermal cracking and associated failure mechanisms still remain a challenge [43–45].

As asphalt binder or mixture is cooled, it transitions from a brittle-ductile to a quasi-brittle physical state. In the quasi-brittle state, resistance to fracture is generally very low, allowing cracks to propagate readily. To complicate matters, asphalt concrete tends to become oxidized, leading to material embrittlement with time. The amount of aging, or the aging rate, has been found to vary significantly depending upon crude source, refining techniques, additives, climate, and

**Table 4.11** Classification grid of the acoustic emission technique

| Technique/device's short name | Acoustic emission technique | |
|---|---|---|
| Measurement principle(1) | (3) | |
| Categories | Yes | No |
| Field system | | X |
| Laboratory system | X | |
| Specific setup dependent measurement system | | X |
| Measurement system associable with different test setup | X | |
| Crack initiation | X | |
| Crack propagation | X | |
| Crack distribution | | X |
| Damage characterization | | X |
| Short description of system(2) (A), (B), (G), (K), (L), (M), (N) | | |

**Fig. 4.46** Thermal cracking in asphalt pavement

characteristics of the mixture (asphalt content, air voids). As a consequence, one of the great challenges for the pavement engineer is to determine when a pavement surface has become vulnerable to embrittlement for a given climate. Currently, pavement engineers lack practical tools to determine the proper time for preventive maintenance, such as the application of a surface treatment, or to determine when a more substantial rehabilitation is required, such as pavement milling and resurfacing. Another challenge facing the pavement engineer is the proliferation of new materials and the lack of field data to assess their suitability for a given climate. Such examples are Recycled Asphalt Pavement (RAP) mixtures, and Warm mixtures (WMA) of which several varieties exist. Considerable interest exists for testing methods capable of predicting the cracking temperature of asphalt material. The Acoustic Emission (AE) testing method described here addresses the current shortage of accurate and practical testing procedures for asphalt binders and mixtures [43].

Acoustic emission is a term used to describe the spontaneous release of localized strain energy in the form of transient elastic waves within a stressed material. Most of the sources of acoustic emissions are damage related. Emission of transient elastic waves results from sudden internal micro-displacements in the stressed material. The AE technique has been proven to be a powerful tool for examining the behavior of materials deforming under stress and has been extensively applied for the condition assessment and damage detection for many materials such as steel, concrete, wood, etc. AE provides comprehensive information on the origin of a flaw in a stressed component and can also provide information pertaining to the development of this flaw as the component is subjected to stress. It is a very sensitive indicator of crack propagation. AE waves from a growing flaw travel within the material and are detected by sensitive surface-mounted sensors. Figure 4.47 schematically illustrates crack nucleation and propagation and corresponding AE wave transmission and detection for material under stress [45, 46].

Recent studies have utilized the AE technique to evaluate the low temperature cracking performance of asphalt mixtures. The "Embrittlement" temperatures of several different asphalt mixtures are presented in following sections.

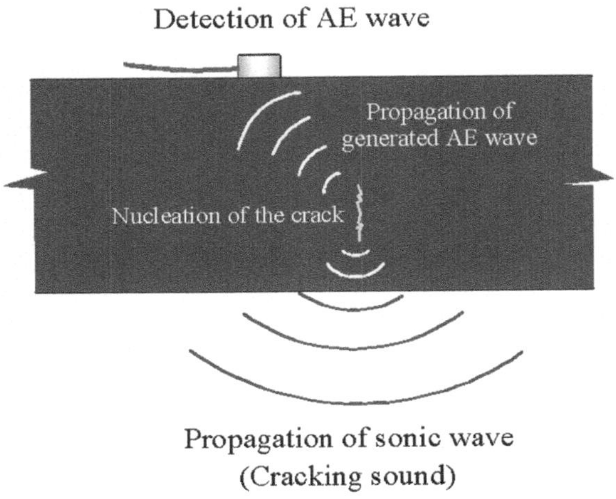


**Fig. 4.47** Nucleation, propagation and detection of AE waves

### *4.10.1 Experimental Procedure and Theoretical Basis*

In this example, specimens of 150 mm diameter semicircular shape with 50 mm thickness were prepared for AE testing. AE specimens were positioned inside the metal cooling chamber surrounded by dry ice as the coolant. The AE testing setup specimen dimensions are schematically shown in Fig. 4.48a, b, respectively [43, 44].

Specimen temperature was continuously recorded using a K-type thermocouple placed on the specimen surface. Wideband AE piezoelectric sensors (Digital Wave, Model B1025) with a nominal frequency range of 50 kHz to 1.5 MHz were utilized in order to monitor and record acoustic activities of the sample during the test. High-vacuum grease was used to couple the AE sensors to the specimen surface. Since by nature the acoustic signals are of low energy, the sensor data is immediately fed into a preamplifier to minimize noise interference and prevent signal loss. Signals from AE sensors were pre-amplified by 20 dB using broad-band pre-amplifiers. Then, the signal was further amplified by 21 dB (for a total of 41 dB) and filtered using a 20 kHz high-pass double-pole filter using the Fracture Wave Detector (FWD) signal condition unit. The signals were then digitized using a 16-bit analog-to-digital converter (ICS 645B-8) using a sampling frequency of 2 MHz and a length of 2048 points per channel per acquisition trigger. The outputs were stored for later processing using Digital Wave software (WaveExplorerTM V7.2.6). A number of parameters were extracted from the AE test. Analysis of thermally induced AE activity of asphalt mixture specimens was conducted on calculated AE energy, event counts and specimen temperature. The emitted energy associated with each event can be calculated by using Eq. (4.17), where $E_{AE}$ is AE energy of an event ($V^2$-μs) with duration of time t (μs) and recorded voltage of V(t) [43, 44].

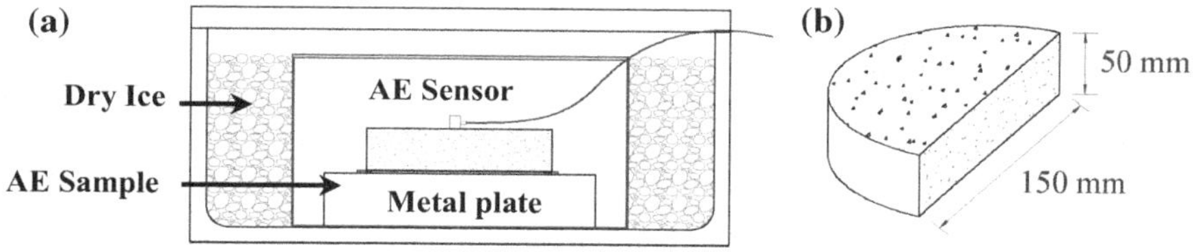


**Fig. 4.48** **a** AE testing set-up; **b** AE testing specimen [43, 44]

$$E_{AE} = \int_{0}^{t} V^2(t)dt \tag{4.17}$$

In order to minimize the amount of extraneous data including electronic noise and to obtain the desirable signals, all AE events with energy lower than 100 $V^2$-μs were filtered out. In addition, AE sensors were conditioned in the cooling chamber prior to starting the test. All results presented hereafter are based on the above filtering procedures. A typical plot of event count versus temperature is shown in Fig. 4.49.

It was observed that for each AE test, there are three distinct regions exist in AE event count versus temperature plot namely: Pre-cracking, Transition and Stable cracking regions. During the so-called pre-cracking region, due to differential thermal contraction between the asphalt mastic and aggregate, thermal stresses build up inside the specimen prior to the onset of material fracture. There were no AE activities recorded in "Pre-cracking region". Progressively higher thermal stresses in the specimen result in the formation of thermal micro cracks in the asphalt mastic, which is accompanied by the release of elastic waves. This manifests itself as a cluster of high amplitude waves during the test. "Transition region" starts when thermal microcracking in the specimen reveals itself as several high-energy events occurring immediately after the pre-cracking period. The temperature corresponding

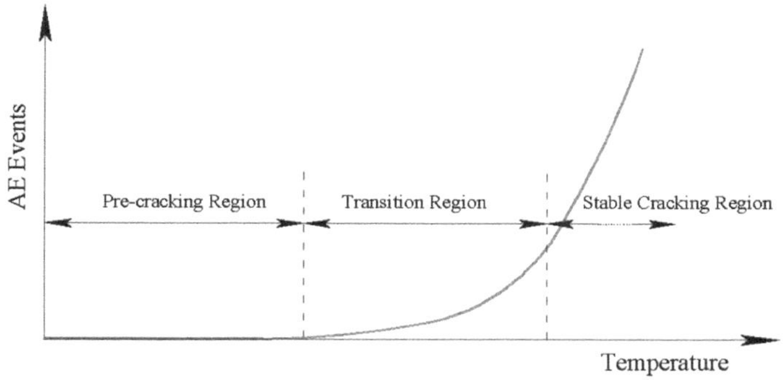


**Fig. 4.49** Typical AE event count versus temperature plot regions [44]

to the event with the first peak energy level has been termed the "Embrittlement temperature" as illustrated in Fig. 4.50. It is hypothesized that the embrittlement temperature represents a fundamental material state, which will be independent of material constraint, sample size (as long as a statistically representative volume or larger is used), and sample shape [44].

The "Transition region" can be considered as the region where material behavior gradually changes from quasi-brittle to brittle state and where resistance to fracture is generally very low, allowing cracks to propagate readily. The "Stable cracking region" usually starts at very low temperature when the material is brittle and generates a significant amount of AE activity. The AE event count versus temperature plot in this region usually has a steep slope that remains relatively constant.

The microcracking results primarily from a combination of asphalt mastic brittleness and the action of tensile thermal stresses within the material. Figure 4.51

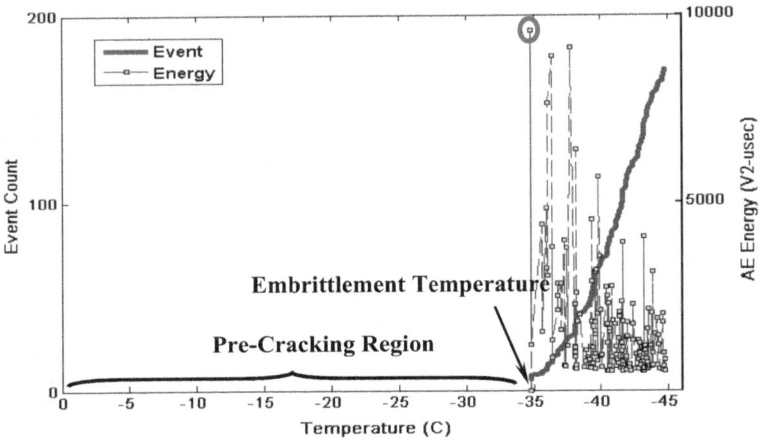


**Fig. 4.50** Typical plot of event count and AE energy versus temperature [44, 45]

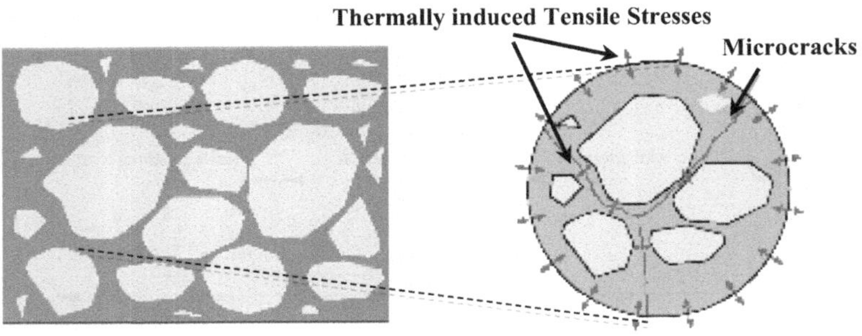


**Fig. 4.51** Thermally induced stresses within the asphalt mastic [44]

schematically shows the thermally induced stresses inside the asphalt mastic, which results in thermal microcracks in the specimen.

### *4.10.2 Benefits, Limitations and Future Extensions*

The newly developed AE-based test is an innovative approach that addresses the current shortage of rapid, practical asphalt material tests available to the pavement engineer for material formulation, selection, design, and QC/QA. This testing method would yield significant payoff to practice for both up-stream and down-stream suppliers and producers. It can be used to evaluate modern materials (highly modified, recycled, warm-mix). This method is extendable to in situ pavement evaluation, capable of testing thin mixture specimens, and is designed specifically to aid in the timing and selection of preventive maintenance and rehabilitation treatments. To this end, the technique is currently being extended to characterize the vertically graded properties of small diameter, field-extracted pavement cores.

## 4.11 An Overview on the Passive and Active Seismic NDT in Asphalt Pavements

NDT is defined as the technical method to examine structures, materials or components in ways that do not impair future usefulness and serviceability. In civil engineering, the NDT is an invaluable tool for structural diagnostics (in-service, in situ or field NDT) and materials characterization (laboratory NDT). According to the proposed classified grid (Table 4.1), this section concerns Table 4.12.

Recently, a NCHRP report mentioned a number of NDT techniques which have been significantly improved. Some techniques have shown promise for potential employment in quality control and acceptance of flexible pavement construction: ground-penetrating radar, falling weight deflectometer, penetrometers, and infrared and seismic technologies. The major advantages of NDT methods are their economy and efficiency. Seismic NDT techniques, which are based on stress wave propagation, are interesting methods for structural evaluation of pavement and asphalt materials characterization. These techniques can be divided into two approaches, the passive approach and the active approach:

- In the passive NDT techniques, signals emanate from the material using proper physical fields reflecting the internal energy of the inspected material. In this context, the Acoustic Emission (AE) method should be considered to be a "passive" nondestructive technique, because it usually identifies defects only while they develop during the test (mechanical, physical and chemical loading).

**Table 4.12** Classification grid of the seismic and acoustic NDT

| Technique/device's short name | Seismic and acoustic NDT | |
|---|---|---|
| Measurement principle$^{(1)}$ | (3), (5) | |
| Categories | Yes | No |
| Field system | X | |
| Laboratory system | X | |
| Specific setup dependent measurement system | | X |
| Measurement system associable with different test setup | X | |
| Crack initiation | X | |
| Crack propagation | X | |
| Crack distribution | | X |
| Damage characterization | | |
| Short description of system$^{(2)}$ (A), (B), (G), (K), (L), (M), (N) | | |

- Active techniques are those in which a material is interrogated by creation of a "forced" physical field with specified orientation. The signal contents of the interrogating medium are changed by way of interaction with the material. In Seismic NDT, potential of active techniques (sonic/ultrasonic methods) in pavement applications has been clearly demonstrated in the literature review. These techniques are based on the propagation of three types of waves: Body Waves (BW); Surface Waves (SW) and Structural Waves.

## *4.11.1 Passive Seismic NDT (AE)*

### 4.11.1.1 AE Principle

Understanding failure mechanisms of construction materials as well as their damage evolution are two key factors to improve design tools for structures. Depending on failure modes to be highlighted and studied, several test methods, and analysis tools have been developed, in particular AE. This latter is an experimental tool well suited for characterizing material behaviour by monitoring fracture process. Despite the wide use of AE technique to characterize and monitor damage evolution of composite materials (often with a brittle behavior at low temperatures), few research studies focused on using AE technique to characterize asphalt concrete behaviour.

Acoustic emission may be defined as transient elastic waves generated by the rapid release of strain energy in a material. A number of micro and macro processes contribute to both the deformation and the deterioration of a material under strain, resulting in a series of acoustic events. Thus, the events released by the material contain information regarding the general deformation. A large amount of literature is available on acoustic emission studies on rock and concrete materials under a variety of loading conditions. Several approaches have been adopted to understand the various aspects of micro-cracking and fracture processes, for example: (i) recording the events by expressing them in the form of event rate or cumulative events; (ii) determination of source locations and the study of mechanism of individual events; (iii) study of amplitude distributions; (iv) investigation of frequency characteristics of emission events.

So, depending upon the application and type of phenomenon being studied, different approaches can be adopted for analyzing AE signals. In the next sections, a number of experimental laboratory studies are described and demonstrate the potential use of AE techniques for asphalt concrete characterization.

#### 4.11.1.2 Cold Cracking: Embrittlement Temperature and Thermal Fatigue

Cracking of asphalt concrete pavements owing to cold temperatures or temperature cycling can occur in structures constructed in cold climates. Cracking that results from cold temperatures is generally referred to as low temperature cracking; cracking that results from thermal cycling is generally referred to as thermal fatigue cracking. Some studies on the application of the AE method for monitoring cold cracking of asphalt mixtures are reported in the following paragraphs.

- Embrittlement: Kim et al. used AE technique to analyze the damage mechanism of asphalt concrete under low temperature (−25 °C). The AE results confirmed occurrence of localized damages as a result of cooling alone. In addition, analysis of AE data showed the temperature where such damage initiated and later propagated under further exposure to lower temperatures. Hesp et al. showed that the temperatures at which acoustic emission starts, or at which the maximum stress in a restrained cooling test (−32 to −20 °C) on a notched sample is reached, are better low temperature failure criteria than the temperature at which the maximum stress is reached in a conventional test on a regular sample. Recently, the NCHRP Project 144, which summaries number of experimental studies, involved research aimed at improving and standardizing the AE technique as a rapid and practical testing method to assess the low temperature cracking behavior of asphalt binder mixtures. During the course of this study, AE tests were conducted to evaluate the concept of an embrittlement temperature through the testing of a wide range of asphalt binders and mixtures, including lab and field specimens. The authors showed that the scientific basis for the material parameters measured in the proposed test, namely, the embrittlement temperature and max event energy temperature, have been solidly supported through lab and field sample testing and comparison to existing test methods.
- Thermal Fatigue: Valkering and Jongeneel used the AE technique to monitor temperature cycling tests with restrained asphalt concrete specimens at low

temperatures (10 to −40 °C). They observed that the repeatability of AE measurements is good, that the AE activity (number of events) correlates with thermal fracture temperatures, and that the AE activity in restrained specimens at low temperatures is caused by defect initiation in the binder. In study reported by Sinha, acoustic emission events were detected in unrestrained asphalt mixture samples exposed to thermal cycling at low temperatures. This indicates that micro cracking occurs, most likely due to significant difference in thermal contraction coefficients of aggregates and binders, which can explain the reduction in stiffness with storage time. Another explanation is provided by the inherent variability present in asphalt mixture samples, which may require testing a larger number of specimens to obtain statistically significant results.

#### 4.11.1.3 Static and Dynamic Loading: Fracture Process and Fatigue Behavior

One of the most important properties of an asphalt concrete in paving is its ability to resist permanent deformation and cracking under dynamic and/or static load. This behavior depends not only on the composition of the concrete and the paving conditions, but also on the mechanical history it has experienced. Many laboratory test methods have been employed for the characterization of asphalt concretes, and also associated to AE measurements. Chang used AE-MTS testing system, with the purpose to first demonstrate that the deformation behavior and microfracture of asphalt concrete under different loading conditions are similar to those of asphalt thin films; and secondly to show the utility of AE technique in assisting the researchers with monitoring the microfracture in asphalt concretes. The author concluded that the micro-damage process in asphalt concretes can be reliably detected by AE. The results from both single cycle and multicycle tests indicate that the deformation of a specimen relates closely to AE event which is the measurement of damage state of the specimen.

- Fracture Process: Li et al. were interested in the location of acoustic events within AC specimens. For purposes of their experimental campaign, pre-notched specimens were tested in three-point bending using a Semi-Circular Bend (SCB) device. Monotonic tests were performed inside a climatic chamber at −20 °C. The authors determined that the acoustic events are located on the fracture path. Moreover, a comparison between the load versus displacement curve and the cumulative AE events versus testing time curve showed an increase of the acoustic activity when the force values (post peak stage) range from 70 to 90% of the strength. AE events located before the peak load define the Fracture Process Zone (FPZ) which is a key factor in the determination of material fracture process. Li and Marasteanu also investigated particularly the size variation of the FPZ based on various parameters. The results showed that the size of the FPZ is mainly affected by different parameters such as temperature, type of aggregates and air voids. It also showed that the binder percentage does not affect the FPZ size. In the energetic approach, more than 50% of events are little energetic and account for about 5% of total energy, whereas only 3% of events are great energetic and account for 40% of total energy. Thus, in the pre-peak stage (load lower than the strength), the energy level of AE events is globally low, and is related to the FPZ as well as the micro cracking. Nesvijki and Marasteanu, investigated the utility of both spectral analysis and wavelet transform in study of AE caused by cold temperature cracking under compression loads.

- Fatigue Behavior: fatigue tests were conducted by Seo and Kim and showed the influence of rest period between loading cycles on mechanical behavior of AC. In fact, both the cumulative AE energy and cumulative AE counts increase when the fatigue test includes rest periods between loading cycles. It indicates that rest period increases the fatigue lifetime which is related to the healing phenomenon. In addition, the Kaiser effect is commonly related to damage state in the material. However, the fatigue tests without rest periods did not allowed for this effect to be systematically observed. Diakhaté et al. reported an experimental study where AE technique is used to characterize AC behavior under a double shear cyclic loading. The results showed that a critical threshold can be identified as a 49% drop in initial value of the specimen stiffness modulus, since it corresponds to a highly and significant increase in acoustic energy levels of the located events. Moreover, this critical threshold also coincides with the conventional failure criterion of asphalt concrete mixes (drop by one-half in the initial value of the specimen stiffness modulus). In addition, by studying cumulative events, several stages were defined in the fatigue test which correspond to different mechanisms such as micro or macro cracking.

## *4.11.2 Active Seismic NDT*

### 4.11.2.1 Seismic NDT Principle

The principle of active seismic methods is based on the generation and detection of stress waves. The raw data collected on the surface of a pavement with seismic methods are simply the time histories of the impact source and the surface deformation. The time record from each sensor contains information about the pavement structure which can be described in terms of two kinds of body waves (compression waves and shear waves), one kind of surface wave (Rayleigh waves), and bending modes. The most commonly used seismic methods are:

- Ultrasonic methods: Ultrasonic Pulse Velocity (UPV); Ultrasonic Echo (UE);
- Sonic methods: Impact Echo (IE);
- Dynamic methods: Impulse Response (IR); Spectral Analysis of Surface Waves (SASW)

#### 4.11.2.2 Laboratory Applications

- Mix Parameters: several laboratory experiments are reported in literature with the objective of investigating the sensitivity of seismic measurements to asphalt concrete mix proprieties. In works presented in CHMR report, the parameters investigated were the gradation, viscosity of the asphalt, voids in total mix (VTM), and temperature. The laboratory tests were performed using a V-Meter and generate compression waves through AC briquettes, and the seismic modulus was calculated by considering wave propagation theory in elastic material. Therefore, the report gives a number of comments and conclusions on the effect of the mixture parameters on the seismic modulus. Similar results were reported on field cored asphalt mixtures samples, and laboratory prepared samples. Jiang et al. investigated three signal processing techniques (Fourier transform, short-time Fourier transform, and the wavelet transform) which were used to analyze wave attenuation and frequency content to develop a suitable condition index for HMA mixtures based on wave characteristics. The results of UPV tests indicated that the condition index based on the amplitude measurement has the potential to provide a reliable assessment of the quality of HMA mixtures. The characterization and quantification of anisotropy in asphalt specimens was also investigated. Biligiri and Kaloush adopted the UPV testing method to assess the dampening properties of pavement materials. Detection of aggregate segregation in asphaltic concrete is also possible when combining analysis of the both phase velocity and amplitude attenuation of ultrasonic Rayleigh surface waves.
- Cracking and Delamination: Underwood and Kim used surface wave and ultrasonic methods to measure crack depths of asphalt slabs. To generate surface waves, three impact methods were developed and tested. Both test methods show promising results for crack analysis. Khazanovich et al. presented results of an explanatory study aimed at examining the applicability of the ultrasonic technology for evaluation of cracks and longitudinal joints in flexible pavements. The authors used self-compensating ultrasonic technique, which involves simultaneous measurements of the wave propagation through cracked and non-cracked portions of the structure. The laboratory measurements were performed on four AC beams. A strong correlation between the depth of the notch and the apparent reduction of the wave velocity was observed for all four beams. Jiang et al. reported a laboratory investigation with the objective of examining the suitability of using a wave attenuation parameter (Transmission Coefficient) for assessing the condition of the joints between two asphalt pavement surfaces. The wave attenuation is measured in terms of two parameters and by using both ultrasonic and hammer testing. The results showed that the considered attenuation parameters were able to differentiate between the fair, weak and poor construction joints built in the laboratory.
- Complex Modulus, High-frequency Modulus: a number of experimental laboratory studies were described in literature and demonstrate the potential use of stress wave propagation techniques for complex modulus measurements. These

techniques are impulse echo method, free-free resonant testing, spectral analysis of surface wave and multichannel analysis of surface wave, flexural wave propagation technique and forced longitudinal vibration technique, and ultrasonic body wave propagation techniques. The testing temperature ranges from −19 to 42 °C, while selected temperatures remained quite limited. Structural, surface, and P-wave have been used, however it should be noted that group velocity and elastic wave propagation hypotheses are most frequently used when measuring |E*|. The viscoelastic approach is used less frequently, and requires combination with a rheological model to identify viscous parameters, such as phase angle or viscous attenuation.

- Fatigue Damage: ultrasonic wave propagation can be used to monitor fatigue damage of asphalt material under mechanical loading. Thus, Arnaud and Houel combined measurements of P-wave velocity and damping coefficient to calculate the complex modulus at each instant during fatigue test on cylindrical samples. Then the damage evolution curve was plotted as a function of the material. Tigdemir et al. presented a study on the possibilities of applying ultrasonic methods in asphalt concrete specimen testing in order to estimate fatigue life. The ultrasonic method was first used in order to obtain the seismic characteristics of specimens. Then, the specimens were tested for fatigue lives using the repeated-loading indirect tensile test equipment. To be able to model the fatigue lives, in addition to conventional fatigue model parameters, new parameters from this ultrasonic analysis were taken into consideration.
- Moisture Damage: Birgisson et al. showed that the small strain Young's modulus obtained with the quick monitoring device ultrasonic pulse wave velocity test appeared to be sensitive to changes in the integrity of hot-mix asphalt mixtures due to moisture conditioning, including the effects of porewater and saturation levels and the effects of aggregate type on mixture conditioning. Barnes et al. tested the feasibility in the laboratory-conditioned specimens of asphalt concrete of multichannel analysis of surface wave in determining the variation of seismic modulus versus depth as a potential indicator of asphalt concrete moisture damage, considering air void content and moisture conditioning. The surface wave method appeared to provide a good sensitivity in detecting stripping damage effects for samples with low air-void contents of 2%.

#### 4.11.2.3 Field Applications

- Thickness and Stiffness: when applied in the field, the main objective of seismic NDT is estimating the unknown structural properties (thickness and stiffness) of different layers in a pavement construction. Baker et al. developed the portable seismic pavement analyzer (PSPA) for monitoring Portland cement concrete (PCC) pavements. Three different techniques are utilized: the ultrasonic body wave (UBW) method for determining Young's modulus of top pavement layers, the ultrasonic surface wave (USW) method for shear modulus of top pavement layer, and the impact echo (IE) method for determining the thickness of pavement layer or depth of a delaminated layer. Techniques of seismic surface wave propagation are most commonly used when evaluating asphalt concrete moduli. They include such methods as: Ultrasonic Surface Wave (USW); Spectral Analysis of Surface Waves (SASW); Multichannel Analysis of Surface Waves (MASW) and Multichannel Simulation with One Receiver (MSOR). The Poisson's ratio (ν) of the concrete layer can be evaluated from combination of USW and UBW measurements. Barnes and Trottier suggested surface-based modulus testing as a field method of evaluating moisture damage-induced stiffness loss.

- Cracking and Delamination: Khazanovich et al. presented results of field evaluation of non-cracked areas, transverse (thermal) cracks and construction longitudinal joints by using self-compensating ultrasonic technique (see paragraph 3.2). The results demonstrated a good potential of the technology. Jiang et al. used two different NDT procedures for condition assessment of the asphalt pavement longitudinal joints in the field (Ultrasonic testing "UPV" and impact hammer testing). In the experiments, sources and receivers were located at both sides of the longitudinal joint, and a wave attenuation parameter (Transmission Coefficient) was considered to assess the relative strength of the joint. The field test clearly identified the deteriorated joints from the newly constructed joints. As well the test is sensitive enough to distinguish between the joints constructed using the traditional method and the echelon paving method. Du Tertre et al. used field MASW measurements to evaluate the attenuation of surface waves due to the joint. This technique provided information regarding the stiffness and attenuation properties of the surface layer which can be used for quality assessment of longitudinal joints. Celaya et al. reported examples on the use of NDT seismic methods for detection of delamination in HMA pavements. A pavement section was specifically constructed with various levels of debonding at different depths and with different asphalt mixes. The utilization of the Impulse Response method show temperature at the time of testing greatly affected the outcome of the results in terms of FFT amplitudes. However, the method seems promising for the detection of the fully-debonded areas both shallow and deep and some of the partially-debonded areas. The authors also showed that the USW method, as implemented in the PSPA, might be able to identify delaminated areas reasonably well, especially the shallow ones. The PSPA was evaluated as a preventive maintenance tool for identifying the early stages of load associated cracking in hot mix asphalt layers.

### 4.11.3 Experimental Device Proposed by GEMH-GCD-Limoges

The laboratory system proposed by the GEMH laboratory allows measurements of both "Ultrasonic Wave Propagation Parameters" and "Acoustic Emission Activities":

Table 4.13 Piezoelectric transducers for "US" and "AE" measurements

| Sensor model | Dim. D × H (mm) | Operating temp. °C | Resonant freq. (kHz) | Operating freq. range (kHz) | Quantity | Wave type |
|---|---|---|---|---|---|---|
| R45 | 29 × 41 | −45 to 150 | 7.3 | 1–30 | 2 | P-Wave |
| R15a | 19 × 22,4 | −65 to 175 | 75 (150) | 50–200 | 4 | P-Wave |
| Nano 30 | 8 × 8 | −65 to 170 | 140 (300) | 125–750 | 4 | P-Wave |
| SH10 | 29 × 41 | −45 to 150 | 70 | 10–120 | 2 | S-Wave |
| WD | 18 × 17 | −65 to 177 | 125 (650) | 100–1000 | 2 | P-Wave |
| P500 | 31 × 33 | −45 to 150 | 500 | 250–800 | 2 | P-Wave |

- The ultrasonic test device is composed of an arbitrary waveform generation card (sinus, dirac, sweep…) with frequency modulation, piezoelectric transducers with different resonant frequencies (see Table 4.13), a 20–40–60 dB analog pre-amplifier, a 40 Ms/s acquisition and sampling card; and software for treatment and analysis of the waveforms.
- The acoustic emission activities, during mechanical or thermal loading, can be recorded using four piezoelectric transducers.

## 4.12 Summary

Among all the 10 contributions received in MCD TG3 and presenting advanced measurement techniques for crack analysis, DIC is extensively used with different algorithms. Table 4.14 provides an overview of the reviewed measurement systems.

It seems that an interesting new field of investigation may be the combination of these technique that are mainly based on "on-surface" analyses with other techniques, i.e., acoustic emissions, that may give information on what happens in the material's internal structure. In principle, X-Ray computed tomography may also aid in this type of analyses, but today, it is hampered by the long detection time in comparison to the speed of crack initiation and propagation. In addition, a common open research area amongst all advanced measurement techniques is the matter of proper calibration and validation. Many of these techniques also lend themselves quite nicely to the fields of intelligent infrastructure and smart cities. For instance, DIC and AE techniques can be applied to fixed optical sensors (overhead cameras) and in-pavement and in-vehicle sensor data collected in a smart city application, to track the pavement system state-of-repair, to allow an entire pavement network to serve as a research testbed, and to provide owner-agencies with optimized recommendations for pavement maintenance and rehabilitation.

**Table 4.14** Synoptic overview of proposed measurement systems

| System's short name | Proposer | System's base method | Obtainable results | Field of application |
|---|---|---|---|---|
| HS-SVS | M. E. Kutay | DIC | Full-field strain | Strain measurement on specimens during laboratory test |
| Optical full-field techniques + F.E. M. | O. Pop, A. Millien, C. Petit | DIC | Full-field strain | Damage process on laboratory specimens for complex loading |
| 4P.B. test + D.I.C. | M. Hun, A. Chabot, F. Hammoum | DIC | Surface displacement, surface strain, rack opening displacement, crack sliding displacement and crack length | Characterization of interface behaviors of double layer specimens |
| Optical strain measurement system | A. Montepara, E, Romeo, R. Roncella, G. Tebaldi | DIC | Full-field strain | Strain measurement on specimens during laboratory test |
| X-Ray CT + tension/ compression testing system | A. Guarin, D. Jelagin, N. Kringos, B. Birgisson | Digital X-Ray imaging | Material characterization and location of internal damages | Analysis on laboratory specimens |
| Colibri | J. M. Simonin | Measurement of vertical surface response made by a dynamic solicitation | Location of internal damages | Pavement analysis |
| Deflection devices | J. M. Simonin, B. Picoux | Deflection measurements | Location of cracks | Pavement analysis |
| Radar analysis | J. M. Simonin, V. Balthazard, X. Dérobert | Imaging of dielectric contrasts trough electromagnetic waves | Location of internal damages | Analysis on laboratory specimens and pavements |
| Acoustic emissions based test | B. Behnia, W. G. Buttlar, H. Reis | Acoustic emission detection | Analysis of crack propagation | Analysis on laboratory specimens and pavements |
| Passive and active seismic NDT | M. Takarli, Angellier | Analysis of stress wave propagation | Material characterization | Analysis on laboratory specimens and pavements |

**Acknowledgements** Chapter made with contributions provided by: G. Tebaldi, M. E. Kutay, O. Pop, A. Millien, C. Petit, M. Hun, A. Chabot, F. Hammoum, A. Montepara, E. Romeo, R. Roncella, J. M. Simonin, B. Picoux, V. Balthazard, X. Dérobert, B. Behnia, W. G. Buttlar, H. Reis, M. Takarli, N. Angellier, A. Guarin, D. Jelagin, N. Kringos, B. Birgisson.

## References

1. Wang L (2010) Mechanics of asphalt: Microstructure and micromechanics. McGraw-Hill ISBN: 0071498540
2. Ketcham RA, Carlson WD (2001) Acquisition, optimization and interpretation of X-ray computed tomographic imagery: application to the geosciences
3. Kutay ME, Guler M, Aydilek AH (2006) Analysis of factors affecting strain distribution in geosynthetics. J Geotech Geoenviron Eng ASCE 132(1):1–11
4. Gautama T, Van Hulle MA (2002) A phase-based approach to the estimation of the optical flow field using spatial filtering. IEEE Trans Neural Netw 13(5):1127–1136
5. Post D (1993) Moiré interferometry. In: Kobayashi AS (ed) Handbook on experimental mechanics. VCH, New York
6. Wells A, Post D (1958) The dynamic stress distribution surrounding a running crack—a photoelastic analysis. Proc Soc Exp Stress Anal 16:69–92
7. Sutton MA, Wolters WJ, Peters WH, Ranson WF, McNeil SR (1983) Determination of displacements using an improved digital correlation method. Image Vis Comput 1(3):133–139
8. Sutton MA, Cheng MQ, Peters WH, Chao YJ, McNeill SR (1986) Application of an optimized digital correlation method to planar deformation analysis. Image Vis Comput 4 (3):143–151
9. Bretagne N, Valle V, Dupre JC (2005) Development of the marks tracking technique for strain field and volume variation measurements. NDT&E Int 38:290–298
10. Orteu JJ (2009) 3-D computer vision in experimental mechanics. Opt Lasers Eng 47:282–291
11. Henshell RD, Shaw KG (1975) Crack tip elements are unnecessary. Int J Numer Meth Eng 9:495–507
12. Barsoum RS (1976) On the use of isoparametric finite elements in linear fracture mechanics. Int J Numer Meth Eng 10:25–37
13. Réthoré J, Roux S, Hild F (2008) Noise-robust stress intensity factor determination from kinematic field measurements. Eng Fract Mech 75:3763–3781
14. Zienkiewicz OC, Taylor RL (2005) The finite element method for solid and structural mechanics, Elsevier Butterworth Heinemann
15. Yoneyama S, Morimoto Y, Takashi M (2003) Automatic determination method of stress intensity utilizing digital image correlation and nonlinear least squares. In: Wu Z, Abe M (eds) Structural health monitoring and intelligent infrastructure. Swet & Zeitlinger, pp 1357–1424
16. Hun M (2012) Influence de l'eau sur le décollement d'une interface par flexion d'un bicouche de chaussée urbaine. ECN PhD thesis, France. Retrieved from https://tel.archives-ouvertes.fr/tel-00851093
17. Chabot A, Hun M, Hammoum F (2013) Mechanical analysis of a mixed mode debonding test for "composite" pavements. Construction and Building Materials 40:1076–1087. https://doi.org/10.1016/j.conbuildmat.2012.11.027
18. Chabot A, Hammoum F, Hun M (2017) A 4pt bending bond test approach to evaluate water effect in a composite beam. European J Environ Civil Eng 11(s1):54–69. https://doi.org/10.1080/19648189.2017.1320237
19. Easley C, Faber KT, Shah SP (2001) Moire interferometry analysis of fiber debonding. J Eng Mech 127(6):625–629
20. Lawler JS, Shah SP (2002) Fracture processes of quasi-brittle materials studied with digital image correlation. In: Recent advances in exp mech. Kluwer, Netherlands, pp 335–344

21. Roux S, Rethore J, Hild F (2009) Digital image correlation and fracture: an advanced technique for estimating stress intensity factors of 2D and 3D cracks. J Phys D Appl Phys 42:214004
22. Sutton M, Mingqi C, Peters W, Chao Y, McNeill S (1986) Application of an optimized digital correlation method to planar deformation analysis. Image Vis Comput 4(3):143–150
23. Sutton M, Wolters W, Peters W, Ranson W, McNeill S (1983) Determination of displacements using an improved digital correlation method. Image Vis Comput 1(3):133–139
24. Bruck H, McNeill S, Sutton MA, Peters W (1989) Digital image correlation using newton-raphson method of partial differential correction. J Exp Mech 29:261–267
25. Touchal SM, Morestin F, Brunet M (1997) Various experimental applications of digital image correlation method. In: International conference on computational methods and experimental measurements. Rhodes, Greece, pp 45–58
26. Eberl C, Thompson R, Gianola D (2006) Digital image correlation and tracking with Matlab. http://www.mathworks.com/matlabcentral/fileexchange/12413
27. Chu TC, Ranson WF, Peters WH, Sutton MA (1985) Applications of digital-image-correlation techniques to experimental mechanics. Exp Mech 25(3):232–245
28. Abanto-Bueno J, Lambros J (2002) Investigation of crack growth in functionally graded materials using digital image correlation. Eng Fract Mech 69(14–16):1695–1711
29. Birgisson B, Montepara A, Romeo E, Roque R, Roncella R, Tebaldi G (2007) Determination of fundamental tensile failure limits of mixtures. J Assoc Asphalt Pavement Technol 76:303–344
30. Birgisson B, Montepara A, Napier JA, Romeo E, Roncella R, Tebaldi G (2006) Micromechanical analyses for measurement and prediction of HMA fracture energy. J Transp Res Rec 1970:186–195
31. Birgisson B, Montepara A, Romeo E, Roncella R, Napier JA, Tebaldi G (2008) Determination and prediction of crack patterns in Hot Mix Asphalt (HMA) mixtures. J Eng Fract Mech 75:664–673 (Elseiver)
32. Ewins DJ (2000) Modal testing: theory, practice and application, 2nd edn. Research studies press LTD, Letchworth
33. Simonin JM, Lièvre D, Dargenton J-C (2009) Structural roadway assessment with frequency response function. In: Proceedings of international conference on bearing capacity of roads and airfields
34. Burmister DM (1943) The theory of stress and displacement in layered systems and application of the design of airport runways. In: Proceedings of the highway research board, vol 23
35. Duhamel D, Chabot A, Tamagny P, Harfouche L (2005) Viscoroute: visco-elastic modeling for asphalt pavements. Bulletin des Laboratoires des Ponts et Chaussées 258–259:89–103
36. El Ayadi A (2010) <<Apport de l'analyse dynamique pour le diagnostic de chaussées>>, PhD thesis, In French, University of Limoges
37. Simonin JM, Cottineau LM, Muzet V, Heinkele C, Guillard Y (2009) Deflection measurement: the need of a continuous and full view approach. In: Proceedings of international conference on bearing capacity of roads and airfields, pp 467–476
38. Chew CW (1990) Waves and fields in inhomogeneous media. Van Nostrand Reinhold, New York
39. Dérobert X (2004) Step-frequency radar technique applied on very-thin layer pavements. In: Daniels DJ (ed) Surface-penetrating radar. Instit. Electrical Engineers, London (UK), pp 386–394
40. Le Bastard C, Baltazart V, Wang Y, Saillard J (2007) Thin pavement thickness estimation using GPR with high and super resolution methods. IEEE Trans Geosci Remote Sens 45 (8):2511–2519
41. Spagnolini U (2007) IEEE Trans Geosci Remote Sens 35(2)
42. Fauchard C, Dérobert X, Côte Ph (2003) GPR performance for thickness calibration on road test sites. NDT&E Int 36:67–75

43. Behnia B, Dave EV, Ahmed S, Buttlar WG, Reis H (2011) Investigation of effects of the Recycled Asphalt Pavement (RAP) amounts on low temperature cracking performance of asphalt mixtures using Acoustic Emissions (AE). J Transp Res Board 64–71 (2208, Washington, D.C.)
44. Behnia B, Buttlar WG, Apeagyei AK, Reis H (2010) Determining the embrittlement temperature of asphalt binders using an acoustic emission approach, NDE/NDT for highways and bridges: Structural Materials Technology (SMT), New York
45. Dave EV, Behnia B, Ahmed S, Buttlar WG, Reis H (2011) Low temperature fracture evaluation of asphalt mixtures using mechanical testing and acoustic emissions techniques. J Assoc Asphalt Paving Technol 80:193–226
46. Apeagyei AK, Buttlar WG, Reis H (2009) Estimation of low-temperature embrittlement for asphalt binders using an acoustic emission approach. In INSIGHT—nondestructive testing and condition monitoring. J Br Inst Non-Destr Test 51(3):129–136

# Chapter 5
# Summary

**William G. Buttlar and Armelle Chabot**

## 5.1 Introduction

This chapter presents a summary of the key findings of this state-of-the-art report on the mechanisms of cracking and debonding in asphalt and composite pavements. Some of these key findings have been presented during the 8th Rilem International Conference on Mechanisms of cracking and Debonding in Pavements in Nantes (MCD2016) [1] and in the three supplementary issues that have been published thereafter [2–4]. It also contains the key recommendations developed by the TG2 group [5], which produced a RILEM recommendation document as a culminating activity in the final year of the RILEM Technical Committee TC241-MCD. Remaining challenges in the area of asphalt and composite pavement cracking and debonding and recommendations for further study are then summarized.

With the exception of permanent deformation (rutting), which appears to be well under control in highly developed countries, the primary challenges in the design and preservation of modern asphalt pavements relate to the prevention and control of various cracking modes. This includes thermal cracking, reflective cracking, block cracking, and fatigue cracking, which can emanate from the bottom, top, or interior of the bound asphalt layers in a flexible pavement system. Interface (partial or total) debonding and interface sliding behaviour are also important topics that need to be analysed in pavement cracking studies, as this form of deterioration can lead to subsequent tearing (slippage cracking) and/or accelerated fatigue cracking of the asphalt pavement surface.

W. G. Buttlar (✉)
University of Missouri, Columbia, USA
e-mail: buttlarw@missouri.edu

A. Chabot
Department of Materials and Structures (MAST), IFSTTAR, Nantes, France
e-mail: armelle.chabot@ifsttar.fr



W. G. Buttlar et al. (eds.), *Mechanisms of Cracking and Debonding in Asphalt and Composite Pavements*, RILEM State-of-the-Art Reports 28,
https://doi.org/10.1007/978-3-319-76849-6_5

Rutting and cracking are often thought to represent opposite sides of the spectrum in asphalt mixture design. A simple rule-of-thumb in asphalt mix design suggests that lower asphalt content mixtures produce less rutting but more cracking in asphalt mixtures, and vice versa. Some pavement engineers describe this dichotomy in terms of a swinging pendulum, where the pendulum has currently swung in the direction of less rutting and more cracking. This may be a result of the focus in modern mix design methods to prevent rutting through more aggressive asphalt compactive effort in design (Superpave gyratory compactor usage, along with relatively high design gyration targets). In addition, some pavement engineers point to higher recycling rates and changes in binder supplies. For instance, modern refining techniques are squeezing more light ends out of asphalt, rendering it less ductile and/or leading to the heterogenization of the binder system through the blending of soft and hard binder components, the use of acids, rejuvenators, recycled engine oil bottoms, and/or other modifiers. These changes may alter the binder's ability to resist cracking and fatigue, to heal microdamage, and to bond with aggregates in the mixture or with other pavement layers. Moreover, the aforementioned rule-of-thumb misses a number of important considerations necessary for a more comprehensive flexible and composite pavement engineering approach, including: consideration of the role of the entire pavement structure (all layers and interfaces) in the development of cracking and debonding in the asphalt layers, the possibility for the development of highly crack and rut resistant surface materials (such as stone-mastic asphalt), and the intentional use of preventive maintenance as a strategy to achieve economical rut and crack resistant pavement system designs. To this end, the development of advanced tests and comprehensive models and field studies continue to be a pressing need in the pavement engineering community. These contemporary pavement cracking issues were the basis for the activities conducted under RILEM TC-241MCD, and have been addressed, at least in part, in this state-of-the-art report.

## 5.2 TC-241 MCD Star Document Summary

The organization of a large group of researchers into teams to work on modern pavement cracking challenges can take one of several logical forms. In the precursor to RILEM TC241-MCD, the RILEM Cracking in Pavements (CAP) technical committee divided researchers into three teams, focused on field studies (CAP TG-01), testing (CAP TG-02), and modelling (CAP TG-03). RILEM TC241-MCD sought to re-shuffle research teams to invoke new synergies, by reorganizing technical committees into the categories of cracking in bulk materials (TG-01), cracking at pavement interfaces (TG-02), and advanced measurement techniques for cracking studies (TG-03), whereby each TG could consider contemporary tests, models, and field studies to reach their stated objectives. Common sets of field and lab data, and generalized modelling schemes, and researchers working across the various committees, served to connect the three technical committees in an integrated fashion.

### 5.2.1 Summary of STAR Chapter 1

As developed in TC-CAP, and summarized in this STAR (Chapter 1, Section 10), scientific progress in complex 'mechanistic-empirical' fields such as pavement engineering necessarily involve a cyclic progression of 'QVCV,' or Qualification, Verification, Calibration, and Validation. Chapter 1 of this STAR document provided background information on field observations and accelerated pavement testing (APT) campaigns, which were used by researchers in the test and model qualification process. In other words, the observed cracking patterns and their hypothesized driving mechanisms on modern pavements and APT facilities across the world served to define the requisite physics to be included in tests and models, and to be studied using advanced measurement techniques by TC241-MCD researchers in the three aforementioned technical committees.

Among the information presented in Chapter 1, the following key field and APT observations provided motivation for the development of better tests and models to understand and describe cracking mechanisms in modern pavement structures:

- Examination of more than a dozen French motorways over the past decade has highlighted the importance of layer debonding in the development of fatigue and reflective cracking within asphalt paving layers. Layer debonding was evident in pavement cores taken on these facilities and could also be detected through in situ deflection testing, as indicated by higher pavement deflections and wider deflection basins as compared to pavements with well-bonded layers. Techniques for detecting debonding using ground penetrating radar have also shown promise.
- In the US, recent studies have recognized the need to better understand and predict thermal, block, and reflective cracking mechanisms via discrete fracture testing and modelling approaches, leading to fracture-mechanics inspired tests and models in the asphalt pavement research and practice communities. For instance, in a recent Pooled Fund study sponsored by ten states residing in colder climates in the US and the Federal Highway Administration, the disk-shaped compact tension test (DC(T)), which is now specified in ASTM D7313, and a cohesive zone fracture model were implemented in a thermal cracking specification, involving fracture energy requirements and an optional finite element based thermal cracking simulation model, called ILLI-TC.
- Field studies involving ultra-thin bonded wearing courses and stress absorbing membrane interlayers also highlighted the need to better understand the mechanics-related phenomena associated with pavement layer bonding and bond degradation, as conflicting viewpoints regarding proper tack coat (adhesive) materials and application rates and the relative contribution of pavement grinding/milling seem to prevail in the current literature.
- In accelerated pavement testing (APT) studies conducted in France, fatigue prediction models appear to provide reasonable estimates for fatigue life of asphalt pavements, but greatly under-predict the fatigue life of thicker asphalt pavements. This highlighted the need to develop better models for damage and crack initiation and propagation and to introduce a very thin bitumen interface

layer or to include discontinuities such as vertical discontinuities or partial debonding between layers into pavement models. The findings also suggested limitations which need to be addressed in developing and interpreting future APT studies, such as differences between APT and actual pavement performance in terms of asphalt aging, healing, and climate. The studies also highlighted the need for improvements in sensors and other methods to collect data related to pavement cracking mechanisms in the field and on APT experiments.

- In reflective cracking APT studies in France carried out using the FABAC device, clear evidence of the role of interface debonding between Portland cement concrete (PCC) and asphalt layers was observed and modelled. A flow-chart type schema describing numerous manifestations of reflective crack development and propagation was developed as part of the study.
- In another APT cracking study carried out at IFSTTAR, geotextile grids were shown to significantly retard the rate of cracking in asphalt paving surfaces, and differences between layer bonding and subsequent cracking rates in asphalt overlays were reported when comparing the various grids investigated.
- Similarly, in an APT reflective cracking study conducted in the US, the role of debonding in the specific type of reflective cracking observed, and its rate of development, was studied. In additional, the phenomena of 'crack jumping' over single-layered stress absorbing membrane interlayer (SAMI) systems was described, and dual-layer SAMI systems were shown to be more effective in mitigating reflective cracking in asphalt overlays placed over PCC.
- An overarching strategy for the continued evolution of cracking tests and models was presented, termed the QVCV approach. Building on the V&V technique used in the experimental mechanics community, a 'Qualification, Verification, Calibration, and Validation,' or QCVC approach was introduced for use in the pavement mechanics community. In addition to model verification (checking vs. closed-form solutions) and validation (checking model efficacy vs. real-world cracking phenomena and/or cracking data outside of the calibration data set), the importance of model qualification (motivating/informing/refining models with field observations) and proper distinction between model calibration and validation is highlighted in the QVCV approach.

Following the motivation and test/model qualification information presented in Chaps. 1–4 of this state-of-the-art report described the recent and ongoing work of the three RILEM TC241-MCD technical groups. The major findings presented in these chapters are now summarized.

### *5.2.2 Summary of STAR Chapter 2: Cracking in Asphalt Material*

Chapter 2 contains a number of contributions from RILEM TC241-MCD, TG-01 members pertaining to tests and models developed to better understanding and characterize cracking within the asphalt 'bulk' material. In light of these contributions, continued research needs were identified. These include:

- A unified laboratory test and model that can characterize and capture the response of asphalt mixtures under all the modes that affect its properties (temperatures, loading rates, aging effect etc.) has been identified as an important research need by a number of groups. Although the present report demonstrates that the current state-of-the-art in characterization and modelling of asphalt bulk fracture is capable of capturing material response under wide-range of conditions, the present approaches are still incapable of capturing a sufficient combination of variables that affect the fracture behaviour in asphalt paving layers.
- The major scientific shortcomings in the field of fracture in bulk asphalt materials identified through this review include:
  - Limited ability of current models to capture both static and cyclic crack growth in a seamless, universal manner
  - Current models and tests still lack a complete fundamental understanding of the healing processes present in asphalt
  - Most laboratory tests and analysis procedures focus on the use of global specimen responses; more research is needed on the incorporation of local material responses from laboratory tests in models, such as those being measured in new, advanced asphalt cracking characterization techniques (Chap. 4)
  - The development of universal criteria from model qualification, verification, calibration and validation are still needed

### *5.2.3 Summary of STAR Chapter 3: Interface Debonding Behavior*

Chapter 3 contains contributions from RILEM TC241-MCD, TG-02 members pertaining to tests and models developed to better understand and characterize interface debonding behaviour in layered flexible and composite pavement systems. In light of these contributions, continuing research needs were identified. Members of the TG-02 group also developed and submitted formal recommendations to RILEM [5]. The group reported that, for further development of interlayer bond tests and models, it is crucial to improve the phenomenological understanding of the mechanisms and influencing parameters associated with interlayer bonding and bond deterioration. The recommendations of the TG-02 group included:

- Moisture/water effect infiltration through cracks and other pavement defects have a significant influence on the quality of interlayer bond. Moisture was found to reduce bond strength drastically. Its effect should be investigated in more detail and moisture should be integrated in the standard methods and considered in pavement design tools.

- Improved phenomenological understanding of the mechanisms associated with interlayer debonding in terms of the geometrical surface conditions of the interface are needed. Advanced contact interface constitutive models, finite element modeling to directly consider local interface topology, as well as roughness measurements at the interface should be further developed. Improved, simplified models should then be developed and introduced into standard pavement design tools.
- The size of test specimens bears great influence on the outcome of interlayer bond tests. As demonstrated by an interlaboratory test program initiated by RILEM, specimens with larger diameters were found to achieve higher bond values compared to specimens with smaller diameters. Further, it was demonstrated that larger nominal maximum aggregate size mixtures require larger test specimens, when testing pavements with reinforcing interlayers). In some larger test assemblies, the dead weight contribution of the specimen might have to be taken into account to properly interpret results.
- Temperature effects on bond behaviour should be studied in more detail. For instance, on one hand shear bond strength decreases with increasing temperature, while on the other hand, increased temperatures may positively affect self- and traffic-induced healing effects. It was further observed that interlayer bond defects or weaknesses seem to appear more obvious when testing is performed at very low temperatures, such as −20 °C.
- There is a need to conduct cyclic and static round robin tests to compare the outcome of different devices and methods in more detail and to develop harmonized testing procedures and requirements.
- There is a need to focus more on comparing laboratory results with in situ measurements and observations, as significant differences existing between laboratory results and field performance have been identified. Such comparisons will likely lead to adjustments and improvements in laboratory bond tests and specification requirements. For instance, the technique of testing actual field specimens and comparison to full-scale accelerated loading experiments have been shown to be effective in improving the understanding of debonding mechanisms and also lead to an improved interpretation of laboratory test data.
- There is a need to develop energy-based approaches for modelling interface damage, with the goal of quantifying damage evolution during bond tests. The introduction of rest periods during cyclic testing can be performed to assess the healing mechanism.
- The importance of testing a sufficient number of test specimens, which depends on the test method used and the applied testing conditions (i.e., static vs. cyclic), was identified.
- There is a need to implement advanced measurement systems in the study of pavement debonding, such as:
    - NDT (Non-destructive testing) in field experiments;
    - Advanced sensors for the measurement of in situ interface behaviour under traffic loading, and

– DIC (Digital image correlation) techniques for advanced interpretation of laboratory interface debonding tests.

Building on this identified scientific lack, other future research topics were identified by TG-02, including:

- Advances in mechanical modelling, such as debonding energy and surface temperature gradient, should be developed.
- As observed in the construction process, less-than-ideal moisture and dust/debris conditions existing prior to asphalt layer placement and more exact pavement texture information should be included as parameters in advanced debonding models.
- Further investigation on different tack coat types, a wider range of application rates, and different HMA mixtures under different temperature and moisture conditions and different modes of failure including mode I, mode II and mixed mode is recommended.
- Data coming from interface bonding tests should also be utilized in numerically-based pavement simulation models, in order to gain additional insight towards the mechanisms of interface debonding and its prevention.
- Further research should consider predicting results for interlayer bond problems using Artificial Neural Network (ANN) modelling, utilizing combined data sets from laboratories and field installations worldwide.

### *5.2.4 Summary of STAR Chapter 4: Advanced Measurement Techniques*

Chapter 4 contains a number of contributions from RILEM TC241-MCD, TG-03 members pertaining to advanced measurements techniques for asphalt cracking studies. This included many emerging techniques that are new (or relatively new) to the pavement engineering community, such as: digital image correlation (DIC) [6], optical strain measurement systems, X-ray computed tomography + tension/compression testing systems, deflection devices and analysis of stress wave propagation, radar analysis, imaging of dieletric contrasts through electromagnetic waves, acoustic emissions (AE) based tests, and passive and active seismic NDT testing.

In light of these contributions, continued research needs were identified. These include:

- Continued development of advanced, in situ tests, possibly combining two or more of the new advanced measurement techniques reviewed herein.
- Further development of acoustic emissions techniques, which provide considerable new insight towards the microcracking behaviour in modern, heterogeneous paving materials.

- Improvements in x-ray CT techniques, particularly in terms of scanning speed, so that more useful 3D cracking process zone development and crack propagation information can be obtained.
- Common to virtually all emerging, advanced measurement systems is the need for continued calibration, validation and standardization of the techniques.
- The reviewed techniques lend themselves nicely to implementation in intelligent infrastructure and smart city systems, but integration into these systems is an open research topic.

## 5.3 Assessment of Research Needs

As summarized in the previous section, significant progress has been made over the past decade in understanding and characterizing the mechanisms of cracking and debonding in asphalt pavement systems. That notwithstanding, significant challenges still remain in the quest for a deeper understanding of these mechanisms, leading to better tests, better specifications, and better asphalt durability. Many of the remaining challenges, which are summarized below, are driven by the complexity of asphalt concrete and composite pavements, the stochastic nature of material, traffic and environment, the scarcity and expense associated with detailed field data, and the challenges associated with measuring and imaging cracking mechanisms in asphalt concrete. Some of the key remaining *research needs*, as outlined and/or informed by this compilation of recent works and as identified by RILEM TC-241 MCD members include:

- More resolute and less costly approaches for 3-D microstructural imaging of asphalt materials, particularly to capture crack various cracking stages and mechanisms (nucleation, initiation, propagation, fracture process zone behavior, etc.)
- More accurate and efficient micromechanical simulations of cracking in asphalt
- Advancing mixed-mode crack tests and models
- Advancing interface crack measurements and models
- Transfer of developed technologies into practical tests and specifications
- Better and more accessible field data, which is required for test and model development and validation
- Cracking evaluation of sustainable asphalt materials

At the time of this writing, the aforementioned research needs were being discussed and prioritized by RILEM TC 241-MCD members. These recommendations can be used to form a work plan for a new Technical Committee, which can serve to help advance the state-of-the-art in understanding, testing, and designing crack-resistant asphalt pavement materials and systems through collaboration across the international pavement engineering community.

## References

1. Chabot A, Buttlar B, Dave E, Petit C, Tebaldi G (Eds) (2016) 8th RILEM International Conference on Mechanisms of Cracking and Debonding in Pavements, Nantes. Springer Series: RILEM Bookseries, vol 13. ISBN 978-94-024-0866-9, https://doi.org/10.1007/978-94-024-0867-6
2. Chabot A, Petit C (2017) Mechanisms of cracking and debonding in pavements: debonding mechanisms in various interfaces between layers (Guest Eds). Eur J Environ and Civ Eng 11 (sup1):1–2. https://doi.org/10.1080/19648189.2017.1361649
3. Buttlar W, Tebaldi G (2017) Special section on advanced measurements for asphalt concrete characterization (Guest Eds). J Test Eval 45(6)
4. Dave E, Buttlar W (2018) Mechanisms of cracking: characterisation and modelling (Guest Eds). Road Materials and Pavement Design19(3):495. https://doi.org/10.1080/14680629.2017.1417759
5. Petit C, Chabot A, Destrée A, Raab C (2018) Recommendation of RILEM TC 241-MCD on interface debonding behavior in pavements. Mater Struct (MAAS-D-17-01269 in press)
6. Buttlar WG, Hill BC, Kim YR, Kutay ME, Millien A, Montepara A, Paulino GH, Petit C, Pop IO, Romeo E, Roncella R, Safavizadeh SA, Tebaldi G, Wargo A (2014) Digital image correlation techniques to investigate strain fields and cracking phenomena in asphalt materials. Mater Struct 47(8):1373–1390. https://doi.org/10.1617/s11527-014-0362-z

The manufacturer's authorised representative in the EU is Springer Nature Customer Service Centre GmbH, Europaplatz 3, 69115 Heidelberg, Germany. If you have any concerns regarding our products, please contact ProductSafety@springernature.com

Printed and bound by CPI Group (UK) Ltd, Croydon, CR0 4YY
05/10/2026
02228135-0007